GTPase Protocols

METHODS IN MOLECULAR BIOLOGY™

John M. Walker, Series Editor

216. **PCR Detection of Microbial Pathogens:** *Methods and Protocols,* edited by *Konrad Sachse and Joachim Frey, 2003*
215. **Cytokines and Colony Stimulating Factors:** *Methods and Protocols,* edited by *Dieter Körholz and Wieland Kiess, 2003*
214. **Superantigen Protocols,** edited by *Teresa Krakauer, 2003*
213. **Capillary Electrophoresis of Carbohydrates,** edited by *Pierre Thibault and Susumu Honda, 2003*
212. **Single Nucleotide Polymorphisms:** *Methods and Protocols,* edited by *Piu-Yan Kwok, 2003*
211. **Protein Sequencing Protocols, 2nd ed.,** edited by *Bryan John Smith, 2003*
210. **MHC Protocols,** edited by *Stephen H. Powis and Robert W. Vaughan, 2003*
209. **Transgenic Mouse Methods and Protocols,** edited by *Marten Hofker and Jan van Deursen, 2002*
208. **Peptide Nucleic Acids:** *Methods and Protocols,* edited by *Peter E. Nielsen, 2002*
207. **Recombinant Antibodies for Cancer Therapy:** *Methods and Protocols.* edited by *Martin Welschof and Jürgen Krauss, 2002*
206. **Endothelin Protocols,** edited by *Janet J. Maguire and Anthony P. Davenport, 2002*
205. ***E. coli* Gene Expression Protocols,** edited by *Peter E. Vaillancourt, 2002*
204. **Molecular Cytogenetics:** *Protocols and Applications,* edited by *Yao-Shan Fan, 2002*
203. ***In Situ* Detection of DNA Damage:** *Methods and Protocols,* edited by *Vladimir V. Didenko, 2002*
202. **Thyroid Hormone Receptors:** *Methods and Protocols,* edited by *Aria Baniahmad, 2002*
201. **Combinatorial Library Methods and Protocols,** edited by *Lisa B. English, 2002*
200. **DNA Methylation Protocols,** edited by *Ken I. Mills and Bernie H, Ramsahoye, 2002*
199. **Liposome Methods and Protocols,** edited by *Subhash C. Basu and Manju Basu, 2002*
198. **Neural Stem Cells:** *Methods and Protocols,* edited by *Tanja Zigova, Juan R. Sanchez-Ramos, and Paul R. Sanberg, 2002*
197. **Mitochondrial DNA:** *Methods and Protocols,* edited by *William C. Copeland, 2002*
196. **Oxidants and Antioxidants:** *Ultrastructural and Molecular Biology Protocols,* edited by *Donald Armstrong, 2002*
195. **Quantitative Trait Loci:** *Methods and Protocols,* edited by *Nicola J. Camp and Angela Cox, 2002*
194. **Posttranslational Modifications of Proteins:** *Tools for Functional Proteomics,* edited by *Christoph Kannicht, 2002*
193. **RT-PCR Protocols,** edited by *Joseph O'Connell, 2002*
192. **PCR Cloning Protocols, 2nd ed.,** edited by *Bing-Yuan Chen and Harry W. Janes, 2002*
191. **Telomeres and Telomerase:** *Methods and Protocols,* edited by *John A. Double and Michael J. Thompson, 2002*
190. **High Throughput Screening:** *Methods and Protocols,* edited by *William P. Janzen, 2002*
189. **GTPase Protocols:** *The RAS Superfamily,* edited by *Edward J. Manser and Thomas Leung, 2002*
188. **Epithelial Cell Culture Protocols,** edited by *Clare Wise, 2002*
187. **PCR Mutation Detection Protocols,** edited by *Bimal D. M. Theophilus and Ralph Rapley, 2002*
186. **Oxidative Stress Biomarkers and Antioxidant Protocols,** edited by *Donald Armstrong, 2002*
185. **Embryonic Stem Cells:** *Methods and Protocols,* edited by *Kursad Turksen, 2002*
184. **Biostatistical Methods,** edited by *Stephen W. Looney, 2002*
183. **Green Fluorescent Protein:** *Applications and Protocols,* edited by *Barry W. Hicks, 2002*
182. **In Vitro Mutagenesis Protocols, 2nd ed.,** edited by *Jeff Braman, 2002*
181. **Genomic Imprinting:** *Methods and Protocols,* edited by *Andrew Ward, 2002*
180. **Transgenesis Techniques, 2nd ed.:** *Principles and Protocols,* edited by *Alan R. Clarke, 2002*
179. **Gene Probes:** *Principles and Protocols,* edited by *Marilena Aquino de Muro and Ralph Rapley, 2002*
178. **Antibody Phage Display:** *Methods and Protocols,* edited by *Philippa M. O'Brien and Robert Aitken, 2001*
177. **Two-Hybrid Systems:** *Methods and Protocols,* edited by *Paul N. MacDonald, 2001*
176. **Steroid Receptor Methods:** *Protocols and Assays,* edited by *Benjamin A. Lieberman, 2001*
175. **Genomics Protocols,** edited by *Michael P. Starkey and Ramnath Elaswarapu, 2001*
174. **Epstein-Barr Virus Protocols,** edited by *Joanna B. Wilson and Gerhard H. W. May, 2001*
173. **Calcium-Binding Protein Protocols, Volume 2:** *Methods and Techniques,* edited by *Hans J. Vogel, 2001*
172. **Calcium-Binding Protein Protocols, Volume 1:** *Reviews and Case Histories,* edited by *Hans J. Vogel, 2001*
171. **Proteoglycan Protocols,** edited by *Renato V. Iozzo, 2001*
170. **DNA Arrays:** *Methods and Protocols,* edited by *Jang B. Rampal, 2001*
169. **Neurotrophin Protocols,** edited by *Robert A. Rush, 2001*
168. **Protein Structure, Stability, and Folding,** edited by *Kenneth P. Murphy, 2001*
167. **DNA Sequencing Protocols,** *Second Edition,* edited by *Colin A. Graham and Alison J. M. Hill, 2001*
166. **Immunotoxin Methods and Protocols,** edited by *Walter A. Hall, 2001*
165. **SV40 Protocols,** edited by *Leda Raptis, 2001*
164. **Kinesin Protocols,** edited by *Isabelle Vernos, 2001*
163. **Capillary Electrophoresis of Nucleic Acids, Volume 2:** *Practical Applications of Capillary Electrophoresis,* edited by *Keith R. Mitchelson and Jing Cheng, 2001*
162. **Capillary Electrophoresis of Nucleic Acids, Volume 1:** *Introduction to the Capillary Electrophoresis of Nucleic Acids,* edited by *Keith R. Mitchelson and Jing Cheng, 2001*
161. **Cytoskeleton Methods and Protocols,** edited by *Ray H. Gavin, 2001*
160. **Nuclease Methods and Protocols,** edited by *Catherine H. Schein, 2001*
159. **Amino Acid Analysis Protocols,** edited by *Catherine Cooper, Nicole Packer, and Keith Williams, 2001*
158. **Gene Knockoout Protocols,** edited by *Martin J. Tymms and Ismail Kola, 2001*
157. **Mycotoxin Protocols,** edited by *Mary W. Trucksess and Albert E. Pohland, 2001*
156. **Antigen Processing and Presentation Protocols,** edited by *Joyce C. Solheim, 2001*
155. **Adipose Tissue Protocols,** edited by *Gérard Ailhaud, 2000*
154. **Connexin Methods and Protocols,** edited by *Roberto Bruzzone and Christian Giaume, 2001*
153. **Neuropeptide Y Protocols,** edited by *Ambikaipakan Balasubramaniam, 2000*

METHODS IN MOLECULAR BIOLOGY™

GTPase Protocols

The Ras Superfamily

Edited by

Ed Manser

and

Thomas Leung

Glaxo-IMCB Group,
Institute of Molecular and Cell Biology,
Singapore

Humana Press **Totowa, New Jersey**

Softcover reprint of the hardcover 1st edition 2002
999 Riverview Drive, Suite 208
Totowa, New Jersey 07512

www.humanapress.com

This publication is printed on acid-free paper. ∞
ANSI Z39.48-1984 (American Standards Institute) Permanence of Paper for Printed Library Materials.

Cover design by Patricia F. Cleary.
Cover illustration: Images derived from Fig. 1 in Chapter 16 "*Arf6 and its Role in Cytoskeletal Modulation,*" by Julie G. Donaldson.

Production Editor: Kim Hoather-Potter.

For additional copies, pricing for bulk purchases, and/or information about other Humana titles, contact Humana at the above address or at any of the following numbers: Tel: 973-256-1699; Fax: 973-256-8341; E-mail: humana@humanapr.com, or visit our Website at www.humanapress.com

10 9 8 7 6 5 4 3 2 1

Library of Congress Cataloging in Publication Data

Main entry under title: Methods in molecular biology™.

GTPase protocols: the Ras superfamily/edited by Edward J. Manser, Thomas Leung.
p.cm. -- (Methods in molecular biology; v. 189)
Includes bibliographical references and index.
ISBN 978-1-61737-261-2 e-ISBN 978-1-59259-281-4
1. Guanosine triphosphatase--Laboratory manuals. 2. Ras proteins--Laboratory manuals. I. Manser, Edward J. II. Leung, Thomas. III. Methods in molecular biology (Clifton, N.J.); v. 189

QP609.G83 G774 2002
612'.01575--dc21

2002022973

Preface

In the last 10 years researchers have firmly established key roles for Ras-related GTPases in almost every aspect of cell biology. In the 1980s the proto-oncogene Ras itself was the focus of interest, though in the 1990s this shifted to the increasing variety of Ras-related proteins. In this new decade much yet needs to be done to establish the role for all the small GTPases now uncovered by the human genome project. In particular, these GTPases need to be understood in the appropriate biochemical and cellular contexts. In the process of trying to uncover the role of these versatile proteins, a variety of novel techniques and methodologies has been developed. These now enable investigators to move easily within a diversity of fields ranging from structural studies to real-time in vivo analysis of a GTPase.

In recognition of the need for access to key background methodologies, *GTPase Protocols: The Ras Superfamily* is devoted to techniques that are presently widely used and that will continue to be the standard for researchers worldwide. Each chapter is aimed at supplying detailed methodologies to allow reproduction in any laboratory, while also providing the general principles on which the methods are based. Some of the techniques grouped in the first section apply broadly to small GTPases, whereas others in Part II are more applicable within each GTPase subfamily.

We would like to thank the many authors who have spent time in sharing their expertise. It is hoped that *GTPase Protocols: The Ras Superfamily* will bring access to and understanding of these remarkable proteins for all those embarking on new studies.

Ed Manser
Tom Leung

Contents

Preface v
Contributors ix

Part I. General Protocols

1 The GTPase Cycle: *How Dominant Inhibitory Mutants Block the Biological Functions of Small GTPases*
Edward J. Manser *3*

2 Preparation of GTPases for Structural and Biophysical Analysis
Susan J. M. Smith and Katrin Rittinger *13*

3 Using cDNA-Representational Difference Analysis (cDNA-RDA) in Combination with Microarrays to Identify Rac Regulated Genes
Arndt A. P. Schmitz, Robert Lucito, and Linda Van Aelst *25*

4 Fluorescence Methods in the Study of Small GTP-Binding Proteins
Mohammad Reza Ahmadian, Alfred Wittinghofer, and Christian Herrmann *45*

Part II. Specific Methods:
Ras Family GTPases

5 Ras and Rac as Activators of Reactive Oxygen Species (ROS)
Herbert Archer and Dafna Bar-Sagi *67*

6 Interfering with Ras Signaling Using Membrane-Permeable Peptides or Drugs
Hiroshi Maruta, Hong He, and Thao Nheu *75*

7 A Ras-Based Module to Generate ^{32}P-Labeled Fusion Proteins for Blot Overlays
Zhoushen Zhao and Louis Lim *87*

Rho Family GTPases

8 Determination of the Activity of Rho-Like GTPases in Cells
Angeliki Malliri, Jean P. ten Klooster, Christina Olivio, and John G. Collard *99*

9 Introduction of Dominant Inhibitory Proteins Directed Against ROK and MRCK Kinases
Ivan Tan, Louis Lim, and Thomas Leung *111*

10 Effects of Rho Family GTPases on Cell–Cell Adhesion
Masaki Fukata, Masato Nakagawa, Shinya Kuroda, and Kozo Kaibuchi *121*

11 Cell Motility and Invasion Assays
Jackie Banyard and Marc Symons *129*

Rab Family GTPases

12 Isolation of Regulatory Proteins for the Rab3 Subfamily GTPases
Hiroyuki Nakanishi, Takuya Sasaki, and Yoshimi Takai *143*

13 Analysis and Preparation of Stable Complexes between Rab GTPases, Rab Escort Protein, and Rab Geranylgeranyl Transferase
José B. Pereira-Leal, Anita Q. Gomes, and Miguel C. Seabra *157*

Arf Family GTPases

14 Preparation of Myristoylated Arf1 and Arf6 Proteins
Paul A. Randazzo and Henry M. Fales *169*

15 ARF-Directed Guanine-Nucleotide-Exchange (GEP) Proteins
Gustavo Pacheco-Rodriguez, Joel Moss, and Martha Vaughan *181*

16 Arf6 and its Role in Cytoskeletal Modulation
Julie G. Donaldson *191*

17 ARF GTPase-Activating Protein 1
Irit Huber, Edna Cukierman, Miriam Rotman, and Dan Cassel *199*

Ran GTPases

18 The Use of Permeabilized Cell Systems to Study Nuclear Transport
Amy M. Brownawell, James M. Holaska, Ian G. Macara, and Bryce M. Paschal *209*

19 Analysis of Nuclear Protein Import and Export In Vitro Using Fluorescent Cargoes
Ralph H. Kehlenbach and Larry Gerace *231*

20 Characterization of the Effects of RanGTP on the Microtubule Cytoskeleton
Sofia B. Lizarraga, Yixian Zheng, and Andrew R. Wilde *247*

Index *261*

Contributors

MOHAMMAD REZA AHMADIAN • *Abteilung Strukturelle Biologie, Max-Planck-Institut für Molekulare Physiologie, Dortmund, Germany*
HERBERT ARCHER • *Department of Molecular Genetics and Microbiology, State University of New York, Stony Brook, NY*
JACKIE BANYARD • *Department of Surgical Research, Children's Hospital, Harvard Medical School, Boston, MA*
DAFNA BAR-SAGI • *Department of Molecular Genetics and Microbiology, State University of New York, Stony Brook, NY*
AMY M. BROWNAWELL • *Department of Pharmacology, Center for Cell Signaling, University of Virginia, Charlottesville, VA*
DAN CASSEL • *Department of Biology, Technion-Israel Institute of Technology, Haifa, Israel*
JOHN G. COLLARD • *Division of Cell Biology, The Netherlands Cancer Institute, Amsterdam, The Netherlands*
EDNA CUKIERMAN • *Department of Biology, Technion-Israel Institute of Technology, Haifa, Israel*
JULIE G. DONALDSON • *Laboratory of Cell Biology, National Heart, Lung, and Blood Institute, National Institutes of Health, Bethesda, MD*
HENRY M. FALES • *Laboratory of Biophysical Chemistry, National Heart, Lung, and Blood Institute, National Institutes of Health, Bethesda, MD*
MASAKI FUKATA • *Department of Cell Pharmacology, Nagoya University Graduate School of Medicine, Nagoya, Japan*
LARRY GERACE • *Department of Cell and Molecular Biology, The Scripps Research Institute, La Jolla, CA*
ANITA Q. GOMES • *Division of Biomedical Sciences, Imperial College School of Medicine, London, UK*
HONG HE • *Tumor Suppressor Laboratory, Ludwig Institute for Cancer Research, Melbourne, Australia*
CHRISTIAN HERRMANN • *Abteilung Strukturelle Biologie, Max-Planck-Institut für Molekulare Physiologie, Dortmund, Germany*
JAMES M. HOLASKA • *Department of Microbiology, University of Virginia, Charlottesville, VA*
IRIT HUBER • *Department of Biology, Technion-Israel Institute of Technology, Haifa, Israel*
KOZO KAIBUCHI • *Department of Cell Pharmacology, Nagoya University Graduate School of Medicine, Nagoya, Japan*

RALPH H. KEHLENBACH • *Abteilung Virologie, Universität Heidelberg, Heidelberg, Germany*

SHINYA KURODA • *Division of Signal Transduction, Nara Institute of Science and Technology, Ikoma, Japan; Inheritance and Variation Group, PRESTO, Japan Science and Technology, Kyoto, Japan*

THOMAS LEUNG • *Glaxo-IMCB Group, Institute of Molecular and Cell Biology, Singapore*

LOUIS LIM • *Glaxo-IMCB Laboratory, Institute of Molecular and Cell Biology, Singapore; and Department of Neurochemistry, Institute of Neurology, University College London, London, UK*

SOFIA B. LIZARRAGA • *Howard Hughes Medical Institute, Department of Embryology, Carnegie Institute of Washington, Baltimore, MD*

ROBERT LUCITO • *Cold Spring Harbor Laboratory, Cold Spring Harbor, NY*

IAN G. MACARA • *Department of Pharmacology, Center for Cell Signaling, University of Virginia, Charlottesville, VA*

ANGELIKI MALLIRI • *Division of Cell Biology, The Netherlands Cancer Institute, Amsterdam, The Netherlands*

EDWARD J. MANSER • *Glaxo-IMCB Group, Institute of Molecular and Cell Biology, Singapore*

HIROSHI MARUTA • *Tumor Suppressor Laboratory, Ludwig Institute for Cancer Research, Melbourne, Australia*

JOEL MOSS • *Pulmonary-Critical Care Medicine Branch, National Heart, Lung, and Blood Institute, National Institutes of Health, Bethesda, MD*

MASATO NAKAGAWA • *Division of Signal Transduction, Nara Institute of Science and Technology, Ikoma, Japan*

HIROYUKI NAKANISHI • *Department of Molecular Biology and Biochemistry, Faculty of Medicine, Osaka University Graduate School of Medicine, Osaka, Japan*

THAO NHEU • *Tumor Suppressor Laboratory, Ludwig Institute for Cancer Research, Melbourne, Australia*

CHRISTINA OLIVIO • *Division of Cell Biology, The Netherlands Cancer Institute, Amsterdam, The Netherlands*

GUSTAVO PACHECO-RODRIGUEZ • *Pulmonary-Critical Care Medicine Branch, National Heart, Lung, and Blood Institute, National Institutes of Health, Bethesda, MD*

BRYCE M. PASCHAL • *Department of Biochemistry and Molecular Genetics, Center for Cell Signaling, University of Virginia, Charlottesville, VA*

JOSÉ B. PEREIRA-LEAL • *Division of Biomedical Sciences, Imperial College School of Medicine, London, UK*

PAUL A. RANDAZZO • *Laboratory of Cellular Oncology, National Cancer Institute, National Institutes of Health, Bethesda, MD*
KATRIN RITTINGER • *Division of Protein Structure, National Institute for Medical Research, London, UK*
MIRIAM ROTMAN • *Department of Biology, Technion-Israel Institute of Technology, Haifa, Israel*
TAKUYA SASAKI • *Department of Biochemistry, The University of Tokushima School of Medicine, Tokushima, Japan*
ARNDT A. P. SCHMITZ • *Enabling Technologies, Schering, Germany*
MIGUEL C. SEABRA • *Division of Biomedical Sciences, Imperial College School of Medicine, London, UK*
SUSAN J. M. SMITH • *Division of Protein Structure, National Institute for Medical Research, London, UK*
MARC SYMONS • *Center for Oncology and Cell Biology, North Shore-Long Shore Island Jewish Research Institute and Department of Surgery, North Shore-Long Shore Island Jewish Medical Center, Manhasset, NY*
YOSHIMI TAKAI • *Department of Molecular Biology and Biochemistry, Faculty of Medicine, Osaka University Graduate School of Medicine, Osaka, Japan*
IVAN TAN • *Glaxo-IMCB Laboratory, Institute of Molecular and Cell Biology, Singapore*
JEAN P. TEN KLOOSTER • *Division of Cell Biology, The Netherlands Cancer Institute, Amsterdam, The Netherlands*
LINDA VAN AELST • *Cold Spring Harbor Laboratory, Cold Spring Harbor, NY*
MARTHA VAUGHAN • *Pulmonary-Critical Care Medicine Branch, National Heart, Lung, and Blood Institute, National Institutes of Health, Bethesda, MD*
ANDREW R. WILDE • *Department of Medical Genetics and Microbiology, University of Toronto, Toronto, Ontario, Canada*
ALFRED WITTINGHOFER • *Abteilung Strukturelle Biologie, Max-Planck-Institut für Molekulare Physiologie, Dortmund, Germany*
ZHOUSHEN ZHAO • *Glaxo-IMCB Group, Institute of Molecular and Cell Biology, Singapore*
YIXIAN ZHENG • *Howard Hughes Medical Institute, Department of Embryology, Carnegie Institute of Washington, Baltimore, MD*

I

General Protocols

1

The GTPase Cycle

How Dominant Inhibitory Mutants Block the Biological Functions of Small GTPases

Edward J. Manser

1. Introduction

There are two key methods which yield information about the cellular function of a protein. One can introduce into cells or whole organisms a form of the protein that is constitutively active, and which therefore functions independent of the normal regulatory mechanisms. This technique has been used very successfully with many small GTPases and is based largely on the early work on "natural" *Ras* mutants (associated with cancers), which were found to contain amino-acid substitutions predominantly at residues 12 and 61 of *Ras* *(1)*. Although introducing GTPases with such mutations can reveal dramatic effects on cell signaling, membrane trafficking, or cytoskeletal architecture, this method must be complemented by a second technique—studying the loss of protein function. This method can be introduced by genetic manipulation, anti-sense DNA/RNA, specific antibodies, drugs, or expression of so-called "dominant inhibitory" proteins. These proteins have emerged as popular tools to accomplish GTPase inhibition in mammalian cells; mutated proteins can interfere with the function of their normal cellular counterparts or with the proteins that interact with them. This approach has been used extensively in the elucidation of signal-transduction cascades involving *Ras*-superfamily proteins. This chapter examines the mechanism underlying such dominant-inhibitory *Ras* proteins and some potential problems associated with their use. Considering the extensive literature, this chapter is primarily limited to discussion of *Ras* itself, as the prototype for a small GTPase.

From: *Methods in Molecular Biology, vol. 189: GTPase Protocols: The Ras Superfamily*
Edited by: E. J. Manser and T. Leung © Humana Press Inc., Totowa, NJ

Ras proteins came to prominence when the first evidence of their involvement in human oncogenesis was revealed *(2)*. Progress has been made in understanding the normal function of *Ras* proteins in cells through the use of both dominant-active and dominant-inhibitory *Ras* mutants. Following the development of variants that cause suppression of *Ras* activity (cf *Ras*N17), researchers were to investigate the consequences of such shut-off in a variety of cell systems. Analogous dominant-inhibitory versions of other *Ras*-related GTPases, (i.e., the Rho, Rab, Arf and Ran families) are well-documented, and the number of these reports continues to grow. Understanding the mechanisms behind the action of this class of inhibitor is clearly important for researchers who intend to use them.

2. The GTPase Cycle of Ras Proteins

The extracellular signals leading to recruitment of *Ras*-specific guanine-nucleotide-exchange factors (GEFs) are many, but biochemically they promote the same effect—the release of bound guanosine diphosphate (GDP) from Ras and its replacement with guanosine triphosphate (GTP) (**Fig. 1**). Inactivation is achieved by GTP hydrolysis, which is stimulated by GTPase-activating proteins (GAPs). In its GTP-bound state, *Ras* binds with high affinity to a set of cytoplasmic target or "effector" proteins that initiate distinct intracellular signaling cascades. The *Raf* kinases, phosphatidylinositol-3-OH kinases, and Ral-specific GEFs constitute three well-established classes of effector proteins *(3)*. The mechanism by which the *Ras* GTPase interact with its target kinase Raf, involves the formation of a beta-sheet interface and contacts in both "switch I and II" regions of *Ras* *(4)*. The recent structure of *Ras* bound to phosphoinositide 3-kinase gamma suggests *Ras* can also directly modulate catalytic-domain function *(5)*.

Other members of the *Ras* superfamily, such as Rho, Rab, and Ran proteins, exhibit similar dependence on bound GTP for activity, and are often referred to as "molecular switches." With the exception of Ran, all these proteins function at cytosol-membrane interfaces where they interact with the lipid bilayer through myristoyl, farnesyl, or geranylgeranyl moieties. Because of the wide variety of both regulators and downstream target proteins, much work has yet to be done to understand the unique cellular functions of each GTPase.

3. Dominant Inhibitory *Ras* Mutants

3.1. How Inhibitory Mutants Block GTPase Function

The first set of inhibitory *Ras* mutants was identified over 10 years ago as a result of random mutagenesis studies *(6,7)*. The three mutants found contain substitutions at amino-acid positions 15, 16, or 17 and function by virtue of the requirement for GEFs in the GTPase cycle. Of these, a mutant with a

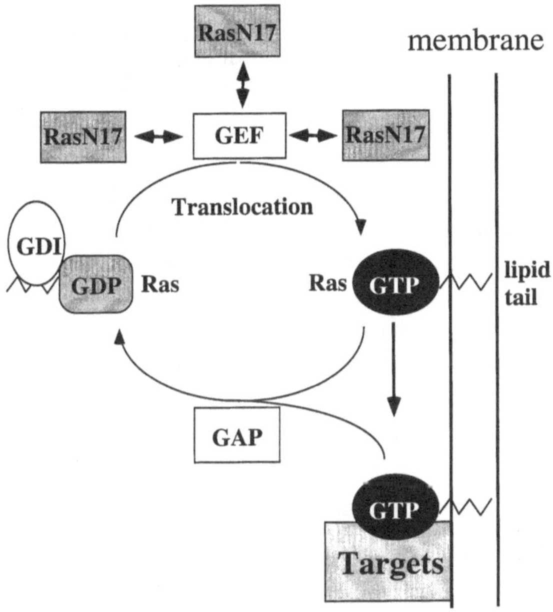

Fig. 1. The Figure shows the activation-inactivation cycle of Ras GTPases, and how this relates to the effects of dominant-inhibitory Ras. Inactive Ras.GDP requires interaction with GEFs to promote exchange of GTP onto Ras. The reverse step involves GTP hydrolysis stimulated by GAPs. Ras.GTP activates effector proteins when it is present at the membrane-cytosol interface. Expression of excess RasN17 in cells leads to preferential reversible formation of Ras-GEF.RasN17 complexes (indicated by double arrows), which prevents endogenous Ras interaction with the GEFs. Because RasN17 itself cannot bind to appropriate target proteins (even in its GTP state), downstream signaling pathways remain dormant.

replacement of serine by asparagine at position 17, *Ras*N17, has been used most extensively to suppress *Ras* function. Subsequently, the *Ras*15A mutant has revealed further details of the underlying mechanism by which these mutants function. The consensus is that dominant-inhibitory mutants work in cells by competing with normal Ras for binding to *Ras*GEFs. The mutants cannot interact with downstream target proteins, so when they are expressed in cells they bind to GEFs and form "dead-end" complexes. This prevents the activation of endogenous *Ras* by *Ras*GEFs. Biological experiments supporting this view have shown that the growth-inhibitory effect of *Ras*N17 expression in mammalian cells ***(6)***, or *Ras*15A expression in yeast ***(7)***, is overcome by increased expression of either a *Ras*-specific GEF or wild-type *Ras*. In addition, mutations within the region of *Ras* that interact with GEFs suppress the inhibitory phenotype of *Ras*N17 ***(8,9)***.

Biochemical characterizations of *Ras*N17 and the closely related inhibitory mutant *Ras*17A showed that they have higher affinities for GDP than for GTP ***(6)***. The mutant proteins were believed to be incapable of binding downstream targets in cells because they cannot bind GTP in vivo. However, subsequent experiments indicated that this is not the case: the alteration in affinity is insufficient to prevent *Ras*N17 or *Ras*17A from binding GTP in vivo ***(10,11)*** and the mutants fail to bind to downstream target proteins even in the GTP state ***(10)***. Crystallographic analysis of normal *Ras* has shown that a serine residue at position 17 binds magnesium ions located in the nucleotide-binding pocket of both GTP- and GDP-bound *Ras* ***(12)***. *Ras*17A exhibits reduced Mg2+ binding ***(11)***. Proper binding of Mg2+ by serine 17 may be necessary for *Ras* to reach an active conformation, because the "effector" domain also binds this tethered Mg2+ (via threonine 35) when *Ras* switches to the active state. It is notable that the functions of amino acids 17 and 35 appear to be conserved in all *Ras* GTPases: comparable inhibitory mutants of other small GTPases can be made in many cases, but not necessarily all (*see* **Note 4.1.**) by a mutation at the equivalent serine.

A different set of inhibitory GTPases have been designed to block the interaction of *Ras* with downstream targets. An example is a constitutively active *Ras* mutant that also includes a mutation which prevents membrane association ***(13)***. Presumably, this inhibitor binds to and traps *Ras* effector proteins in the cytosol, where they are ineffectual. Interestingly, this is not as effective as *Ras*N17 at inhibiting normal *Ras* activity in cells. One explanation is that it also competes for *Ras*GAP, which plays an inhibitory role by enhancing their hydrolysis of bound GTP to GDP. The mutants could thus increase endogenous GTP-bound *Ras* at the plasma membrane while competing for effector proteins. Another method of blocking the interaction of *Ras* with its downstream targets is to express the *Ras*-binding domain of its effector proteins, such as Ral-GDS ***(14)***.

3.2. Choosing Inhibitory GTPase Mutants

Any *Ras* mutant that cannot bind downstream target proteins should, in theory, also compete with normal *Ras* and suppress the activity of *Ras*GEFs. However, the N17 and 15A *Ras* mutants have an extra property that makes them particularly effective at this task: they bind more tightly to *Ras*GEFs in cells than normal *Ras*. A brief description of how RasGEFs promote nucleotide exchange is needed to explain this process. The GDP-bound *Ras* binds to *Ras*GEFs with relatively low affinity (>10 μ*M*). This binding promotes the release of bound GDP, leading to a high-affinity (K_d 4–5 n*M*) transient binary complex between nucleotide-free *Ras* and the RasGEF ***(15)***. This complex would be very stable in cells except for the fact that nucleotides, which are present at millimolar concentrations, have an even higher affinity for nucleotide-free *Ras* (~K_d 10^{-10} *M*) than RasGEF ***(15)***. Because GTP is in excess, GTP quickly binds nucleotide-free *Ras* and displaces the *Ras*GEF. In cells both *Ras*N17 and *Ras*15A proteins bind more strongly to *Ras*GEFs than normal *Ras* (*see* **Fig. 1**), for two reasons. First, *Ras*N17 and *Ras*15A show severely reduced affinities for nucleotides; thus, guanine nucleotides are less likely to displace *Ras*GEFs from these mutants. Second, the nucleotide-free forms of the mutants have higher affinities for *Ras*GEFs than wild-type *Ras*. These properties have been described in vitro, where stability of preformed complexes consisting of *Ras*15A or *Ras*N17 and a *Ras*GEF have been measured as a function of GTP concentration ***(16)***. *Ras*15A, which exhibits the poorest binding to guanine nucleotides of these two mutants, cannot be dissociated from *Ras*GEFs even when physiological (m*M*) concentrations of GTP are added. Free *Ras*15A can bind GTP at these concentrations, indicating that the mutant may have a sharply increased affinity for *Ras*GEFs compared with normal *Ras*, and probably forms an essentially irreversible complex with them in vivo.

*Ras*N17 has a less severe defect in nucleotide binding (K_d for GTP = 250 n*M*) relative to wild-type ***(11)***. Therefore, sub-millimolar concentrations of GTP are required to dissociate *Ras*N17 from a *Ras*GEF. This is consistent with the finding that the *Ras*N17 mutant retains the ability to bind nucleotides in cells, and does not form an irreversible complex with *Ras*GEFs in vivo. Thus, *Ras*N17 is not an extremely efficient competitor of normal *Ras*. In fact, only a threefold excess of normal *Ras* is required to overcome the growth-inhibitory effects of *Ras*N17 in NIH-3T3 cells ***(17)***. Nevertheless, because *Ras*N17 is easy to express at levels much higher than those of endogenous *Ras*, it can be an efficient attenuator of *Ras* function.

4. Notes

4.1. Potential Problems with Ras*N17*

The properties of *Ras*N17 described in **Subheading 3.1.** explain how it can inhibit *Ras*GEF proteins. What are the limitations to the possible interpretations of results obtained using *Ras*N17? A major issue is that this mutant inhibits the catalytic domain of *Ras*GEFs rather than *Ras* itself. This distinguishes it from some dominant interfering mutants of receptor tyrosine kinases or transcription factors which also inhibit their normal counterparts by forming unproductive dimers with them. Strictly speaking, one can attribute a phenotype associated with *Ras*N17 expression to suppression of *Ras* activity only if it is clear that the *Ras*GEFs in a particular cell activate the *Ras* proteins present (H-*Ras*, N-*Ras*, or K-*Ras*) exclusively. This is not always the case. For example, the *Ras*GEF Sos can also activate the related GTPase TC21 ***(18)***. The biological activities of TC21 are apparently similar to those of *Ras* ***(19)***, so if one considers it to be another *Ras* protein, the interpretation of results obtained with *Ras*N17 expression is the same. However *Ras*-GRF1 can also activate the *Ras*-related protein R-*Ras* ***(20)***. R-*Ras* has different biological effects, and in cells that express *Ras*-GRF1, such as neurons of the central nervous system (CNS), expression of *Ras*N17 potentially blocks the activity of this related but distinct GTPase. A control for this potential problem is to demonstrate that co-expression of the normal version of the GTPase reverses the phenotype of its dominant-negative counterpart.

Nucleotide exchange on *Ras* can be induced directly by nitrous oxide ***(21)***, circumventing the need for GEFs. Alternatively, *Ras* may be activated by a decrease in GAP activity ***(22)***. Neither of these mechanisms is influenced by *Ras*N17 expression. Direct measurements of *Ras*-GTP levels can resolve this issue ***(23)***. As the affinity of nucleotide-free *Ras*N17 for GEFs is not dramatically higher than that of normal *Ras*, the mutant should be expressed at levels at least two to three times that of endogenous *Ras* to achieve complete inhibition. In fact, expression of various different amounts of *Ras*N17 has shown that individual biological effects may require different levels of *Ras* activation ***(24)***.

Finally, one cannot always assume that a mutation at this conserved serine (S17) will produce exactly the same effect in all GTPases. A striking example is the Rap1A protein, which is a remarkably close relative of *Ras*. For reasons that are not yet clear, the N17 mutant of Rap1A does not compete well with normal Rap for their GEF, C3G, in vitro ***(25)***. Thus inhibition of Rap1A by Rap1AN17 in cells would be expected to require very high levels of expression of the mutant.

4.2. Potential Problems with Rho GTPases

This issue of specificity is even more of a concern when dominant-inhibitory versions of Rho-family GTPases such as Rho, Rac and Cdc42 are used. Many of the nucleotide-exchange factors for these GTPases activate more than one Rho-family member, at least in vitro ***(26)***. Thus expression of one dominant-inhibitory Rho-family member may be expected to block the activation of other family members. Fortunately, expression of individual inhibitory Rho-family members has yielded remarkably specific phenotypes in cells ***(27)***. This has proven to be true even in the context of whole organisms ***(28)***. Considering how frequently these mutants are being used, it is important to develop the mechanism underlying this specificity. As for *Ras*, there is also evidence that

Rac1 can be activated by phospholipids ***(29)***, and in this case Rac1N17 should be ineffective as an inhibitor.

Acknowledgment

This work is funded by the Glaxo Singapore Research Fund.

References

1. Lowy, D. R. and Willumsen, B. M. (1993) Function and regulation of ras. *Annu. Rev. Biochem.* **62,** 851–891.
2. McCoy, M. S., Toole, J. J., Cunningham, J. M., Chang, E. H., Lowy, D. R., and Weinberg, R. A. (1983) Characterization of a human colon/lung carcinoma oncogene. *Nature* **302,** 79–81.
3. Katz, M. E. and McCormick, F. (1997) Signal transduction from multiple Ras effectors. *Curr. Opin. Genet. Dev.* **7,** 75–79.
4. Nassar, N., Horn, G., Herrmann, C., Scherer, A., McCormick, F., and Wittinghofer, A. (1995) The 2.2 A crystal structure of the Ras-binding domain of the serine/threonine kinase c-Raf1 in complex with Rap1A and a GTP analogue. *Nature* **375,** 554–560.
5. Pacold, M. E., Suire, S., Perisic, O., Lara-Gonzalez, S., Davis, C.T., Walker, E. H., et al. (2000) Crystal structure and functional analysis of *Ras* binding to its effector phosphoinositide 3-kinase gamma. *Cell* **103,** 931–943.
6. Feig, L. A. and Cooper, G. M. (1988) Inhibition of NIH 3T3 cell proliferation by a mutant Ras protein with preferential affinity for GDP. *Mol. Cell. Biol.* **8,** 3235–3243.
7. Powers, S., O'Neill, K., and Wigler, M. (1989) Dominant yeast and mammalian Ras mutants that interfere with the CDC25-dependent activation of wild-type RAS in Saccharomyces cerevisiae. *Mol. Cell. Biol.* **9,** 390–395.
8. Quilliam, L. A., Kato, K., Rabun, K. M., Hisaka, M. M., Huff, S. Y., Campbell-Burk, S., et al. (1994) Identification of residues critical for Ras(17N) growth-inhibitory phenotype and for Ras interaction with guanine nucleotide exchange factors. *Mol. Cell. Biol.* **14,** 1113–1121.
9. Crechet J. B., Bernardi, A., and Parmeggiani, A. (1996) Distal switch II region of Ras2p is required for interaction with guanine nucleotide exchange factor. *J. Biol. Chem.* **271,** 17,234–17,240.
10. Farnsworth, C. L. and Feig, L. A. (1991) Dominant inhibitory mutations in the Mg(2+)-binding site of RasH prevent its activation by GTP. *Mol. Cell. Biol.* **11,** 4822–4829.
11. John, J., Rensland, H., Schlichting, I., Vetter, I., Borasio, G. D., Goody, R. S., et al. (1993) Kinetic and structural analysis of the Mg(2+)-binding site of the guanine nucleotide-binding protein p21H-ras. *J. Biol. Chem.* **268,** 923–929.
12. Milburn, M. V., Tong, L., deVos, A. M., Brunger, A., Yamaizumi, Z., Nishimura, S., et al. (1990) Molecular switch for signal transduction: structural differences between active and inactive forms of protooncogenic ras proteins. *Science* **247,** 939–945.

13. Stacey, D. W., Feig, L. A., and Gibbs, J. B. (1991) Dominant inhibitory Ras mutants selectively inhibit the activity of either cellular or oncogenic Ras. *Mol. Cell. Biol.* **11,** 4053–4064.
14. Peterson, S. N., Trabalzini, L., Brtva, T. R., Fischer, T., Altschuler, D. L., Martelli, P., et al. (1996) Identification of a novel RalGDS-related protein as a candidate effector for Ras and Rap1. *J. Biol. Chem.* **271,** 29,903–29,908.
15. Lenzen, C., Cool, R. H., Prinz, H., Kuhlmann, J., and Wittinghofer, A. (1998) Kinetic analysis by fluorescence of the interaction between Ras and the catalytic domain of the guanine nucleotide exchange factor Cdc25Mm. *Biochemistry* **37,** 7420–7430.
16. Lai, C. C., Boguski, M., Broek, D., and Powers, S. (1993) Influence of guanine nucleotides on complex formation between Ras and CDC25 proteins. *Mol. Cell. Biol.* **13,** 1345–1352.
17. Farnsworth, C. L., Marshall, M. S., Gibbs, J. B., Stacey, D. W., and Feig, L. A. (1991) Preferential inhibition of the oncogenic form of RasH by mutations in the GAP binding/"effector" domain. *Cell* **64,** 625–633.
18. Graham, S. M., Vojtek, A. B., Huff, S. Y., Cox, A. D., Clark, G. J., Cooper, J. A., et al. (1996) TC21 causes transformation by Raf-independent signaling pathways. *Mol. Cell. Biol.* **16,** 6132–6140.
19. Rosario, M., Paterson, H. F., and Marshall, C. J. (1999) Activation of the Raf/MAP kinase cascade by the Ras-related protein TC21 is required for the TC21-mediated transformation of NIH 3T3 cells. *EMBO J.* **18,** 1270–1279.
20. Gotoh, T., Niino, Y., Tokuda, M., Hatase, O., Nakamura, S., Matsuda, M., et al. (1997) Activation of R-Ras by Ras-guanine nucleotide-releasing factor. *J. Biol. Chem.* **272,** 18,602–18,607.
21. Mott, H. R., Carpenter, J. W., and Campbell, S. L. (1997) Structural and functional analysis of a mutant Ras protein that is insensitive to nitric oxide activation. *Biochemistry* **36,** 3640–3644.
22. Downward, J., Graves, J. D., Warne, P. H., Rayter, S., and Cantrell, D. A. (1990) Stimulation of p21ras upon T-cell activation. *Nature* **346,** 719–723.
23. Marais, R., Light, Y., Mason, C., Paterson, H., Olson, M. F., and Marshall, C. J. (1998) Requirement of Ras-GTP-Raf complexes for activation of Raf-1 by protein kinase C. *Science* **280,** 109–112.
24. Szeberenyi, J., Cai, H., and Cooper, G. M. (1990) Effect of a dominant inhibitory Ha-ras mutation on neuronal differentiation of PC12 cells. *Mol. Cell. Biol.* **10,** 5324–5332.
25. van den Berghe, N., Cool, R. H., Horn, G., and Wittinghofer, A. (1997) Biochemical characterization of C3G: an exchange factor that discriminates between Rap1 and Rap2 and is not inhibited by Rap1A(S17N). *Oncogene* **15,** 845–850.
26. Cerione, R. A. and Zheng, Y. (1996) The Dbl family of oncogenes. *Curr. Opin. Cell Biol.* **8,** 216–222.
27. Ridley, A. J. and Hall, A. (1992) The small GTP-binding protein rho regulates the assembly of focal adhesions and actin stress fibers in response to growth factors. *Cell* **70,** 389–399.

28. Harden, N., Ricos, M., Ong, Y. M., Chia, W., and Lim, L. (1999) Participation of small GTPases in dorsal closure of the Drosophila embryo: distinct roles for Rho subfamily proteins in epithelial morphogenesis. *J. Cell Sci.* **112,** 273–284.
29. Missy, K., Van Poucke, V., Raynal, P., Viala, C., Mauco, G., Plantavid, M., et al. (1998) Lipid products of phosphoinositide 3-kinase interact with Rac1 GTPase and stimulate GDP dissociation. *J. Biol. Chem.* **273,** 30,279–30,286.

2

Preparation of GTPases for Structural and Biophysical Analysis

Susan J. M. Smith and Katrin Rittinger

1. Introduction

Members of the *Ras* superfamily of small GTPases (p21) are involved in the regulation of a large variety of key cellular processes, including cell differentiation and proliferation, membrane trafficking, and nuclear import and export. Based on sequence homology, this superfamily can be divided into the Ras, Rho, Ran, Arf, Rab, and Rad subfamilies, which all have distinct biological activities. All members of this superfamily act as molecular switches and become activated and capable of transducing a signal upon binding to GTP, while guanosine triphosphate (GTP) hydrolysis returns them to the inactive state. Most members of this superfamily are post-translationally modified and carry isoprenoids at their C-termini, which anchors them to the membrane.

Significant progress has been made over the last few years in the structural and biophysical characterization of many members of this superfamily. In particular, the structures of complexes of small GTPases with regulatory proteins, such as GTPase-activating proteins (GAPs) ***(1,2)***, guanine-nucleotide exchange factors (GEFs) ***(3,4)***, guanine-nucleotide dissociation inhibitors (GDIs) ***(5,6)***, and downstream effectors ***(7–11)*** has immensely increased our understanding of the regulation and function of these molecular switches (reviewd in ***12,13***). With the exception of GDI complexes with Rac and Cdc42 (which have been expressed in insect cells), all the GTPases employed in these studies have been expressed and purified from bacterial sources.

A prerequisite for the biophysical and structural characterization of a single protein or multiprotein complex is the availability of significant amounts (>10 mg) of highly pure, homogeneous protein solutions. This chapter describes

From: *Methods in Molecular Biology, vol. 189: GTPase Protocols: The Ras Superfamily*
Edited by: E. J. Manser and T. Leung © Humana Press Inc., Totowa, NJ

the bacterial expression, purification and characterization of small GTPases for biophysical and structural studies. Although this chapter focuses on the expression and purification of members of the Rho subfamily, all the protocols described here are also applicable, with suitable modifications, to members of other subfamilies.

Ras was the first member of this superfamily to be extensively characterized by biophysical and structural methods. For these studies, *Ras* was often expressed from plasmids such as pTac and purified by classical methods, such as ion-exchange chromatography and hydrophobic-interaction chromatography ***(14,15)***, which can be highly time-consuming. Over the last couple of years, systems that enable expression of the protein of interest fused to a removable affinity tag, such as glutathione S-transferase (GST) or hexahistidine, have become widely available. These expression systems offer the advantage of a quick, one-step purification, which often yields high amounts of up to 90% pure protein, as judged by sodium dodecyl sulfate-polyacrylamide gel electrophoresis (SDS-PAGE) and ultraviolet (UV)-spectra. The protein can subsequently be cleaved from the affinity tag by a variety of proteases (e.g., thrombin, factor Xa, or TEV protease), and separated from the tag and "polished" by gel filtration, resulting in a 95–99% pure protein. Expression and purification methods based on affinity tags are now widely used to produce proteins for biophysical and structural studies. This chapter describes the use of glutathione S-transferase and hexahistidine-tagged systems ***(16,17)***.

Because of their intrinsic GTPase rate, most wild-type small GTP-binding proteins are bound to GDP after purification. However, the activated GTP-bound form of the protein is often of interest, particularly when studying the interaction with regulatory proteins or down-stream effectors. Various methods exist to circumvent the problem of GTP hydrolysis; the bound GDP can be exchanged against non-hydrolyzable GTP analogs such as guanylylimido-diphosphate (GMPPNP) or guanosine 5′-(3-O-thio)triphosphate (GTPγS), techniques which are described later in this chapter. Alternatively, mutant GTPases can be expressed with an intrinsic hydrolysis rate that is reduced enough to purify these proteins in the GTP-bound form. The most commonly used mutants are G12V and Q61L in the Rho or Ras numbering scheme. The use of mutant, GTPase-deficient proteins is the method of choice for co-expression and copurification of GTPase/effector complexes. The structure of the complex between Rab and its downstream effector Rabphilin is a very interesting example of this method, as in this particular case co-expression was the only way to prevent degradation of the effector protein in *Escherichia coli* (*E. coli*) ***(18)***. The nucleotide content of small GTPases can be analyzed by HPLC using ion-exchange or reversed-phase chromatography techniques

(19,20). Conditions for both types of chromatography are described at the end of this chapter.

Many small GTPases carry a polybasic sequence at their C-terminus. This can be a potential problem during expression in *E. coli* due to the proteolytic degradation of this sequence, which results in a mixture of proteins with heterogeneous C-termini. The Rho-family proteins Rac and Cdc42 are particularly susceptible to degradation of up to eight residues. We routinely carry out mass spectrometric analysis of the purified proteins to confirm their molecular weight and to detect possible proteolytic degradation. Most interactions of small GTPases with regulatory proteins and downstream effectors do not involve the C-terminus of the GTPase, and in these cases it is advisable to re-engineer the protein to remove those amino acids which are susceptible to proteolysis.

2. Materials

2.1. Expression of Recombinant p21s

Preparation of expression clones for small GTPases are carried out by standard cloning procedures. However, for the use of proteins in structural studies, certain points which are outlined in the following section should be taken into account when deciding on a cloning strategy.

2.1.1. Expression of GST-Fusion p21s

Recombinant Rho-family proteins are expressed as glutathione S-transferase (GST) fusion proteins in pGEX-2T or pGEX-4T (Amersham Pharmacia Biotech). For structural studies, it is advantageous to produce a protein with as few additional residues N-terminal to the initiating methionine as possible after cleavage from the GST-tag. Cloning into the vector BamH I site leaves only two additional amino acids after cleavage, and should be done whenever possible. However, we and others have experienced difficulties in thrombin cleavage of the fusion protein under these conditions. We have found that a single amino-acid insertion, introduced by PCR, can be sufficient to restore cleavage (the insertion of a histidine has worked well for us in many cases). This is preferable to the use of cloning sites further 3′ to the *Bam*H I site or vectors with a "spacer" sequence, which will add five or more amino acids. The *E. coli* strain BL21 should be used for expression of small GTPases in pGEX vectors because it lacks the bacterial proteases *lon* and *ompT*, which can contribute to the degradation of the polybasic C-termini of Rho-family proteins.

2.1.2. Expression of Hexahistidine-Tagged p21s

Various vectors are available for the expression of hexahistidine-tagged proteins. The major differences between these vectors are the specificity of the protease cleavage site (thrombin, Factor Xa, enterokinase, or TEV protease) and the restriction sites present in the multiple cloning site. As discussed previously, the choice of vector and cloning strategy should leave as few extra residues as possible after protease cleavage. pET vectors containing a thrombin site are the vectors most widely used for the expression of small GTPases (e.g., ***3,21–23***), and require expression hosts that contain a chromosomal copy of the gene for T7 RNA polymerase, such as the *E. coli* strain BL21(DE3).

2.2. Purification

The pH of all buffers used during purification depends on the p*I* of the protein of interest and should be in the order of 0.5 pH units above or below the p*I* to maintain solubility (*see* **Note 1**). We routinely calculate the theoretical p*I* from the amino-acid composition (as well as mol-wt and extinction coefficient) using the "ProtParam" program on the Expasy server http://www.expasy.ch/tools/protparam.html.

Glutathione Sepharose 4B and the S-75 Hiload 26/60 column are from Amersham Pharmacia Biotech, and Ni-NTA superflow is from Qiagen. We usually pack our matrices in XK-16 columns and carry out all chromatographic steps using an FPLC or Gradifrac system (Amersham Pharmacia Biotech). However, any other chromatographic equipment (columns and pumps) can be used. Imidazole is from Fluka (>99.5% purity. If imidazole of lower purity is used, UV-absorption of the buffer may become a problem during purification). Human thrombin is from Calbiochem. The lyophilized thrombin is solubilized in H_2O (at 2 U/μL) and stored in aliquots of 100 U at –70°C. Thrombin adsorbs to glass surfaces and should only be stored in plastic containers.

2.2.1. Purification of GST-Fusion p21s

We use the following buffers, at pH 7.5, for the purification of Rho and Cdc42. For the purification of other GTPases, the pH of the buffer must be changed according to the p*I*.

1. Lysis buffer: 50 m*M* Tris-HCl, pH 7.5, 300 m*M* NaCl, 10 m*M* $MgCl_2$, 4 m*M* DL-dithiothreitol (DTT), 4 m*M* benzamidine, 1 m*M* phenylmethylsulfonyl fluoride (PMSF). Chilled previously to 4°C (*see* **Notes 2–5**).
2. Wash buffer: 50 m*M* Tris-HCl, pH 7.5, 1.0 *M* NaCl, 10 m*M* $MgCl_2$, 4 m*M* DTT, 4 m*M* benzamidine, 1 m*M* PMSF.
3. Thrombin-cleavage buffer: 50 m*M* Tris-HCl, pH 7.5, 50 m*M* NaCl, 10 m*M* $MgCl_2$, 4 m*M* DTT, 2.5 m*M* $CaCl_2$ (*see* **Notes 6** and **7**).
4. Gel-filtration buffer: 50 m*M* Tris-HCl, pH 7.5, 50 m*M* NaCl, 10 m*M* $MgCl_2$, 2 m*M* DTT.

2.2.2. Purification of His-Tagged p21s

1. Lysis buffer: 50 m*M* Tris-HCl, pH 7.5, 500 m*M* NaCl, 10 m*M* imidazole, 5 m*M* $MgCl_2$, 4m*M* β-mercaptoethanol, 4 m*M* benzamidine, 1 m*M* PMSF. Chilled previously to 4°C (*see* **Notes 2–5** and **8**).
2. Elution buffer: (*see* **Note 9**).
   a. 50 m*M* Tris-HCl, pH 7.5, 300 m*M* NaCl, 10 m*M* imidazole, 5 m*M* $MgCl_2$, 4 m*M* β-mercaptoethanol.
   b. 50 m*M* Tris-HCl, pH 7.5, 300 m*M* NaCl, 300 m*M* imidazole, 5 m*M* $MgCl_2$, 4m*M* β-mercaptoethanol.
3. Thrombin-cleavage buffer: 50 m*M* Tris-HCl, pH 7.5, 50 m*M* NaCl, 10 m*M* $MgCl_2$, 2 m*M* DTT, 2.5 m*M* $CaCl_2$ (*see* **Note 7**).
4. Gel-filtration buffer: 50 m*M* Tris-HCl, pH 7.5, 50 m*M* NaCl, 10 m*M* $MgCl_2$, 2 m*M* DTT.

2.3. Nucleotide Exchange

Various methods exist to exchange the protein-bound GDP against non-hydrolyzable analogs. EDTA-facilitated exchange reactions are often used for the exchange against radioactive nucleotides. However, this method requires a large excess of nucleotide (50–100-fold) for a quantitative exchange. A less nucleotide-consuming exchange protocol uses alkaline phosphatase, which degrades GDP but leaves GMPPNP, GTPγS, and GMPPCP intact. The affinity of GMP for small GTPases is so low that a 10-fold excess of the nucleotide analog is sufficient for a complete exchange. We use alkaline phosphatase coupled to agarose beads (Sigma P-0762), which allows removal of the enzyme by centrifugation. PD10 Sephadex G25 columns are from Amersham Pharmacia Biotech. Centricon 10 concentrators are from Amicon Inc.

1. Exchange buffer: 40 m*M* Tris-HCl, pH 7.5, 200 m*M* $(NH_4)_2SO_4$, 10 μ*M* $ZnCl_2$, 5 m*M* DTT. When preparing the reaction mixture, $(NH_4)_2SO_4$ should be added last to avoid precipitation of the protein.
2. PD10 buffer: 20 m*M* Tris-HCl, pH 7.5, 2 m*M* $MgCl_2$, 2 m*M* DTT.

2.4. Analysis of Bound Nucleotide

The completeness of the nucleotide-exchange reaction can be monitored by HPLC, using anion exchange or reverse-phase chromatography. Both chromatography methods work well for the separation of GTP, GDP, GMP, and GMPPNP, and it is up to the user to choose between the two. Reverse-phase chromatography has the disadvantage of containing acetonitrile, which is toxic in the mobile phase, while a high-salt buffer is used for ion-exchange chromatography which can lead to salt deposits in the HPLC pumps.

The ion-exchange Partisil 10 SAX column is from Whatman and the reverse-phase ZORBAX SB-C18 column is from Rockland Technologies, Inc. KH_2PO_4 and K_2HPO_4 are from BDH, "HiperSolv for HPLC" quality.

1. Ion-exchange buffer: 0.6 *M* $NH_4H_2PO_4$, pH 4.0.
2. Reverse-phase buffer; 100 m*M* KH_2PO_4/K_2HPO_4, pH 6.5, 10 m*M* tetrabutylammonium bromide, 8.5% acetonitrile.

3. Methods

3.1. Expression of Recombinant p21s

3.1.1. Expression of GST-Fusion p21s

BL21 cells are transformed using standard procedures. A single colony is picked and spread on a plate containing 100 µg/mL ampicillin, in order to form a lawn, and incubated at 37°C overnight (one plate per flask is prepared). The next morning, 5 mL of broth (LB or terrific broth) containing 100 µg/mL ampicillin are withdrawn from 750 mL in a 2-L flask, used to resuspend the bacterial lawn and are returned to the flask. We usually grow 6 × 750 mL for a standard protein preperation. Cells are grown at 37°C to an A_{600} of 1.0, and protein expression is induced by addition of 0.35 m*M* IPTG. The incubation temperature is lowered to 28°C, and the cells are harvested after 3 h of growth by centrifugation at 4500*g* for 20 min at 4°C in a Beckman J6-MC centrifuge. 750 mL of culture (using terrific broth) generally yields 6–8 g of wet cell pellet, which can be stored at –70°C for several months.

3.1.2. Expression of His-Tagged p21s

Antibiotic selection depends on the specific vector used. All other procedures are carried out as described in **Subheading 3.1.1.**

3.2. Purification

All subsequent steps are performed at 4°C. We recommend removing aliquots at each step of this protocol to follow the success of the purification by SDS-PAGE analysis. Cell lysis can be carried out using a variety of techniques, including treatment with lysozyme, homogenization (liquid extrusion), or sonication. The latter method is routinely used in our group.

3.2.1. Purification of GST-Fusion p21s

1. About 40 g of cell pellet is resuspended in 280 mL of lysis buffer (7 mL per g of cell pellet).
2. Cells are lysed by sonication on ice in 6 × 30 second bursts, separated by 1-min breaks to prevent overheating of the sample. The lysate is clarified by

centrifugation at 25,000*g* for 1 h at 4°C in a Beckman L8-70 m Ultracentrifuge using a 45 TI rotor.

3. The glutathione Sepharose matrix is equilibrated with 10 column volumes of lysis buffer. The manufacturer recommends the use of 1 mL of glutathione Sepharose per 5 mg of fusion protein. However, we have previously recovered up to 10 mg of cleaved protein from 1 mL glutathione Sepharose (*see* **Note 10**).
4. The supernatant from **step 2** is loaded onto the column at a flow rate of 1 mL/min.
5. The column is washed with 10-column volumes of wash buffer containing 1 *M* NaCl at a flow rate of 1 mL/min to remove nonspecifically bound proteins.
6. The column is washed with 10 column volumes of thrombin cleavage buffer (*see* **Note 6**).
7. Cleavage of the GTPase from the GST tag is carried out by addition of 200 U of thrombin to 25 mL of cleavage buffer (use approx 4 U/mg protein). This buffer is recirculated through the column at a flow rate of 0.3 mL/min for 12–16 h.
8. The column is washed with cleavage buffer at 1 mL/min until all cleaved protein is eluted (*see* **Note 11**).
9. Generally, we have found it necessary to carry out a further purification step. When the protein appears pure by SDS-PAGE, we have quite often found significant contamination by DNA, observable spectrophotometrically as an absorbance peak at 260 nm.
10. The protein is concentrated in an Amicon pressure cell, using a regenerated cellulose membrane with a mol-wt cutoff of 10 kDa (YM 10).
11. A maximum of 40 mg of protein are loaded in a volume not exceeding 6 mL onto an S-75 Hiload 26/60 gel filtration column, which is run at a flow rate of 1.5 mL/min. Rho family proteins elute after 160–170 mL on this column.
12. Peak fractions are concentrated to 20–30 mg/mL in an Amicon pressure cell.
13. The concentrated protein is characterized by SDS-PAGE (*see* **Note 12**), UV spectro photometry (*see* **Note 13**), and mass spectrometry, flash-frozen in liquid nitrogen as 100-µL aliquots, and stored at –70°C.
14. The glutathione Sepharose can be regenerated by a wash with 5 column volumes of 10 mM reduced glutathione in 50 m*M* Tris, pH 7.5, followed by 5 column volumes of 3.0 *M* NaCl. Glutathione Sepharose is stored in 0.02% sodium azide at 4°C (*see* **Note 10**).

3.2.2. Purification of His-Tagged p21s

1. Approximately 40g of cell pellet are resuspended in 280 mL of lysis buffer (7 mL of buffer per g of wet cell pellet).
2. Cells are lysed by sonication on ice in 6 × 30 second bursts, separated by 1-min breaks to prevent overheating of the sample. The lysate is clarified by centrifugation at 25,000*g* for 1 h at 4°C in a Beckman L8-70 m Ultracentrifuge using a 45 TI rotor.
3. A 25-mL colum of Ni-NTA Superflow (capacity 5–10 mg/mL) is equilibrated in lysis buffer.

4. The supernatant from **step 2** is loaded onto the Ni-NTA column at a flow rate of 1 mL/min.
5. The column is washed back to baseline with elution buffer A at a flow rate of 2 mL/min.
6. Bound fusion protein is eluted with an increasing gradient from 10–300 m*M* imidazole (elution buffer B) over 5–10 column volumes at a flow rate of 2 mL/min.
7. Peak fractions are pooled, $CaCl_2$ is added to 2.5 m*M*, 200 U of thrombin are added, and the solution is dialyzed twice (using a 8–10-kDa cutoff membrane) against 2 L of thrombin cleavage buffer overnight.
8. The mixture is incubated with 2 mL of Ni-NTA Superflow to remove uncleaved fusion protein, and is concentrated in an Amicon pressure cell.
9. The subsequent purification by gel filtration is carried out as described in **Subheading 3.2.1.**
10. The Ni-NTA colum is regenerated according to the supplier's instructions.

3.3. Nucleotide Exchange

Generally, we perform the nucleotide exchange on approx 10 mg of the GTPase in a 1-mL reaction mixture. The time-course of the exchange reaction can be followed by nucleotide analysis on HPLC as described in **Subheading 3.4.** We found that the exchange reaction is normally complete after 1 h.

1. The concentrated protein is incubated in exchange buffer in the presence of a 10 *M* excess of GMPPNP and 2–4 U of alkaline phosphatase per mg GTPase. Alkaline phosphatase beads are washed twice with 20 m*M* Tris-HCl, pH 7.5 prior to addition.
2. Incubation is for 1 h at room temperature with agitation.
3. The reaction mixture is centrifuged in a benchtop microfuge to pellet the alkaline phosphatase beads. In order to remove the excess unbound nucleotide from the reaction mixture, the sample is applied to a PD10 Sephadex G25 column equilibrated in PD10 buffer, according to the manufacturer's instructions. Next, 300 μL fractions are collected and analyzed for protein content by spotting 2 μL of each fraction onto a filter paper and staining the filter in Coomassie blue SDS gel-staining solution. Fractions containing protein will appear as blue dots.
4. Protein-containing fractions are pooled, concentrated to 20–30 mg/mL using a Centricon 10, flash-frozen in liquid nitrogen, and stored at –70°C.

3.4. Analysis of Bound Nucleotide

To analyze the nature of the nucleotide bound to small GTPases by HPLC, the protein must be denatured and removed from the sample. This is achieved by incubation with perchloric acid and subsequent centrifugation to pellet the denatured protein.

1. Sufficient volume to contain a minimum of 1 nmol of protein should be withdrawn from the reaction mixture and brought up to a volume of 40 µL with PD10 buffer. Next, 2.5 µL 10% perchloric acid is added to this solution to denature the protein. The pH is raised by the addition of 1.75 µL 4 *M* CH_3COONa (sodium acetate) pH 4.0.
2. Precipitated protein is spun out in a benchtop microfuge for 2 min, and the supernatant containing the nucleotide is loaded onto the HPLC column.

3.4.1. Ion-Exchange HPLC

1. The column is run in 0.6 *M* $NH_4H_2PO_4$, pH 4.0, at 1 mL/min, and nucleotide absorbance is measured at 254 nm. The HPLC should be calibrated with solutions of known nucleotide composition.
2. Retention times: GDP: 4.8 min; GTP: 13.2; GMPPNP: 9.8 min.

3.4.2. Reverse-Phase HPLC

1. Better results are achieved using this method if the nucleotide solution is diluted 1:1 with reverse-phase buffer minus acetonitrile prior to injection onto HPLC (*see* **Note 14**).
2. The column is run isocratically in reverse-phase buffer at a flow rate of 1 mL/min, and nucleotide absorbance is measured at 254 nm.
3. Retention times: GDP: 5.3 min, GTP: 8.1 min, GMPPNP: 6.0 min.

4. Notes

1. The p*I* of a GST-fusion *p21* may be significantly different from that of the cleaved protein, making it necessary sometimes to alter the pH of the buffers during the course of purification in order to compensate for this. For example, glutathione S-transferase (GST) has a p*I* of 6.2, and Rac1 has a p*I* of 8.8, resulting in a p*I* of 8.1 for the fusion protein. We carried out the purification of Rac at a pH of 7.5, which seemed adequate. However, we regularly experienced large losses of protein because of precipitation after cleavage and concentration. Mass spectrometric analysis revealed that the protein was proteolytically degraded by eight residues at the C-terminus, resulting in a protein with a p*I* of 7.5. When the pH of the cleavage buffer was adjusted to 7.0 or 8.0, 100% of the cleaved protein was retained in solution. However, when selecting an appropriate pH for the cleavage buffer, the pH range at which thrombin is maximally active (pH of 7.5–8.4) should also be considered.
2. The pH of Tris-HCl changes significantly with temperature, and the pH at 4°C is more than 0.5 pH U higher than that of the same buffer at 25°C. For this reason, the pH should be adjusted at the temperature at which the buffer will be used.
3. All buffers are chilled to 4°C before use. DTT or β-mercaptoethanol should be included in all buffers to keep cysteines reduced. DTT and protease inhibitors are made freshly and added to buffers immediately before use. Use of the protease

inhibitor 4-(2-aminoethyl)-benzenesulfonyl fluoride (AEBSF, commercially available as Pefabloc SC) should be avoided because the protein can become irreversibly modified by this reagent (unpublished data).

4. Having a relatively high ionic strength in the buffer helps to prevent unspecific binding of proteins to the column. If necessary, the salt concentration can be increased up to 2 *M* for purification of His-tagged proteins on Ni-NTA.
5. $MgCl_2$ should be kept in all buffers, as it increases the affinity of the nucleotide for the protein. Nucleotide-free GTPases are often unstable.
6. It is essential to remove all traces of protease inhibitors from the column before thrombin cleavage in order to obtain maximal thrombin activity.
7. Thrombin is optimally active in the presence of physiological concentrations of $CaCl_2$.
8. DTT cannot be used for the purification on Ni-NTA, as it reduces the nickel ions. β-mercaptoethanol can be used at concentrations up to 20 m*M*.
9. The pH of the elution buffer must be checked and possibly adjusted to pH 7.5 with HCl before use (a 0.5 *M* solution of imidazole has a pH of 10–11).
10. To prevent cross-contamination between different GTPases, particularly when purifying mutants, we re-use 1 aliquot of glutathione Sepharose only for the purification of the same protein.
11. Thrombin can be removed after cleavage by passing the sample through a 1-mL column of either p-aminobenzamidine agarose or antithrombin III agarose.
12. When the SDS-PAGE gel is not freshly poured, we sometimes observe that the GTPase appears as a double band, even though mass spectrometric analysis of the same sample showed that only a single species is present.
13. The absorbance maximum for guanine nucleotides lies at 254 nm, and their spectrum partially overlaps with that of proteins. To account for this, the extinction coefficient for guanine nucleotides at 280 nm, 7765 $cm^{-1}M^{-1}$, is added to the extinction coefficient of the GTPase (calculated using the "ProtParam" tool on the Expasy server).
14. Tetrabutylammonium ions act as counter ions to the phosphate groups, and neutralize their charge. They also increase the interaction with the hydrophobic column, and thereby increase the retention time with an increasing number of phosphates.

Acknowledgment

We thank all colleagues in the Divisions of Protein Structure and Physical Biochemistry for stimulating discussions and advice.

References

1. Scheffzek, K., Ahmadian, M. R., Kabsch, W., Wiesmuller, L., Lautwein, A., Schmitz, F., et al. (1997) The Ras-RasGAP complex: structural basis for GTPase activation and its loss in oncogenic Ras mutants. *Science* **277,** 333–338.
2. Rittinger, K., Walker, P. A., Eccleston, J. F., Smerdon, S. J., and Gamblin, S. J. (1997) Structure at 1.65 A of RhoA and its GTPase-activating protein in complex with a transition-state analogue. *Nature* **389,** 758–762.

3. Goldberg, J. (1998) Structural basis for activation of ARF GTPase: mechanisms of guanine nucleotide exchange and GTP-myristoyl switching. *Cell* **95,** 237–248.
4. Boriack-Sjodin, P. A., Margarit, S. M., Bar-Sagi, D., and Kuriyan, J. (1998) The structural basis of the activation of Ras by Sos. *Nature* **394,** 337–343.
5. Hoffman, G. R., Nassar, N., and Cerione, R. A. (2000) Structure of the Rho family GTP-binding protein Cdc42 in complex with the multifunctional regulator RhoGDI. *Cell* **100,** 345–356.
6. Scheffzek, K., Stephan, I., Jensen, O. N., Illenberger, D., and Gierschik, P. (2000) The Rac-RhoGDI complex and the structural basis for the regulation of Rho proteins by RhoGDI. *Nat. Struct. Biol.*, **7,** 122–126.
7. Nassar, N., Horn, G., Herrmann, C., Scherer, A., McCormick, F., and Wittinghofer, A. (1995) The 2.2 A crystal structure of the Ras-binding domain of the serine/threonine kinase c-Raf1 in complex with Rap1A and a GTP analogue. *Nature* **375,** 554–560.
8. Huang, L., Hofer, F., Martin, G. S., and Kim, S. H. (1998) Structural basis for the interaction of Ras with RalGDS. *Nat. Struct. Biol.* **5,** 422–426.
9. Abdul-Manan, N., Aghazadeh, B., Liu, G. A., Majumdar, A., Ouerfelli, O., Siminovitch, K. A., et al. (1999) Structure of Cdc42 in complex with the GTPase-binding domain of the "Wiskott-Aldrich syndrome" protein. *Nature* **399,** 379–383.
10. Mott, H. R., Owen, D., Nietlispach, D., Lowe, P. N., Manser, E., Lim, L., et al. (1999) Structure of the small G protein Cdc42 bound to the GTPase-binding domain of ACK. *Nature* **399,** 384–388.
11. Vetter, I. R., Nowak, C., Nishimoto, T., Kuhlmann, J., and Wittinghofer, A. (1999) Structure of a Ran-binding domain complexed with Ran bound to a GTP analogue: implications for nuclear transport. *Nature* **398,** 39–46.
12. Sprang S. R. (1997) G protein mechanisms: insights from structural analysis. *Annu. Rev. Biochem.* **66,** 639–678.
13. Geyer, M. and Wittinghofer, A. (1997) GEFs, GAPs, GDIs and effectors: taking a closer (3D) look at the regulation of Ras-related GTP-binding proteins. *Curr. Opin. Struct. Biol.* **7,** 786–792.
14. Tucker, J., Sczakiel, G., Feuerstein, J., John, J., Goody, R. S., and Wittinghofer, A. (1986) Expression of p21 proteins in Escherichia coli and stereochemistry of the nucleotide-binding site. *EMBO J.* **5,** 1351–1358.
15. John, J., Frech, M., and Wittinghofer, A. (1988) Biochemical properties of Ha-ras encoded p21 mutants and mechanism of the autophosphorylation reaction. *J. Biol. Chem.* **263,** 11,792–11,799.
16. Amersham Pharmacia Biotech (2000) The Recombinant Protein Handbook.
17. Novagen (2000) pET System Manual.
18. Ostermeier, C. and Brunger, A. T. (1999) Structural basis of Rab effector specificity: crystal structure of the small G protein Rab3A complexed with the effector domain of rabphilin-3A. *Cell* **96,** 363–374.
19. Scherer, A., John, J., Linke, R., Goody, R. S., Wittinghofer, A., Pai, E. F., et al. (1989) Crystallization and preliminary X-ray analysis of the human c-H-ras-oncogene product p21 complexed with GTP analogues. *J. Mol. Biol.* **206,** 257–259.

20. Graham, D. L., Eccleston, J. F., and Lowe, P. N. (1999) The conserved arginine in rho-GTPase-activating protein is essential for efficient catalysis but not for complex formation with Rho.GDP and aluminum fluoride. *Biochemistry* **38,** 985–991.
21. Guo, W., Sutcliffe, M. J., Cerione, R. A., and Oswald, R. E. (1998) Identification of the binding surface on Cdc42Hs for p21-activated kinase. *Biochemistry* **37,** 14,030–14,037.
22. Leonard, D. A., Satoskar, R. S., Wu, W. J., Bagrodia, S., Cerione, R. A., and Manor, D. (1997) Use of a fluorescence spectroscopic readout to characterize the interactions of Cdc42Hs with its target/effector, mPAK-3. *Biochemistry* **36,** 1173–1180.

3

Using cDNA-Representational Difference Analysis (cDNA-RDA) in Combination with Microarrays to Identify Rac Regulated Genes

Arndt A. P. Schmitz, Robert Lucito, and Linda Van Aelst

1. Introduction

Small GTPases of the Ras superfamily are molecular switches which cycle between an active guanosine triphosphate (GTP)-bound and an inactive guanosine diphosphate (GDP)-bound state. They integrate signals from the cell surface to the nucleus, regulating important cellular activities. For example, Ras itself is activated when extracellular growth factors such as platelet derived growth factor (PDGF) or epidermal growth facor (EGF) bind to their receptors at the cell surface. This activation of Ras ultimately leads to changes in the transcriptional activity of the cell, e.g., via the canonical mitogen activated protein kinase (MAPK) cascade. Constitutively activated, mutant forms of Ras such as RasV12 are found frequently in human tumors, and it is widely assumed that this oncogene acts via transcriptional activation of growth and proliferation pathways.

Although the Rho family members—including Rho, Rac, and Cdc42—are best known for control of the actin cytoskeleton, they have also been linked to transcriptional activation. For example, activation of Rac triggers the activation of p38 and JNK MAPKs, as well as NFκB pathways *(1)*. Furthermore, numerous studies support a role for Rac in the proliferation, invasion, and the control of cell adhesion. All these events occur over a longer time-scale compared to the short-term changes of the actin cytoskeleton, and it is often assumed that these events require transcriptional activation. However, although pathways leading from Rac to the nucleus have been identified, information about genes regulated by Rac (or other members of the Rho GTPase family) remains scarce.

Within the last decade, several methods have been developed to identify changes in gene expression. These include serial analysis of gene expression

From: *Methods in Molecular Biology, vol. 189: GTPase Protocols: The Ras Superfamily*
Edited by: E. J. Manser and T. Leung © Humana Press Inc., Totowa, NJ

(SAGE) *(2)*, differential display (DD) *(3,4)*, representational difference analysis of cDNAs (cDNA-RDA) *(5,6)*, and suppression subtractive hybridization (SSH) *(7,8)*. Each of these techniques has its limitations. For example, Harris et al. identified differentially expressed genes in Aflatoxin B1-treated hepatocytes, using in parallel DD, cDNA-RDA, and SSH, and each of these three methods found a small, nonoverlapping set of differentially expressed genes *(9)*. More recently, expression analysis utilizing microarray technology has become available *(10)*. This technique depends on the availability of reliable cDNA clones that can be arrayed, whereas the former methods allow the identification of novel sequences in incompletely characterized organisms. One major advantage of cDNA-RDA compared to other methods is its low level of false-positives, because RDA eliminates those fragments which are present in both DNA populations. Furthermore, cDNA-RDA does not require sophisticated equipment.

As an example, we describe protocols for using cDNA-RDA to identify genes differentially expressed between cell lines with or without induction of Rac1 expression (under the control of an inducible promoter). cDNA-RDA is a modified form of RDA, a polymerase chain reaction (PCR)-based differential cloning method (**Fig. 1**) *(11,12)*. With this technique, one DNA population

Fig. 1. *(see opposite page)* Principle of cDNA-RDA. Flowchart of cDNA-RDA. (**A**) From cells either induced or left uninduced for expression of the gene of interest, representations are prepared by isolation of mRNAs, synthesis of cDNAs, restriction digest with *Dpn*II, ligation to adaptors, and amplification by PCR. The adaptors are subsequently removed by digestion in order to use the representation as the driver in a cDNA-RDA experiment. To use the representation as the tester, the adaptors are cut off and replaced by newly ligated adaptors of different sequence. Then, the tester from one sample is hybridized to an excess of driver from the other sample and sequences enriched in the tester are selectively amplified by PCR to obtain the difference product one (DP1). Finally, the DP1 is used as the tester in a new round of hybridization and amplification to result in difference product two (DP2). Note that by performing two sets of reactions in parallel, using cells from each of the two samples once as tester and once as driver against the other sample, up- as well as downregulated genes can be identified. (**B**) The three possible outcomes of the tester/driver hybridization. If a sequence is unique to the tester or present at a higher molar ratio in the tester than in the driver, it will be exponentially amplified. If a sequence is found in both driver and tester to equal amounts, only the strand from the tester population bears the adaptor and the sequence will be linearly amplified. If the sequence is found only in the driver, neither strand contains the adaptor sequence and the sequence will not be amplified. (**C**) A typical difference product after two rounds of cDNA-RDA visualized by agarose gel electrophoresis. A difference product consists of a series of visible bands superimposed on a "smear." As indicated, each band may contain fragments of several genes whose different sizes can not be resolved on an agarose gel. Furthermore, candidate genes may be contained in the "smear." Finally, different fragments of the same gene can occur at different positions in the gel, since cDNA-RDA is based on digested cDNA.

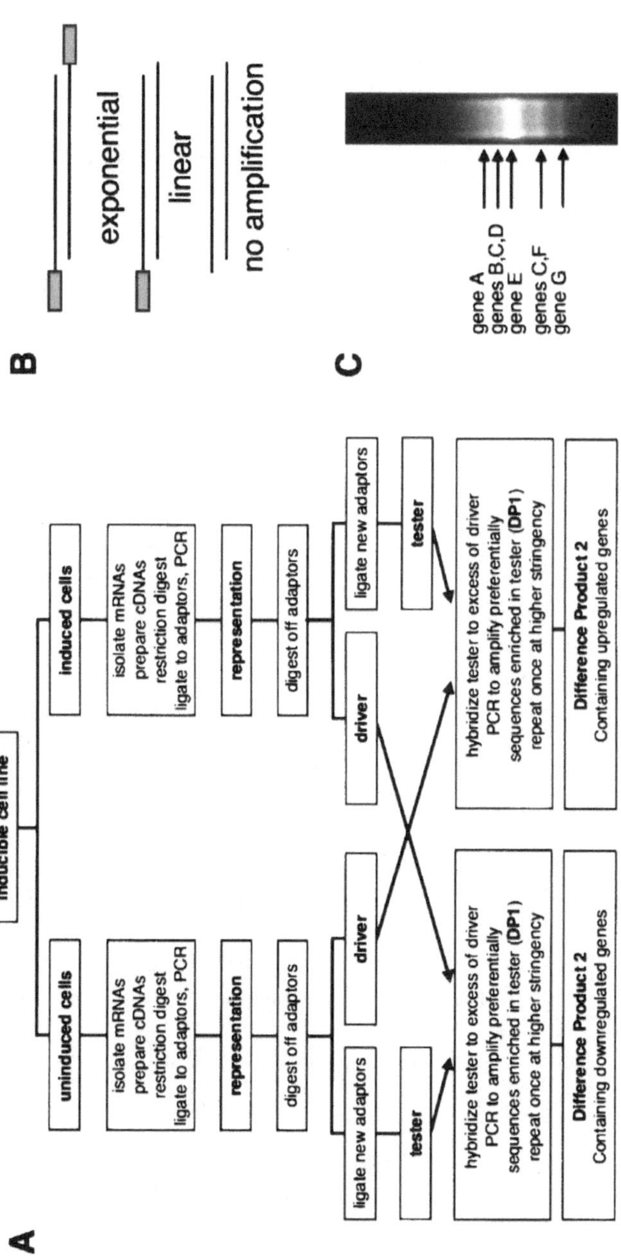
A
inducible cell line
uninduced cells
induced cells
isolate mRNAs
prepare cDNAs
restriction digest
ligate to adaptors, PCR
representation
digest off adaptors
ligate new adaptors
tester
driver
hybridize tester to excess of driver
PCR to amplify preferentially
sequences enriched in tester (DP1)
repeat once at higher stringency
Difference Product 2
Containing downregulated genes
Containing upregulated genes
B
exponential
linear
no amplification
C
gene A
genes B,C,D
gene E
genes C,F
gene G

Fig. 1.

(the driver) is hybridized in excess against a second population (the tester) to remove common hybridizing sequences, thereby enriching target sequences unique to the tester. RDA relies on the use of representations of the DNAs of interest. In brief, a representation is prepared by restriction digestion of the DNA (e.g., *Dpn*II in the case of cDNA-RDA), ligation of adaptor oligonucleotides, and subsequent PCR amplification.

One major task is the analysis of the difference product resulting from the cDNA-RDA experiment (**Fig. 1C**). To identify the most promising candidate genes, we and others ***(13)*** used the microarray technique as detailed here (**Fig. 2**). Only the clones with the highest differential expression as judged by microarray analysis were sequenced and further pursued. By doing so, we were able to significantly increase the number of clones screened per cDNA-RDA experiment.

2. Materials

2.1. Equipment

1. PCR machine capable of handling 0.5 mL polymerase chain reaction (PCR) tubes to perform cDNA-RDA (e.g., Perkin Elmer DNA Thermal Cycler 480).
2. PCR machine which can hold 96-well plates (e.g., Perkin Elmer GeneAmp 9600) or up to four 96-well plates simultaneously (e.g., MJ Research PTC-225) to process the samples for microarraying.
3. Speedvac (e.g., Savant DNA 100).
4. 8-channel pipetters such as Labsystems Finnpipette 4510000 (0.5–10 µL) and 4510020 (5–50 µL) and a repeat pipet (e.g., Brinkmann Eppendorf 22260006) are very helpful for working with 96-well plates.
5. 96-well PCR plates are obtained from Perkin Elmer (N801-0560), 96-well plates for dilutions of DNA samples are retrieved from Nunc, and 96-well plates for microarraying (with V-shaped bottom) are obtained from Corning-Costar. Plates are covered either with caps (during PCR, Perkin Elmer N801-0535) or sealing film (Sigma Z36, 968-3).
6. Ultraviolet (UV) crosslinker (Stratalinker 2400, Stratagene).
7. Minifold I dot-blotting apparatus (Schleicher and Schuell SRC096/0).
8. Hybond-N+ nylon membrane (Amersham Pharmacia Biotech RPN203B).
9. Microcon YM-30 ultrafiltration columns (Amicon) to purify and concentrate labeling reactions.
10. Vacuum oven.
11. Microarrayer, e.g., Cartesian PixSys 5500 (Cartesian Technologies, Irvine, CA).
12. Pins for the arrayer (Chipmaker 2, Telechem International).
13. Silanated glass slides (Corning).
14. Humidified hybridization chamber (Telechem International).
15. Scanner suitable for microarray fluorescence detection, such as GSI Lumonics ScanArray3000 or Axon GenePix4000.

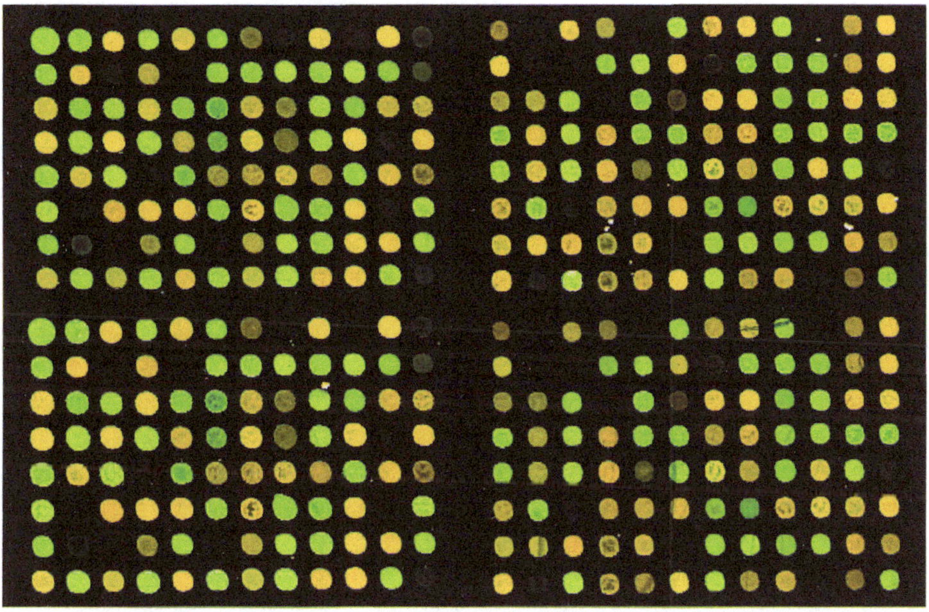

Fig. 2. Analysis of cDNA-RDA products by microarray analysis. 190 clones derived from a cDNA-RDA experiment using cells induced for RacV12 expression as the tester and uninduced cells as the driver were arrayed in duplicate. The array was then hybridized simultaneously to a representation from RacV12 induced cells (labeled in green) and to a representation from uninduced cells (labeled in red). The clones with different shades of green represent genes whose expression is upregulated as a result of RacV12 expression, whereas "yellow clones" are transcripts unaffected by RacV12 expression. Genes downregulated by RacV12 would be identified as red spots, but are absent as expected from the design of this particular experiment. Such microarraying allows rapid screening of a difference product for the clones with the highest differential expression ratio, which are then further pursued.

16. Analysis software: We made use of ScanAlyze (Stanford University) or Axon GenePix to determine features and for the quantitative analysis of the resulting TIFF files.

2.2. Enzymes and Reagents

1. Organic solvents such as 70% and 100% ethanol (EtOH), isopropanol (iPrOH), chloroform, dimethyl sulfoxide (DMSO), and phenol/chloroform/isoamylalcohol (25/24/1) saturated with TE pH 8.0 (Sigma P 2069).
2. FastTrack mRNA isolation kit (Invitrogen K1593-02).
3. CopyKit cDNA synthesis kit (Invitrogen L1311-03).
4. *Dpn*II restriction enzyme and 10 x DpnII buffer (New England Biolabs 543L).

5. 10 mg/mL tRNA (Sigma R 8759) is used as a carrier during precipitations of small amounts of DNA.
6. Sheared salmon-sperm DNA (Stratagene 201 190, diluted to 50 ng/μL).
7. DNA mol-wt marker ϕ174 *Hae*III digest (New England Biolabs 3026L).
8. T4 ligase and 10 × T4 buffer (New England Biolabs 202S).
9. The primers listed in **Table 1**.
10. AmpliTaq polymerase, 25 m*M* $MgCl_2$, and 10 × Taq buffer without $MgCl_2$ (5 U/μL, Perkin Elmer N8080-153).
11. Deoxynucleotide 5′ triphosphate (dNTPs) (100 m*M*, Roche Molecular Biochemicals 1 969 064).
12. Mung bean nuclease (MBN) and 10 × MBN buffer (New England Biolabs 250S).
13. Qiaquick gel-extraction kit (Qiagen 28704).
14. *Bam*HI restriction enzyme, 10 × *Bam*HI buffer, and 10 mg/mL bovine serum albumin (BSA) (New England Biolabs 136S).
15. Calf intestinal phosphatase (CIP) and 10 × CIP buffer (New England Biolabs 290S).
16. Qiaquick PCR purification kit (Qiagen 28104).
17. DNA polymerase I Klenow fragment and 10 × reaction buffer (Amersham Pharmacia Biotech E2141).
18. Cy3-deoxycytidine 5′ triphosphate (dCTP) and Cy5-dCTP for probe labeling (Amersham Pharmacia Biotech PA53021 and PA55021).
19. dRhodamine dye terminator kit for nonradioactive DNA sequencing (Perkin Elmer 403045).

2.3. Buffers

1. 3 *M* sodium acetate (NaAc), pH 5.2.
2. Elution buffer (EB): 10 m*M* Tris-HCl pH 8.5.
3. 6 × GLB (Gel-loading buffer): 30% glycerol and 0.25% Bromophenol blue.
4. 500 m*M* Tris-HCl pH 8.9 (It is important to use Tris base adjusted to pH with HCl instead of Tris hydrochloride adjusted to pH with NaOH.)
5. 100 m*M* $MgCl_2$.
6. 200 m*M* $(NH_4)_2SO_4$, sterile-filtered.
7. TE buffer: 10 m*M* Tris-HCl, 1 m*M* ethylenediaminetetraacetic acid (EDTA), pH 8.0.
8. 10 *M* ammonium acetate (NH_4Ac).
9. EE buffer: 30 m*M* *N*-(2-Hydroxyethyl) piperazine-*N*′-3-propane sulfonic acid (EPPS) (Sigma E 1894) pH 8.0, 3 m*M* EDTA pH 8.0, sterile-filtered.
10. 5 *M* NaCl.
11. 20 × SCC: 3 *M* NaCl, 0.3 *M* sodium citrate, pH 7.0.

2.4. PCR Buffers

PCR buffers are prepared from autoclaved or sterile-filtered stock solutions and autoclaved water (*see* **Note 1**). All PCR buffers are made immediately before use and stored on ice.

Table 1
Sequences of the Used Primers

Name	Sequence	Use	Ref.
RBgl24	AGCACTCTCCAGCCTCTCACCGCA	Representation	*11*
RBgl12	GATCTGCGGTGA	Representation	*11*
NBgl24	AGGCAACTGTGCTATCCGAGGGAA	1st round of cDNA-RDA	*11*
NBgl12	GATCTTCCCTCG	1st round of cDNA-RDA	*11*
IBgl24	TCAGCATCGAGACTGAACGCAGCA	2nd round of cDNA-RDA	[a]
IBgl12	GATCTGCTGCGT	2nd round of cDNA-RDA	[a]
SP6R	GGTGACACTATAGAATACTCAAGC	Single-colony PCR	[b]
T7E	TGTAATACGACTCACTATAGGGC	Single-colony PCR	[b]
Random nonamer	NNNNNNNNN	Probe labeling	[a]

[a]Robert Lucito, unpublished results.
[b]This report.

1. PCR buffer 1 is used in **Subheading 3.2.1.3.** for the preparation of representations. This buffer consists of 67 m*M* Tris-HCl, pH 8.9, 4 m*M* $MgCl_2$, 10 m*M* 2-mercaptoethanol, 0.1 mg/mL BSA, 16 m*M* $(NH_4)_2SO_4$, 320 μ*M* each dNTP, 1.25 μ*M* primer RBgl24, and 0.04 U/μL (15 U/tube) Taq.
2. PCR buffer 2 is identical to buffer 1, except that the primer and the Taq polymerase are added later. This buffer is used for the first PCR in each round of cDNA-RDA described in **Subheadings 3.2.2.3.** and **3.2.3., item 4**.
3. PCR buffer 3 is also identical to buffer 1, except that primer NBgl24 (in **Subheading 3.2.2.4.**) or IBlg24 (in **Subheading 3.2.3., item 4**) are used. This buffer is employed during the second PCR in each round of cDNA-RDA.
4. PCR buffer 4 is used for the single-colony PCR described in **Subheading 3.3.3.** This buffer consists of 12% DMSO in 1 × Taq buffer supplemented with 2 m*M* $MgCl_2$, 200 μ*M* each dNTP, 0.5 μ*M* primer SP6R, 0.5 μ*M* primer T7E, and 0.025 U/μL (0.625 U/well) Taq polymerase.
5. PCR buffer 5 is a standard PCR buffer used in **Subheading 3.3.5.** for reamplification. Buffer 5 is identical to buffer 4, except that dimethyl sulfoxide (DMSO) is omitted.

3. Methods

A detailed description of the various steps in cDNA-RDA and the analysis of the difference product by microarray analysis is given. More common techniques used throughout our study, such as Western, Northern, and Southern blot analysis, are not described here in detail.

3.1. Preparation of cDNAs

In order to obtain meaningful results by cDNA-RDA, the samples used for the preparation of tester and driver must be selected carefully. It is best to differ in only one parameter and to compare two samples matched as close as possible. For this reason, we made use of an inducible system for the expression of the target gene ***(14–16)***.

1. Perform time-courses and titrate the concentration of inducer to establish optimal induction conditions for the selected inducible clones.
2. Grow approx 5 × 10 ^ 7 cells of one selected clone in one incubator. Split cells into two halves and induce one-half of the population. For example, in the case of MDCK cells with expression of RacV12 under the control of the tetracycline-off system, 2 d in the absence of doxycycline were sufficient to induce expression of RacV12 to the level of endogenous Rac ***(17)***. Continue to use the same batch of medium for tester and driver populations. Harvest approx 1 × 10 ^ 8 each of uninduced and induced cells by scraping into ice-cold PBS. Process immediately, or store the cell pellet at –80°C. Before continuing, check by Northern or Western blot analysis to see whether induction was successful.
3. Isolate mRNA from cells induced and not induced for expression of the gene of interest. Several commercial kits are available. We used the Fast Track kit for mRNA isolation and the CopyKit for cDNA preparation (both from Invitrogen). A yield of approx 3 µg mRNA/10 ^ 7 cells can be expected, depending on the cell type. Use 5 µg mRNA to generate approx 7 µg cDNA (*see* **Note 2**).

3.2. cDNA-RDA

3.2.1. Preparation of Representations

Representations are obtained by digestion of the cDNA sample of interest with a frequently cutting restriction enzyme such as *Dpn*II (GATC), followed by ligation to short-adaptor oligonucleotides and amplification by PCR, using the same sequences as primers. Representations to be compared to each other by cDNA-RDA must be prepared simultaneously.

3.2.1.1. Digestions of the cDNAs

1. Digest the two cDNA samples derived from the induced and the uninduced cells in parallel. Each digest consists of 1.2 µg of cDNA in 10 µL 10 × *Dpn*II buffer, 10 µL 10 U/µL *Dpn*II, and H_2O to a total vol of 100 µL overnight (16 h) at 37°C (*see* **Note 3**).
2. Add 1 µL 10 µg/µL tRNA, vortex, and extract twice with 100 µL phenol/chloroform. Add 10 µL 3 *M* NaAc pH 5.2 and 330 µL 100% EtOH (kept at –20°C), vortex, and incubate at –70°C for 10 min.
3. Spin 10 min at 15,000*g* at 4°C, and wash the pellets with 500 µL 70% EtOH (kept at –20°C). Spin again and remove wash solutions as completely as possible by pipetting. Dry in a speedvac at medium heat and resuspend carefully in 12 µL EB (*see* **Note 4**).

4. Transfer 2 µL to a tube containing 8 µL H_2O and 2 µL 6 × GLB. Load the samples onto a 2% agarose gel, side-to-side with 200 ng undigested cDNA (to check digest), a dilution series of 50–100–200 ng sheared salmon-sperm DNA (to check recovery and concentration), and 200 ng of ϕ174 *Hae*III as the marker.

3.2.1.2. Ligation

The ligation is preceded by an incubation step ensuring optimal conditions for the adaptors to hybridize to the cDNA fragments as follows:

1. To the remaining 10 µL of each cDNA digest, add 7.5 µL 62 µ*M* RBgl24, 7.5 µL 62 µ*M* RBgl12, 3 µL of 10 × T4 buffer, and 2 µL H_2O.
2. Mix well and transfer to a heatblock (holes filled with glycerol for optimal thermal contact) at 55°C.
3. Transfer the block with the two tubes to the cold room, and allow the temperature to decrease to 10–15°C over the next 90 min (check).
4. Add 1 µL T4 ligase to each tube, mix by pipetting, and incubate for 16 h at 12°C.

3.2.1.3. Amplification by PCR

To ensure equal conditions during preparation of representations, these PCRs must be performed in parallel in the same PCR machine using the same master mix for both sets of reactions.

1. Add 970 µL of TE supplemented with 200 µg/mL (20 µL of 10 µg/µL) tRNA to each of the two ligations, vortex, and store on ice.
2. Label 2× ten 0.5 mL PCR tubes, a set each for the induced and the uninduced sample, respectively. Prepare 8 mL of PCR buffer 1 and aliquot 360 µL into each of the 20 PCR tubes. Add 40 µL from each of the two diluted ligations to the ten corresponding tubes. Vortex each tube briefly before adding one drop of mineral oil to each tube.
3. Incubate tubes at 72°C for 5 min prior to cycling, to allow extension of adaptor ends. Cycle all 20 tubes together 20× in a two-temperature PCR cycle (1 min at 95°C and 3 min at 72°C). Finish the reaction by a final extension at 72°C for 10 min.
4. To check the outcome of the PCR, run 8 µL of the reaction from each tube + 1.6 µL 6 × GLB on a 2% agarose gel, using ϕ174 *Hae*III as the marker and 200-300-400 ng sheared salmon-sperm DNA to check yield (approx 300 ng per lane can be expected). All ten aliquots prepared from the induced sample should appear identical. Also, all ten aliquots from the uninduced sample should look identical among themselves. Often, a band pattern specific for each of the two samples is observed.

5. Withdraw the PCR reactions under the mineral oil and collect each of the two samples (in total 4 mL from the ten matching tubes) in a 15-mL Falcon tube. Extract both pooled samples twice with 3 mL phenol/chloroform. Add 400 µL 3 *M* NaAc and 8 mL iPrOH. Incubate for 15 min at 4°C (or overnight at –20°C), and spin at 15,000*g* 30 min at 4°C.
6. Wash pellets once with 5 mL 70% EtOH (kept at –20°C), recentrifuge for 5 min, remove supernatants, and dry pellets in a desiccator. Redissolve pellets in 500 µL TE. Check recovery by diluting 2 µL sample 1:10 with TE and running 2, 5, and 10 µL on a 2% agarose gel with 100-250-500 ng salmon-sperm DNA as standards.
7. Estimate the amount of DNA on the gel and use it to calculate the total amount of representation recovered. Approximately 150 µg for each of the two representations prepared can be expected. This corresponds to approx a 400-fold amplification, since the representations were made from 0.4 µg template each (*see* **Note 5**).

3.2.1.4. Removal of the Adaptors by Digestion

Before the representations can be used in a cDNA-RDA reaction, the adaptors at their ends must be cleaved by restriction enzyme digestion.

1. Use 100 µg of each of the two representations prepared in **Subheading 3.2.1.3.** for a digest to remove adaptors. In a 2-mL Eppendorf tube, add 100 µg representation to a reaction mix consisting of 100 µL 10 × *Dpn*II buffer, 100 µL 10 U/µL *Dpn*II, and water to 1 mL final vol. Incubate overnight at 37°C, preferentially in an incubator.
2. Extract twice with 1 mL phenol/chloroform. Add 100 µL 3 *M* NaAc and 1 mL iPrOH. Incubate for 15 min at 4°C (or overnight at –20°C), and spin the two tubes at 15,000*g* at 4°C for 15 min. Wash pellets once with 1 mL 70% EtOH (kept at –20°C), and carefully dry pellets in a speedvac. Redissolve carefully in 100 µL TE by pipetting and vortexing.
3. Check recovery and completeness of digestion by preparing 10 µL of a 1:10 dilution and run 1-, 2.5-, and 5-µL samples on a 2% agarose gel. Include 200 ng samples of the undigested representations (*see* **Subheading 3.2.1.3.**). The band pattern before and after digestion should be identical, except that after digestion it will be shifted slightly downward because of the removal of 48 bp of adaptor sequence. Use 100-250-500 ng salmon-sperm DNA to check yield. Adjust concentrations to 500 ng/µL with TE, pipet, and revortex.

3.2.2. First Round of cDNA-RDA

Since cDNA-RDA identifies only samples upregulated in one sample relative to the other, we perform two cDNA-RDA experiments in parallel (*see* **Fig. 1**). To identify downregulated genes, use the representation derived from induced cells as the driver and the tester derived from uninduced cells. To identify upregulated genes, use the representation derived from uninduced cells as the driver, and the tester derived from induced cells. In this experiment, it is advisable to add the induced gene of interest back to the driver, since this sequence may otherwise constitute a major part of the difference product (*see* **Note 6**).

3.2.2.1. Ligation of New Adaptors to Tester

1. Set up in parallel two ligations to obtain testers from each of the two samples. Each ligation contains 2 µL of 500 ng/µL digested representation (*see* **Subheading 3.2.1.4.**), 7.5 µL 62 µ*M* NBgl24, 7.5 µL 62 µ*M* NBgl12, 3 µL 10 × T4 buffer, and 10 µL H_2O.
2. Mix well and transfer to a heatblock (holes filled with glycerol for optimal thermal contact) at 55°C. Transfer the block with the tubes to the cold room and allow the temperature to fall to 10–15°C during the next 90 min.
3. Add 1 µL T4 ligase, mix by pipetting, and incubate for 16 h at 12°C.
4. Add 70 µL TE supplemented with 20 µg/mL tRNA, and store on ice.

3.2.2.2. Hybridization

1. Adjust one thermoblock with glycerol- or mineral oil-filled holes to 67°C and another one to 98°C.
2. For each of the two reactions, add 80 µL of 500 ng/µL (40 µg) digested representation, corresponding to the driver (*see* **Subheading 3.2.1.4.**), to 40 µL of 10 ng/µL (0.4 µg) representation ligated to new adaptors (*see* **Subheading 3.2.2.1.**; the tester). Vortex the Eppendorf tubes and extract once with 120 µL phenol/chloroform. Add 30 µL 10 *M* NH_4Ac, vortex, and add 380 µL EtOH (kept at –20°C). Incubate at –70°C for 10 min, warm the samples for 2 min at 37°C, and spin at 4°C at 15,000*g* for 20 min. Wash twice with 500 µL 70% EtOH, and dry carefully in a speedvac (up to 2 min, without heating).
3. Resuspend each pellet in 5 µL EE (*see* **Subheding 2.3., item 9**) by pipetting and vortexing 4× for 30 s each. Spin down very briefly and carefully add 35 µL mineral oil. Denature DNA at 98°C for 4 min, add 1.5 µL 5 *M* NaCl to the aqueous phase by pricking through the mineral oil with the pipet tip, and incubate at 67°C for 20 h.

3.2.2.3. First PCR

1. Remove mineral oil from the hybridizations (*see* **Subheading 3.2.2.2.**). Add tRNA (8 µL 5 µg/µL) to the hybridization mix. Mix by pipetting, add 390 µL TE, and vortex.
2. Prepare 1.5 mL of PCR buffer 2, distribute 352 µL each into four PCR tubes. Add 40 µL (4 µg total DNA) from the one hybridization to two of the four tubes and 40 µL from the other hybridization to the remaining two tubes, and place them in a PCR machine kept at 72°C.
3. Add 3 µL 5 U/µL Taq polymerase to each of the four tubes, vortex, and incubate for 5 min at 72°C. During this time, the 3′ ends of the adaptor sequences will be filled in. Add 10 µL of 62 µ*M* primer NBgl24 to each tube, vortex, add mineral

oil, and cycle 10 times in a two-temperature PCR (keep at 95°C for 1 min and at 72°C for 3 min).

3.2.2.4. MBN Digest and Second PCR

The PCR step described in **Subheading 3.2.2.3.** selectively amplifies the sequences enriched in the tester. To improve amplification of tester-specific sequences during the second PCR amplification, the single-stranded driver DNA is digested by MBN treatment.

1. Place a water bath at 30°C and a heatblock at 98°C.
2. Pool the identical samples obtained after the first PCR (*see* **Subheading 3.2.2.3.**). Extract once with 600 µL phenol/chloroform and once with 600 µL chloroform. In a 2-mL Eppendorf tube, add 80 µL 3 *M* NaAc and 1 mL isopropanol (iPrOH).
3. Incubate for 1 h at –20°C, spin 15 min at 15,000*g* at 4°C. Wash pellets once with 500 µL 70% EtOH (kept at –20°C), recentrifuge for 5 min, and remove supernatants.
4. Dry pellets in a speedvac and redissolve carefully in 40 µL EB by pipetting and vortexing. Prepare 50 µL 2 × MBN buffer.
5. Incubate 20 µL from each of the two samples in parallel with 20 µL 2 × MBN buffer and 2 µL 10 U/µL MBN enzyme for 30 min at 30°C.
6. Neutralize with 160 µL 50 m*M* Tris, pH 8.9 and inactivate the MBN for 5 min at 98°C, then store on ice.
7. Prepare 1.5 mL of PCR buffer 3, distribute 360 µL into four PCR tubes. Add 40 µL from each sample from **step 6** to two PCR tubes. Do not combine the samples at this step. Vortex, add mineral oil, and perform 20 PCR cycles of 1 min at 95°C and 3 min at 72°C.
8. To evaluate the outcome of the PCR, run 10 µL of the reaction from each tube + 2 µL 6 × GLB on a 2% agarose gel and use 100–200–300 ng sheared salmon-sperm DNA to check yield (a yield of approx 200 ng/10 µL can be expected). In the case of a low yield, supplement each tube with 3 µL of fresh Taq polymerase and perform three additional PCR cycles.
9. Pool the identical samples and extract them once with 600 µL phenol/chloroform and once with 600 µL chloroform. In a 2-mL Eppendorf tube, add 80 µL 3 *M* NaAc and 1 mL iPrOH. Incubate overnight at –20°C.
10. Centrifuge for 15 min at 15,000*g* at 4°C. Wash pellet once with 500 µL 70% EtOH (kept at –20°C), and dry the pellets in a speedvac.
11. Redissolve the pellets carefully in 100 µL TE by pipetting and vortexing. Prepare 20 µL of a 1:5 dilution and run 2.5, 5, and 10 µL on a 2% agarose gel. Include lanes with 100–200–300–400 ng salmon-sperm DNA to check yield (approx 20 µg/each sample is to be expected). Adjust the concentration to 100 ng/µL and store the difference product 1 (DP1) at –20°C.

3.2.3. Second Round of cDNA-RDA

The second round of cDNA-RDA is basically a repetition of the first round, incorporating the following changes: The product of the first round of cDNA-RDA (DP1) is used as the tester, and the driver is still the digested representation. The driver to tester ratio is increased from 100:1 to 800:1. Also, a new set of primers is used.

1. Remove the NBlg24 adaptors from 5 µg of DP1 with *Dpn*II following the protocol provided in **Subheading 3.2.1.4.** Resuspend after precipitation in 50 µL TE. Check the yield by agarose gel electrophoresis and adjust the concentration to 20 ng/µL with TE.
2. Ligate new adaptors to 100 ng (5 µL) of digested DP1, using the same conditions and concentrations of reagents as described in **Subheading 3.2.2.1.**, but using IBgl24 and IBgl12 instead of NBgl24 and NBgl12. Incubate overnight, add 50 µL TE supplemented with 20 µg/mL tRNA, and store on ice.
3. Mix 40 µL (50 ng) of tester just ligated to new IBgl adaptors and 80 µL of 500 ng/µL (40 µg) driver (prepared in **Subheading 3.2.1.4.**). Hybridize, following the instructions given in **Subheading 3.2.2.2.**
4. Perform first PCR, MBN digest, and second PCR as outlined in **Subheading 3.2.2.3.** and **Subheading 3.2.2.4.**, substituting primer IBgl24 for NBgl24. Adjust the concentration to 100 ng/µL and store the difference product 2 (DP2) at –20°C (*see* **Note 7**).

3.3. Analysis of the Obtained Difference Products

3.3.1. Southern Blot Analysis

After completion of the cDNA-RDA experiment (which will take approx 2–3 wk), the difference products (DP2) are analyzed. A first standard control is performed by Southern blot analysis, using the induced gene (e.g., *Rac*) as a positive control to probe equal amounts of each driver and tester representation, DP1, and DP2 transferred to a nylon membrane. If looking for genes downregulated by *Rac*, the hybridization signal from the Rac probe should be stronger in the driver than in the tester and absent from both DP1 and DP2. If looking for genes upregulated by Rac, the hybridization signal should be stronger in the tester than in the driver and stronger in both DP1 and DP2 are compared to tester. This signal is likely to be detected, although to a lesser extent, even when the induced gene was added back to the driver (*see* **Note 6**).

If the blot indicates that the cDNA-RDA experiment was successful, the individual clones contained in the DP2 are analyzed. The DP2 library is subcloned, individual inserts are amplified and revalidated by microarraying in order to sequence only the most promising candidate genes.

3.3.2. Subcloning

1. Digest 1 µg DP2 (*see* **Subheading 3.2.3.**) with *Dpn*II as described in **Subheading 3.2.1.4.**
2. After redissolving in 20 µL EB, add 4 µL 6 × GLB and run at low voltage on a 2% agarose gel approx 4 cm distance from the wells. With a fresh scalpel, cut out the entire lane from 800–100 bp, leaving the cleaved adaptors behind.
3. Use the Qiaquick gel extraction kit to recover the DNA and ligate the adaptor-free DP2 to a bacterial expression vector (e.g., pGEM7Zf[–]) digested with *Bam*HI and dephosphorylated with CIP as described in **Subheadings 3.2.1.2.** and **3.2.2.1.** Transform into highly competent *E. coli* and plate the library.

3.3.3. Single-Colony PCR

Individual clones are directly amplified from the bacterial colonies by using the bacteria as the template in a PCR reaction with primers directed against the sequences surrounding the multiple cloning site of the vector. We suggest processing four 96-well plates from each DP2 library.

1. Prepare 2.75 mL of PCR buffer 4 (*see* **Note 8**). Pipet 325 µL into each well of the first column of a 96-well PCR plate. With an 8-channel pipetter, distribute 25 µL into each of the wells across the plate.
2. Pick a single white colony from the plate using a sterile pipet tip, and dip the tip into a well of the PCR plate. Repeat 94 times, inoculating each well with an individual white colony picked at random and using fresh tips. Include one well as a negative control without any template.
3. Amplify by PCR using a three-temperature program (5 min at 94°C for initial denaturation; 25 cycles of 1 min each at 94°C, 60°C, and 72°C; additional 10 min at 72°C for final elongation and hold at 4°C). Run 5 µL sample, 5 µL H_2O, and 2 µL 6 × GLB on a 2% agarose gel with ϕ174 *Hae*III as the standard, and photograph. Store remainder of the PCR reactions at –20°C.

3.3.4. Exclusion of cDNA Clones Which Are Identical to the Induced Gene of Interest

It is to be expected that the induced gene (e.g., Rac) will be found as a difference product if the tester is derived from induced cells, even when the induced gene was added back to the driver (*see* **Note 6**). To prevent sequencing these clones, we identified them by dot-blotting experiments.

1. Prepare a 96-well plate with 10 µL TE in each well. Add 1 µL of each of the PCR reactions from **Subheading 3.3.3.** into the corresponding wells.
2. Following the manufacturer's instructions, place a Hybond-N+ nylon membrane in the Minifold I dot-blotting apparatus and apply vacuum. Load your DNA samples in the corresponding wells. Fix the DNA to the membrane, e.g., by UV-crosslinking.

3. Prepare 50 ng of radiolabeled double-stranded DNA probe from your induced gene (e.g., full-length *Rac1*) and hybridize to the dot-blot filter.

3.3.5. Re-Amplification for Microarraying

The amplified clones must be re-amplified in order to obtain sufficient amounts for microarraying.

1. Prepare 10 mL of PCR buffer 5. Add 100 µL each to the wells of a 96-well PCR plate. With an 8-channel pipetter, add 1 µL each from the PCR reactions (*see* **Subheading 3.3.3.**) into the corresponding well. Amplify by PCR using the following program: 1 min at 94°C for initial denaturation; 25 cycles of 1 min each at 94°C, 60°C, and 72°C; additional 10 min at 72°C for final elongation; hold at 4°C).
2. Run 5 µL of each of the PCR products and 1 µL 6 × GLB on a 2% agarose gel with ϕ174 *Hae*III as the standard, and photograph. Transfer the samples to a Corning 96-well plate, supplement with 10 µL 3 *M* NaAc and 200 µL EtOH, and keep for 16 h at –20°C.
3. Centrifuge for 20 min at + 4°C and 2800*g*. Aspirate supernatants, wash with 100 µL 70% EtOH, spin for 5 min, and aspirate supernatants. Dry the pellets in a vacuum oven at maximal 50°C, resuspend in 15 µL of 3 × SSC per well, and store at –20°C.

3.3.6. Microarraying

In our microarraying experiments, the clones derived from a cDNA-RDA experiment are arrayed. Representations from each driver and tester are labeled with a fluorophore, mixed, and then hybridized simultaneously to the arrayed candidate genes.

3.3.6.1. Array preparation

1. Array the re-amplified inserts from **Subheading 3.3.5.** on commercially prepared silanated slides, using a Cartesian PixSys 5500 or equivalent.
2. Place the array in a humidified chamber for 3–5 min to hydrate the spots. Crosslink the slide by ultraviolet (UV) irradiation with 60 mJ in a UV crosslinker. Rehydrate the slide in the humidified chamber and snap dry by heating on the surface of a hot plate for several seconds. Wash the chip in 0.1% SDS for approx 10 s in MilliQ water for approx 10 s and then denature the chip in boiling MilliQ water for approx 1–2 min. After denaturation, immediately immerse the array in ice-cold benzene-free ethanol for several seconds. Remove and allow the chip to dry. Also perform the wash procedure from the SDS to the ice-cold ethanol with the cover slips to be used with the arrays.

3.3.6.2. Sample Preparation

1. In parallel, denature 10 µg each of the representations derived from induced and uninduced cells in the presence of 5 µg of a random nonamer primer in a total

of 100 µL H_2O. Use either representations ligated to adaptors (*see* **Subheading 3.2.1.3.**) or with the adaptors cleaved off (*see* **Subheading 3.2.1.4.**).

2. To each sample, add 12 µL of 10 × Klenow buffer and supplement with 33 µ*M* dNTPs, 10 nmol of either Cy3 or Cy5, and 4 U of Klenow fragment. Incubate the reactions at 37°C for 2 h.
3. Combine the two reactions and remove unincorporated nucleotides by centrifugation through a Microcon YM-30 ultracentrifugation column according to the manufacturer's instructions. Wash 3× with 400 λ of EB. Adjust the eluate containing the labeled sample to a volume of 15 µL and a concentration of 3 × SSC and 0.2% SDS, and denature by heating to 95°C for 5 min.
4. Carefully place the 15 µl of labeled representations on the array prepared in **Subheading 3.3.6.1.** and slowly place a cover slip on the array. Insert into a hybridization chamber according to the manufacturer's instructions and incubate in the dark at 67°C overnight.
5. After the overnight hybridization, disassemble the hybridization chamber and wash the array at room temperature in 0.1% SDS, 0.2 × SSC for 90 s, followed by a 30-s wash in 0.2 × SSC and a final 30-s wash in 0.05 × SSC.
6. After washing, arrays are imaged in a scanner. Images are saved as TIF files and feature definition and quantitative analysis are performed with either ScanAlyze or Axon GenePix. This analysis yields a text file that is imported into Microsoft Excel or Access for further analysis.

3.3.7. Non-Radioactive Sequencing of PCR Products

Sequence all clones that are found to be differentially expressed based on microarraying (e.g., fluorescence ratio greater than 2). However, do not sequence the Rac clones already identified by dot-blotting in **Subheading 3.3.4.**

1. To obtain sufficient material for nonradioactive sequencing, repeat the PCR from **Subheading 3.3.5.** with only the selected clones as the templates. These PCR samples are checked on a gel and then purified using the Qiaquick PCR purification kit. Determine the concentration of the samples by OD_{260}, which should be at least 25 ng/µL.
2. Perform the sequencing reactions using 8 µL rhodamine dye terminator kit, 2 µL 1.6 µ*M* primer SP6R or T7E, 250 ng ds DNA PCR product, and H_2O to 20 µL. Submit the obtained sequences to BLAST searches.

3.4. Further Characterization of Candidate Genes

It is advisable to reconfirm the expression pattern for the genes which showed altered expression in the microarray experiments by an independent method, such as Northern blot analysis. In the case of less abundant genes, quantitative RT-PCR may be required.

Further analysis of interesting clones will be dependent on the nature of the gene. To further obtain information on the Rac-induced genes, we began to explore the effects of various growth factors and drugs on the transcription of these genes. We used reagents known to activate or to interfere with previously characterized signaling pathways of Rac. For example, the transcription levels of candidate genes can be compared between resting cells and cells stimulated with PDGF or EGF. Furthermore, Rac is known to activate the p38 MAP kinases, and the drug SB202190 is known to be a selective inhibitor of these kinases ***(18)***. The transcription of candidate genes can be compared between cells induced for Rac and either left untreated or treated with SB202190. These experiments are also performed using the microarray technique, allowing simultaneous monitoring of the effect of these reagents on the genes identified by cDNA-RDA.

4. Notes

1. A major challenge in cDNA-RDA experiments is the high risk of contamination, since cDNA-RDA is able to amplify very small differences between two representations. Thus, while performing cDNA-RDA, the use of gloves, sterile filter tips, single-use individually wrapped pipets, and sterile single-use plasticware is highly recommended. Be aware that not all types of plastic withstand the phenol/chloroform solutions used. Therefore, use polypropylene intead of polystyrene tubes. Also, all reagents are used only for cDNA-RDA.
2. To prevent losing sequences later during preparation of the representations (*see* **Subheading 3.2.1.**), one can design specific primers for the preparation of cDNAs. Short or unusually composed mRNA sequences may contain either none or only one *Dpn*II site. To avoid their loss, use primers with a *Dpn*II site added at the 5′ end instead of standard random or oligodT primers. Using large amounts of cells derived from an inducible expression system may not be possible in all cases. When starting with smaller amounts of cells (10 ^ 6), the Oligotex kit (Qiagen) is recommended for mRNA isolation. If working with even smaller amounts of defined starting material or tissue samples, it may be necessary to use specialized methods during the preparation of representations, as discussed elsewhere ***(12,6,19)***. When the samples to be used as tester or driver need to be pooled (e.g., progenitor cells of healthy individuals), it is advisable to avoid pooling the cells or the mRNAs, but to prepare in parallel a representation from each individually and then to combine them.
3. Incubate in an incubator rather than in a water bath or a thermoblock. An incubator is preferred to avoid evaporation of the solution and recondensation on the inner surface of the lid.
4. Be careful not to lose minute amounts of dried DNA in handling. This is a major source of low yields in the following PCR amplification steps. It is also important not to over-dry the DNA pellets, because they become more difficult to redissolve.
5. It is essential to reproducibly obtain representations before starting cDNA-RDA. If a typical yield of approx 150 μg from 0.4 μg template is not obtained, this step must be optimized (such as by preparing new buffers, new cDNAs and/or mRNAs).

6. Depending on the level of induction, the induced gene itself may become a major part of the difference product, when the induced sample is employed as the tester to identify upregulated genes. This can be suppressed by adding this particular transcript back to the driver during both rounds of cDNA-RDA. Although others have recommended to substitution of up to 50% of the driver by the gene *(6)*, we would advise more caution in substituting approx 10% of the driver. An optimal ratio can also be determined experimentally by titration. The Southern blot analysis described in **Subheading 3.3.1.** may not be quantitative enough to detect whether the addition had an effect on the composition of the DP2. However, the difference should be notable in the percentage of individual clones identified by dot-blotting (*see* **Subheading 3.3.4.**) as being the induced gene.
7. There are only a few ways to judge the quality of the cDNA-RDA experiment during the procedure. It is essential, however, that the yields and amplification rates described throughout the protocol are achieved. Also, digests should result in downward shifts of band patterns in agarose gels (*see* **Subheading 3.2.1.4., step 3**). Finally, equal aliquots of driver and tester representations, DP1, and DP2 can be run together on a gel (to perform the Southern blot analysis described in **Subheading 3.3.1.**). The observation of appearing or disappearing bands from representation to DP1 and DP2 may indicate that the cDNA-RDA has worked. However, a constant band pattern does not necessarily indicate that the cDNA-RDA has failed. The best method is to proceed with the Southern blot analysis and use its outcome as described in **Subheading 3.3.1.** to judge the quality of the cDNA-RDA experiment.
8. We found that inclusion of 12% DMSO in this PCR buffer significantly increased the yield, and addition of Triton X-100 had no effect on the yield and reduced the clarity of the run on an agarose gel.

Acknowledgment

The authors would like to thank Joan Alexander for sequencing and Chet Cunha (Friends Academy H.S., Glen Cove, NY) for technical assistance.

This research was supported by grants from the U.S. Army and the NIH to Linda Van Aelst, who is also a Kimmel Foundation and Foundation V award recipient. Arndt A. P. Schmitz is supported by a post-doctoral fellowship from the Deutsche Forschungsgemeinschaft (DFG, Bonn, Germany).

References

1. Van Aelst, L. and D'Souza-Schorey, C. (1997) Rho GTPases and signaling networks. *Genes Dev.* **11,** 2295–2322.
2. Velculescu, V. E., Zhang, L., Vogelstein, B., and Kinzler, K. W. (1995) Serial analysis of gene expression. *Science* **270,** 484–487.
3. Liang, P. and Pardee, A. B. (1992) Differential display of eukaryotic messenger RNA by means of the polymerase chain reaction. *Science* **257,** 967–971.

4. Liang, P. and Pardee, A. B., eds. (1997) *Differential Display Methods and Protocols.* Methods Mol. Biol., Vol. 85, Humana Press, Totowa, NJ, pp. 1–320.
5. Hubank, M. and Schatz, D. G. (1994) Identifying differences in mRNA expression by representational difference analysis of cDNA. *Nucleic Acids Res.* **22,** 5640–5648.
6. Hubank, M. and Schatz, D. G. (1999) cDNA representational difference analysis: a sensitive and flexible method for identification of differentially expressed genes. *Methods Enzymol.* **303,** 325–349.
7. Diatchenko, L., Lau, Y. F., Campbell, A. P., Chenchik, A., Moqadam, F., Huang, B., et al. (1996) Suppression subtractive hybridization: a method for generating differentially regulated or tissue-specific cDNA probes and libraries. *Proc. Natl. Acad. Sci. USA* **93,** 6025–6030.
8. Zuber, J., Tchernitsa, O. I., Hinzmann, B., Schmitz, A. C., Grips, M., Hellriegel, M., et al. (2000) A genome-wide survey of RAS transformation targets. *Nat. Genet.* **24,** 144–152.
9. Harris, A. J., Shaddock, J. G., Manjanatha, M. G., Lisenbey, J. A., and Casciano, D. A. (1998) Identification of differentially expressed genes in aflatoxin B1-treated cultured primary rat hepatocytes and Fischer 344 rats. *Carcinogenesis* **19,** 1451–1458.
10. Phimister, B., ed. (1999) The Chipping Forecast. *Nat. Genet.* **21(Suppl.),** 1–60.
11. Lisitsyn, N., Lisitsyn, N., and Wigler, M. (1993) Cloning the differences between two complex genomes. *Science* **25,** 946–951.
12. Lisitsyn, N. and Wigler, M. (1995) Representational difference analysis in detection of genetic lesions in cancer. *Methods Enzymol.* **254,** 291–304.
13. Welford, S. M., Gregg, J., Chen, E., Garrison, D., Sorensen, P. H., Denny, C. T., and Nelson, S. F. (1998) Detection of differentially expressed genes in primary tumor tissues using representational differences analysis coupled to microarray hybridization. *Nucleic Acids Res.* **26,** 3059–3065.
14. Clackson, T. (1997) Controlling mammalian gene expression with small molecules. *Curr. Opin. Chem. Biol.* **1,** 210–218.
15. No, D., Yao, T. P., and Evans, R. M. (1996) Ecdysone-inducible gene expression in mammalian cells and transgenic mice. *Proc. Natl. Acad. Sci. USA* **93,** 3346–3351.
16. Gossen, M. and Bujard, H. (1992) Tight control of gene expression in mammalian cells by tetracycline-responsive promoters. *Proc. Natl. Acad. Sci. USA* **89,** 5547–5551.
17. Jou, T. S. and Nelson, W. J. (1998) Effects of regulated expression of mutant RhoA and Rac1 small GTPases on the development of epithelial (MDCK) cell polarity. *J. Cell Biol.* **142,** 85–100.
18. Lee, J. C., Laydon, J. T., McDonnell, P. C., Gallagher, T. F., Kumar, S., Green, D., et al. (1994) A protein kinase involved in the regulation of inflammatory cytokine biosynthesis. *Nature* **372,** 739–746.
19. Lockhardt, D. J. and Winzeler, E. A. (2000) Genomics, gene expression, and DNA arrays. *Nature* **405,** 827–836.

4

Fluorescence Methods in the Study of Small GTP-Binding Proteins

Mohammad Reza Ahmadian, Alfred Wittinghofer, and Christian Herrmann

1. Introduction

A great variety of small guanosine triphosphate (GTP)-binding proteins are known, and each of these in turn interacts with a variety of regulatory proteins, including guanine-nucleotide dissociation inhibitors (GDIs), GTPase-activating proteins (GAPs) and guaninine nucleotide exchange factors (GEFs). Most importantly, the small GTPases bind to target "effector" proteins, and many of these have yet to be identified. This process generates a network of protein-nucleotide and protein-protein interactions, which underlies the biology of the corresponding transport or signal transduction process. Many of the interactions defined by two-hybrid analysis, blot overlay or glutathione-S-transferase (GST)-pulldown experiments require a more detailed analysis of their kinetics and thermodynamic properties in order to quantify these processes in vivo. It therefore becomes increasingly important to examine these interactions by appropriate real-time methods.

Fluorescence methods are ideally suited to fulfill the criteria here. This requires the presence of a fluorescence label, which can either be intrinsic (e.g., tryptophan residue) or can be introduced into one of the components. The introduced fluorescence label should not grossly disturb the biochemical properties of the proteins or ligands involved. The use of fluorescence also requires that the reporter group is sensitive enough to changes in the local environment to produce a detectable fluorescence change at reasonable protein concentrations.

From: *Methods in Molecular Biology, vol. 189: GTPase Protocols: The Ras Superfamily*
Edited by: E. J. Manser and T. Leung © Humana Press Inc., Totowa, NJ

Tryptophan residues are occasionally used for the study of GTP-binding proteins. An early example is the large fluorescence change observed when the nucleotide-bound state of the Gα subunit of heterotrimeric G proteins changes from the guanosine diphosphate (GDP) to the GTP form *(1)*. This fluorescence change was later found to derive from a single conserved tryptophan residue in the switch II region of Gα that becomes buried in the GTP-bound state *(2,3)*. Mutagenesis of this residue confirmed the structural prediction *(4)*. In *Ras*, a single tryptophan residue has been introduced into the otherwise tryptophan-free protein, allowing sensing of its GTP/GDP conformational change *(5–7)*.

More widely applicable is the introduction of fluorescently labeled methyl-anthraniloyl (mant or m) guanine nucleotides into GTP-binding proteins. The fluorescent reporter group is located on either the 2′ or 3′ oxygens of the ribose of GDP and GTP-analogs (**Fig. 1**), but can also be specifically introduced into one of these oxygens using 2′ or 3′ desoxy derivatives of the guanine nucleotides *(5,8,9)*. The fluorescent reporter group does not disturb the binding of nucleotides, as evidenced from the similar affinities of the labeled vs the nonlabeled nucleotide, yet it is very sensitive to the chemical environment. In all cases, the fluorescent reporter group experiences a large increase in fluorescence intensity on binding to *Ras*-like GTP-binding proteins (*see* **Note 1**). Furthermore, it is often rather sensitive to the interaction with partner proteins that happen to bind in the neighborhood. In some cases, the properties of tryptophan and mant groups can be combined: a tryptophan residue is excited and acts as a donor in a fluorescence resonance energy transfer (FRET) experiment, whereby the excited tryptohan induces fluorescence of the mant group *(10,11)*.

2. Materials

2.1. Preparation of mant Nucleotide Solutions

1. The fluorescent N-methylanthraniloyl (mant) derivatives of nucleotides are prepared according to a protocol established by John et al. *(9)*, which is a modification of the original procedure developed by Hiratsuka *(12)*. The mant group is attached to either the 2′ or 3′ hydroxyl of the ribose moiety. Guanine nucleotides to be labeled, such as GDP, GTP, or the nonhydrolyzable GTP analog GppNHp (guanosine (5′-β,γ-imidotriphosphate), are purchased from Sigma. In cases in which isomerization between 2′- and 3′-derivatives should be avoided because it interferes with the reaction to be measured (as in interactions with GEFs or GAPs), 2′-desoxy nucleotides can be used alternatively *(13,14)*.
2. The nucleotide is dissolved in water at 50–100 m*M* and adjusted to pH 9.6 by the addition of 1 *N* NaOH. In five steps of 15 min intervals each, methylisatoic anhydride (powdered) is added in 1.5-*M* excess overall.
3. The solution is stirred and incubated at 38°C, and the pH is controlled and held at the value of 9.6 throughout the reaction by addition of NaOH. The progress of the reaction can be monitored by subjecting small aliquots to high-pressure

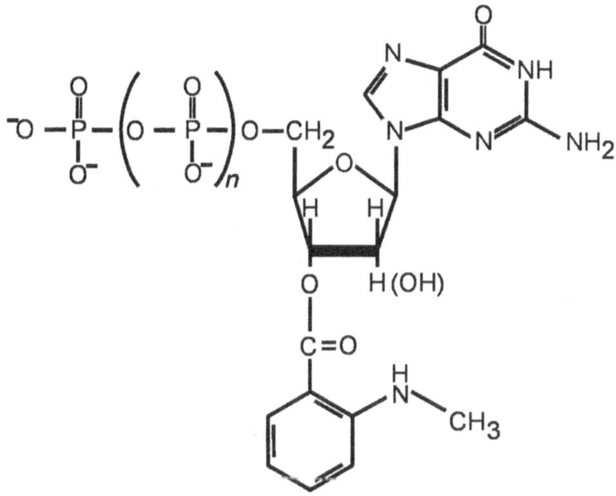

Fig. 1. Structure of mGDP (n = 1) and mGTP (n = 2) with the mant group attached to the 3′ position of the nucleotide. Note that position 2′ can be the desoxy form in order to avoid isomerization of the mant group in 2′ and 3′ position.

liquid chromatography (HPLC) analysis, and after 2 h a yield of 20–30% should be obtained.

4. The solution is centrifuged and extracted twice with 1 vol chloroform each. The pH of the aqueous phase is adjusted to pH 7.5 with acetic acid, and the solution is filtered (0.45 µ*M* pore size).
5. A Q-Sepharose (fast-flow) column is equilibrated with a TBC (triethylamin bicarbonate) buffer which has been produced by gassing an ice-cooled 2 *M* solution of triethylamin with carbon dioxide (evolving from dry ice) overnight (a pH value of 7.5 should be reached). The raw mant nucleotide solution loaded onto this column is fractionated by running a 200–800 m*M* triethylamin bicarbonate (TBC) gradient.
6. Again, samples are analyzed by HPLC showing mant-ethyl ester and unreacted nonmodified nucleotide to elute first, followed by the mant nucleotide the ultraviolet (UV)/VIS spectra displayed in Lenzen et al. ***(15)***. The mant-ethyl ester exhibits an absorption maximum at 330 nm, whereas the corresponding maximum of the mant nucleotide is observed at 350 nm. In order to calculate the concentration of a mant guanine-nucleotide absorption coefficients of 22,600 dm^3/(mol cm) at 252 nm or 5,700 dm^3/(mol cm) at 350 nm are used ***(12)***. The ratio of the absorption values at these wavelengths may also serve as criterion of purity.
7. Pooled fractions heated to 30°C are evaporated in vacuum. The residue is dissolved in methanol, which is evaporated (3–4×). The mant nucleotide is dissolved in water at 20 m*M* and stored at –20°C.

8. Free or protein-bound mant nucleotides and reaction products are analyzed by reverse-phase HPLC using a C-18 column (ODS-Hypersil, 5 μm, Bischoff, Leonberg, Germany) and a prefilter (Nucleosil 100 C18, Bischoff) when applying protein-containing samples. The elution buffer (EB) consists of 100 m*M* potassium phosphate at pH 6.5, 10 m*M* tetrabutylammonium bromide, and 20% acetonitrile. An adequate resolution with nonlabeled nucleotides alone is achieved with 7.5% acetonitrile.

2.2. Preparation of Human Guanylate-Binding Protein 1 (GBP1)

1. This member of the family of large GTP-binding proteins ***(16)***, 67 kDa in size, is synthesized in *Escherichia coli* (*E. coli*) as recombinant protein. The plasmid used is pQE9 (Qiagen, Germany), which encodes a His_6-tag at the amino-terminal end of hGBP1. Therefore, this protein can be purified from the cleared bacterial lysate by means of a Ni-NTA column (Qiagen) and subsequent gel-filtration chromatography.
2. After purification, hGBP1 is nucleotide-free, and nucleotide does not need to be added for stability. The protein can be concentrated up to 1 m*M* by centrifugal concentrators (Vivascience, USA), and can be stored at –80°C after snap-freezing in liquid nitrogen ***(17)***.

2.3. Preparation of Nucleotide-Bound or Free Ras/Exchange of Protein-Bound Nucleotide

1. Recombinant *Ras* is prepared from *E. coli* using the ptac-expression system on a DEAE-Sepharose 6B (anion-exchange chromatography) and subsequent gel-filtration columns ***(18)***. *Ras* protein in GDP-bound form is snap-frozen in liquid nitrogen and stored at –80°C in 64 m*M* Tris pH 7.5, 5 m*M* $MgCl_2$, 5 m*M* dithio-erythritol (DTE).
2. Ras•GTP is prepared by incubating Ras•GDP in the presence of 10 m*M* EDTA and 100-fold molar excess of GTP for 1 h on ice.
3. The separation of unbound nucleotides and EDTA from Ras•GTP is carried out using PD-10 gel-filtration column (Pharmacia, Uppsala, Sweden).
4. Ras•GppNHp is prepared using enzymatic activity of alkaline phosphatase (Boehringer, Mannheim, Germany) to degrade the bound GDP. Add 5 U of the enzyme overnight at 4°C with 1 mg Ras•GDP in the presence of 2-fold excess of GppNHp (a nonhydrolyzable GTP-analog which is resistant to alkaline phosphatase) and 200 m*M* ammonium sulfate and 1 μ*M* $ZnCl_2$. The GDP degradation can be followed by HPLC analysis of small aliquots.
5. After complete GDP degradation, Ras•GppNHp is run over a NAP-5 gel-filtration column (Pharmacia, Uppsala, Sweden).
6. Production of nucleotide-free *Ras* is carried out according to ***(9)*** as described for the preparation of Ras•GppNHp with Gpp(CH_2)p used instead of GppNHp. After GDP is degraded, 1 U snake venom phosphodiesterase (Boehringer, Mannheim, Germany) per mg Ras protein is added to cleave Gpp(CH_2)p between the α- and β-phosphate.

7. In the presence of GMP or guanosine (the products of enzymatic reactions), Ras is quite stable, and can be stored at –80°C for several months.

2.4. Preparation of the Catalytic Domain of Neurofibromin

1. The catalytic domain of NF1, NF1-333, is isolated using the IPTG-inducible pLMM-expression system from *E. coli* ***(19)***. LMM (Light MeroMyosin) is insoluble at low-salt condition (below 100 m*M* KCl), and can be solubilized at high-salt condition (500 m*M* KCl). This property was used to prepare the fusion protein LMM-NF1-333 protein in one step at high purity by dialysis for 2 h against 10 m*M* KP_i pH 6.5, 5 m*M* DTE.
2. The precipitated fusion protein was washed once more in low-salt buffer and dissolved in the same buffer, which included 500 m*M* potassium chloride (KCl). The fusion protein was cleaved off from LMM using IgA-protease digestions at 30°C for 3 h at a substrate:protease ratio of 100:1 (w/w).
3. LMM is separated from NF1-333 just by dialysis of the protein solution against low-salt buffer and subsequent centrifugation to remove the precipitated LMM.
4. The supernatant is concentrated using the Centricon system (Milipore) and NF1-333 protein is applied to a Superdex 75 gel filtration using 30 m*M* Tris pH 7.5, 5 m*M* $MgCl_2$, 3 m*M* DTE. The eluted NF1-333 (>97% purity) fractions are concentrated to 1 m*M* (38 mg/mL) and stored at –80°C by snap-freezing in liquid nitrogen.

2.5. Preparation of Ras-Binding Domain of Raf

1. The cDNA encoding the Ras-binding domain (RBD) of Raf is expressed from the pGEX plasmid. The resulting GST fusion protein is expressed in *E. coli* strain BL21 and purified from the cleared bacterial lysates by glutathione Sepharose chromatography (Pharmacia). Raf is cloned in the pGEX vector.
2. Cleavage of the GST from the RBD part is performed on the column by incubation with 10 U/mL thrombin (Serva) overnight at 4°C.
3. The eluted RBD is concentrated by centrifugal concentrators (Vivaspin 10 kDa cut off, Viva Science) to 20 mg/mL and applied to size-exclusion chromatography (Superdex 75, Pharmacia).
4. The collected fractions are checked for purity by sodium dodecyl sulfate (SDS) polyacrylamide gel electrophoresis and Coomassie blue staining, pooled, and concentrated to 20 mg/mL. Aliquots are snap-frozen in liquid nitrogen and stored at –80°C.

2.6. Preparation of p190DH/PH-Domain and RhoA

1. Recombinant proteins, i.e., DH/PH domains of p190RhoGEF and C-terminal truncated RhoA (amino-acid residues 1-181) were produced as GST-fusion proteins in *E. coli* strain BL21 (DE3) and purified by glutathione-Sepharose affinity chromatography and gel filtration (Pharmacia, Uppsala, Sweden) as described for the preparation of RasRBDs (*see* **Subheading 2.5.**).

2. Nucleotide-free RhoA is prepared as described above for Ras (*see* **Subheading 2.3.**).
3. mGDP-bound RhoA was prepared by mixing nucleotide-free RhoA and mGDP in a molar ratio of 1 : 1.5 and by using prepacked gel filtration column (NAP5, Pharmacia) to remove unbound nucleotides.
4. The concentration of RhoA•mGDP was determined by HPLC (20% acetonitrile). The prepared proteins were snap-frozen in liquid nitrogen and stored at –80°C.

3. Methods

3.1. Assaying Nucleotide Binding, Equilibrium, and Kinetics

3.1.1. Behavior of Guanine Nucleotide-Binding Proteins

Before studying the interaction of GTP-binding proteins with regulatory proteins or effectors it is essential to understand the interaction with the nucleotide. The use of mant nucleotides is exceedingly helpful in this instance, as a large increase in fluorescence intensity upon binding to a protein is consistently observed. Obviously, the environment of a protein-bound mant nucleotide is less polar compared to bulk aqueous solution, and/or the fluorescence experiences less quenching when bound to the protein. For quantifying nucleotide binding, a fluorescence titration experiment seems to suggest itself. Yet, this method is not amenable to most of the GTP-binding proteins as Gα subunits, EF-Tu, or Ras-like proteins, because their nucleotide affinity is too high (i.e., their dissociation constant is too small). The concentration of the mant nucleotide must not be too far above the dissociation constant in order to obtain reasonable accuracy. However, the concentration of the mant nucleotide cannot be taken lower than 10 n*M* because of limited fluorescence detection sensitivity. Therefore, dissociation constants in the subnanomolar range are out of reach of fluorescence titration, and are also difficult to measure with radioactive methods (*see* **ref. *20***). In these cases, affinities for nucleotides are most conveniently determined by measuring the association- and dissociation-rate constants as described for Ras ***(9)*** and Rho ***(21)***.

3.1.2. Deriving Binding Constants from Direct Titration Experiments

In cases in which the affinities are lower, direct titration experiments to determine affinities are convenient for a member of the family of large GTP-binding proteins such as dynamin, Mx, and hGBP1 which have nucleotide dissociation constants in the micromolar range ***(17,22)***. **Figure 2** shows a typical fluorescence titration of mGDP and hGBP1 which includes the fit to the data according to **equation 1** and yielding a dissociation constant of 1.0 μ*M*. Note that an increase of more than threefold in fluorescence intensity occurs.

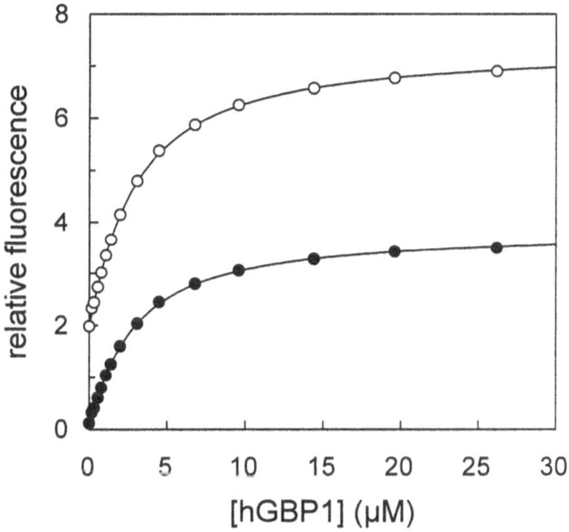

Fig. 2. Titration of hGBP1 with mGDP. From a stock solution of hGBP1 (500 µ*M*) small aliquots amounts are titrated to a solution of 0.5 µ*M* mGDP, producing the concentrations indicated. Signals were monitored either by direct mant fluorescence (•) or through FRET. The curves are calculated according to **Eq. 1**, yielding a K_d value of 1.0 µ*M*.

$$Y = A - (A - B) * \frac{([P1]+[P2]+K_d) - \sqrt{([P1]+[P2]+K_d)^2 - 4 * [P1] * [P2]}}{2 * [P1]} \quad (1)$$

Y corresponds to the observed fluorescence value. A and B represent the minimum and maximum fluorescence. P1 is the binding partner held at constant concentration (here mGDP). P2 is the other partner titrated to the solution (here the protein).

3.1.3. Deriving Binding Constants from Kinetic Data

The equilibrium dissociation constants (K_d) of high-affinity interactions can be obtained by the kinetic approach from the ratio of the dissociation- and association-rate constants. Taking advantage of the large change in fluorescence intensity the time-course of association of the nucleotide-free protein and the mant nucleotide can be followed after rapidly mixing the two components in a stopped-flow apparatus as shown in **Fig. 3A**. The fluorescence trace of increasing intensity corresponds to the association of hGBP1 and mGDP. Using a constant concentration of mGDP and increasing concentrations of the proteins yields a set of primary data. The linear regression (**Fig. 3B**) of these

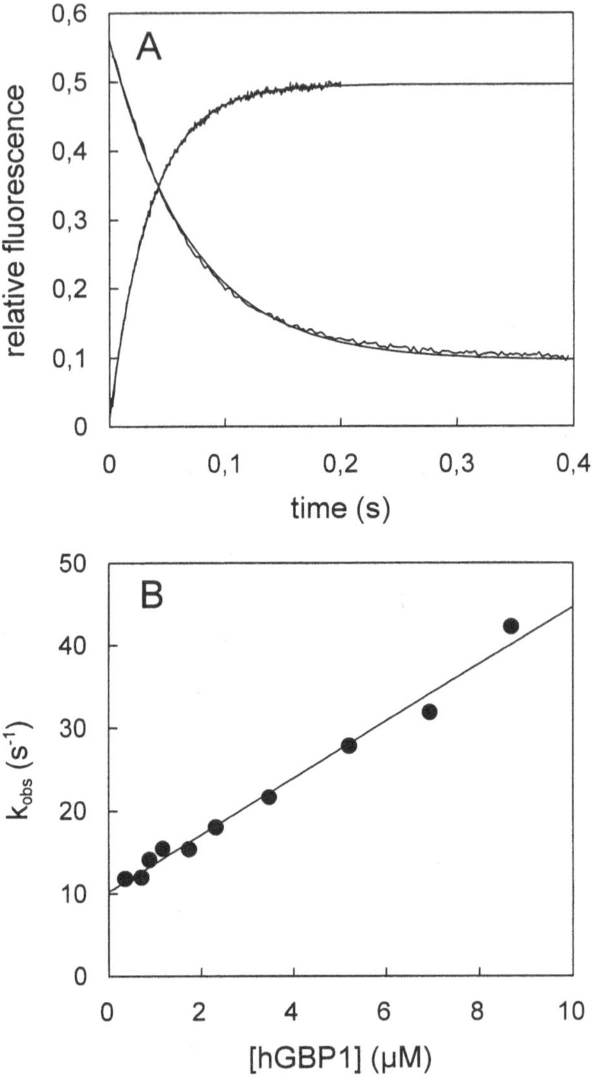

Fig. 3. Kinetics of the interaction of mGDP with hGBP1. (**A**) mGDP (0.2 µ*M*) is mixed with an excess of hGBP1 (10 µ*M*) in a stopped-flow apparatus yielding an exponential fluorescence increase. The decreasing curve corresponds to a displacement experiment in which a solution of 0.2 µ*M* mGDP and 5 µ*M* hGBP1 (preformed complex) is mixed with 200 µ*M* GDP. (**B**) The observed rate constants obtained from the exponential fits to the association data in A are plotted vs the concentration of hGBP1. The slope of the fitted straight line corresponds to k_{on} and the intercept equals k_{off}. The latter is also obtained from the exponential fit to the data of the displacement experiment in A.

observed rate constants plotted vs the concentration of hGBP1 which was in molar excess results in k_{on} (slope) and k_{off} (intercept) according to **Eq. 2**.

$$k_{obs} = k_{on} \text{ [partner in excess]} + k_{off} \quad (2)$$

The displacement of mGDP from its complex with hGBP1 by a large molar excess of GDP resulted in a decrease in fluorescence as shown in **Fig. 3A**. The exponential curve fitted to these data yields the value of k_{off}, which should equal the intercept value of the linear regression of the association rates. Kinetic experiments are particularly useful in the determination of equilibrium constants when a small change in fluorescence intensity does not allow for a direct titration experiment (*see* **ref. *23***). With many GTP-binding proteins it is possible to apply the FRET. This is based on the excitation of tryptophan residues from the protein (290 nm) and detection of the mant fluorescence at 450 nm, as there is an overlap of the tryptophan fluorescence emission and the mant excitation at 360 nm. An example is given in **Fig. 2**, where the excitation at 290 nm (a maximum of optical absorption of tryptophan) of mGDP alone leads to low fluorescence, whereas upon binding of hGBP1 an increase of almost 30-fold in intensity is observed. This method is particularly useful when binding of mant nucleotides must be detected against a high background of nonbinding mant-nucleotide. This is the case when kinetic experiments are done with increasing nucleotide concentrations (i.e., the mant nucleotide is in excess over protein).

3.2. Assaying p190RhoGEF-Catalyzed Nucleotide Dissociation from Rho

1. The intrinsic and p190DH/PH-stimulated dissociation rates of Rho•mGDP were measured on LS50B Perkin-Elmer spectrofluorometer (Norwalk, CT) using an excitation wavelength of 360 nm and an emission wavelength of 450 nm.
2. The reaction was initiated by adding 200-fold excess of unlabeled GDP to 0.1 μ*M* Rho•mGDP and 8 μ*M* DH/PH-domain in 30 m*M* Tris/HCl, 5 m*M* $MgCl_2$, 10 m*M* KPO_4, 3 m*M* DTE, pH 7.5 at 25°C, as originally described for Cdc25 and Ras ***(15,24)***.
3. As shown in **Fig. 4**, intrinsic displacement of the bound mGDP from RhoA by nonlabeled GDP is very slow, but is stimulated by three orders of magnitude in the presence of 8 μ*M* DH/PH-domain. The observed nucleotide dissociation-rate constants of 0.0012 min^{-1} for the intrinsic and 0.987 min^{-1} for the GEF-catalyzed reaction have been evaluated by exponential fitting using the Grafit program (Erithacus software).
4. The kinetics shown here facilitate the quantitative analysis of the GEF-catalyzed nucleotide dissociation from RhoA, and allow the determination of the GEF-specificity within the Rho-protein family ***(25)***.

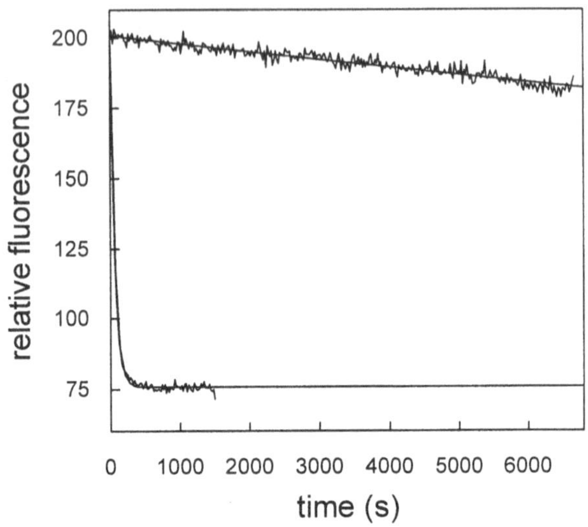

Fig. 4. Kinetics of p190-DH/PH catalyzed nucleotide exchange reaction on RhoA. Addition of 8 μ*M* DH/PH domain of p190RhoGEF to 0.1 μ*M* RhoA•mGDP and 20 μ*M* GDP resulted in a 822-fold stimulation of the nucleotide-dissociation rate (0.987 min^{-1}), measured as the decrease of fluorescence of the mant group, as compared to the intrinsic rate (0.0012 min^{-1}). The rate constants were obtained using a single exponential equation.

3.3. Interaction of Ras Proteins with Effectors

In GTP-binding proteins loaded with a fluorescent derivative of a hydrolysis-resistant form of GTP, such as mGppNHp or mGTPγS, the binding of an effector molecule can often be detected by the extrinsic fluorophore. When the interaction with the effector leads to a large change of the fluorescence intensity, as in the case of the Cdc42/WASP or Ran/RanBP interaction, a direct fluorescence titration is possible ***(26,27)***. But there are other examples like Ras/effector complexes for which the change in fluorescence intensity is less than 10% and is too small to produce an accurate titration curve—particularly when low concentrations have to be used for the titration of high-affinity interactions, as in the case of Ras/Raf. Again, kinetic methods such as stopped-flow can cope with such small signal changes and yield rate constants from which the dissociation constant is calculated as described above in **Subheading 3.1.3.** ***(23)***.

1. To make use of the large fluorescence change upon release of mant nucleotides, another approach has been explored, which is based on the observation that nucleotide dissociation from Ras proteins is inhibited by interaction with

effectors (GDI effect) *(28,29)*. The mant nucleotide bound to the Ras protein is displaced by nonlabeled nucleotide. The observed rate constant is composed of the corresponding fractions of the nucleotide dissociation-rate constants from the Ras protein alone and from the complex with the effector.

2. As shown in **Fig. 5A**, the nucleotide dissociation rate decreases with increasing effector concentration. The dependence of the observed rate constant on the effector concentration can be fitted according to **Eq. 1** described in **Subheading 3.1.2.**
3. Note that A and B, which represent the maximum and minimum rate constants, are reversed in comparison to the titration experiment described in **Subheading 3.1.2.** (**Fig. 2**) in which the curve goes upwards *(29,30)*. The fit in **Fig. 5B** yields a value of 18 nm for the equilibrium dissociation constant of the Ras/RafRBD complex.

3.4. Assaying the Intrinsic GTPase Activity of Ras

1. One generally useful and accurate method is HPLC, by which concentrations of GDP and GTP can be measured from ratio of elution peaks. The relative GTP content determined as the ratio (GTP)/(GTP)+(GDP) is used to describe the reaction progress.
2. The GTPase reaction rates of Ras itself can be more conveniently measured by introducing a tryptophan residue into the otherwise tryptophane-free protein. The mutant has properties similar to wild-type *(5)* and there is large fluorescence change between the GDP- and GTP-bound form. The method can be used to follow the GTPase reaction rates of other mutants as well *(5,6)*.
3. **Figure 6** shows the typical GTP hydrolysis trace with Ras protein (in 30 m*M* Tris/HCl pH 7.5, 3 m*M* DTE, 5 m*M* KP_i buffer and 5 m*M* $MgCl_2$) at 30°C. Fluorescence (excitation 295 nm, emission 350 nm) is followed on a LS50B Perkin-Elmer spectrofluorometer. The spectrometric technique allows an accurate and real-time measurement of the GTPase reaction rates of wild-type (Y32W) and an oncogenic (Y32W, G12V) Ras protein, which are calculated from the exponential fluorescence increase.

3.5. GAP-Stimulated GTPase Reaction

1. Stopped-flow experiments using mGTP and mGppNHp were performed to monitor the interaction of NF1-333 with Ras in a Applied Photophysics SX16MV apparatus similar to that described by two groups (*see* **refs. *13,31–34***).
2. The reactions were followed at 25°C in 40 m*M* HEPES pH 7.4, 5 m*M* DTE, and 5 m*M* $MgCl_2$ using an excitation wavelength of 360 nm and a cut off filter (408 nm) in front of the detector.
3. By mixing Ras•mGTP with GAP-334 (catalytic domain of p120RasGAP), a relatively rapid decrease of fluorescence can be monitored and the rate constant for this change can be fitted as a single exponential *(32)*.
4. Unlike GAP-334, stopped-flow experiments with NF1-333 and Ras•mGTP showed a fast concentration-dependent initial phase corresponding to the binding

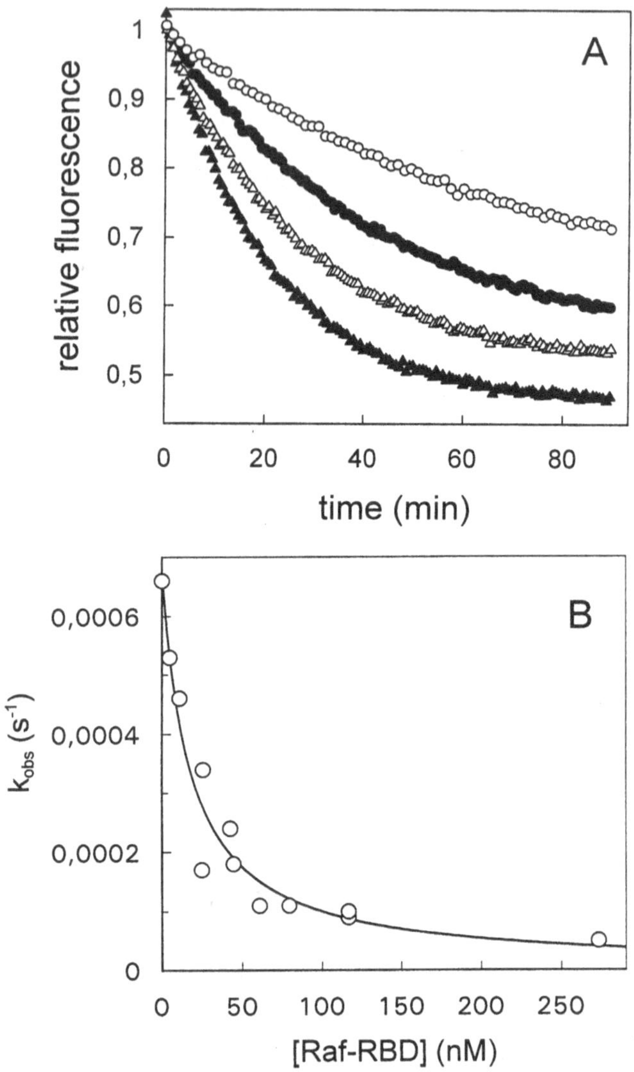

Fig. 5. Effect of GDI on nucleotide exchange. **(A)** The dissociation of mGppNHp from Ras protein (induced by the addition of an excess of nonlabeled nucleotide) is inhibited by the interaction with an effector protein. The dissociation is measured in the presence of increasing Raf-RBD concentrations (▲, 0 nm, △ 20 nm, ●, 50 nm, and ○ 100 nm). **(B)** The observed rate constants obtained from the exponential curves in A are plotted against the concentration of the effector. A fit to the data according to equation 1 yields a K_d value of 18 nm.

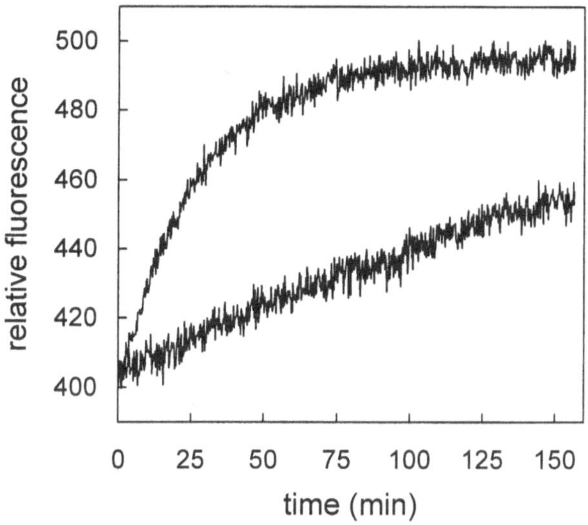

Fig. 6. Kinetics of the GTP-hydrolysis of Ras(Y32W) and Ras(Y32W/G12V) using tryptophan fluorescence. Time-course of tryptophan fluorescence was followed by rapid-mixing of 1.1 μ*M* nucleotide free *Ras*-proteins with 1 μ*M* GTP on a stopped-flow apparatus The data were fitted to a single exponential to obtain rate constants of 0.037 min^{-1} for *Ras*(Y32W) and 0.003 min^{-1} for *Ras*(Y32W/G12V).

reaction followed by a slower phase, which corresponds to the hydrolysis reaction (**Fig. 7A**). The curves through the data points were obtained by fitting with a numerical integration procedure which allowed calculation of $k_{on} = 5.3 \times 10^7\ M^{-1}s^{-1}$ and $k_{cat} = 5.4\ s^{-1}$.

5. With Ras bound to the nonhydrolyzable GTP-analog mGppNHp, only the concentration-dependent first phase was observed, which led to the Ras•mGpp NHp•NF1-333 complex formation. This complex was used to measure the dissociation-rate constants of NF1 from Ras by mixing the complex with 100-fold excess of nonfluorescent Ras•GppNHp, leading to an easily measurable signal for NF1-333 dissociation (**Fig. 7B**).
6. The fitted curve was obtained using a single exponential equation ($k_{off} = 6.8\ s^{-1}$). K_d values (0.13 μ*M*) have been calculated from the association-rate constants of Ras•mGTP and the dissociation-rate constants of Ras•mGppNHp ($K_d = k_{off}/k_{on}$), assuming that the dissociation-rate constants are similar for Ras•mGTP and Ras•mGppNHp.

3.6. Summary

We have presented examples for the analysis of small GTP-binding protein interactions with bound nucleotides, GEFs, GAPs, and effectors. Studies

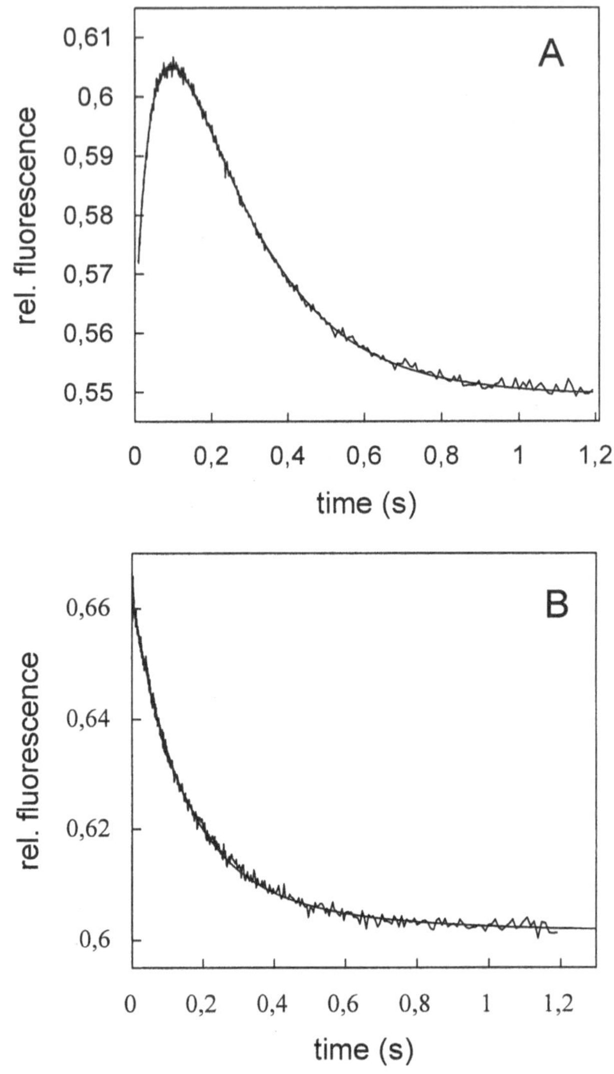

Fig. 7. Kinetics of the GAP-stimulated GTPase reaction of *Ras*. (**A**) Fluorescence transients obtained by mixing 0.1 μ*M* *Ras*•mGTP with 0.2 μ*M* NF1-333 on a stopped-flow apparatus. The curves through the data points were obtained by fitting with a numerical integration procedure as described in **Subheadings 2.** and **3.** (**B**) Displacement of 0.1 μ*M* Ras•mGppNHp from its complex with NF1-333 (1 μ*M*) by mixing with 10 μ*M* Ras•GppNHp. The fitted curve was obtained using a single exponential equation.

involving fluorescent nucleotides and all subfamilies of the Ras subfamily, including Ran, Rab and Arf (which were not mentioned here) can be found elsewhere in the literature. For reference, additional examples involving GTPase regulatory proteins are noted:

1. For Rab7, interactions have been measured with RabGDI ***(35)*** and either Rab Escort Protein (REP) alone, or in ternary complex with Rab geranylgeranyl transferase ***(36–38)***.
2. The mechanism of nucleotide release from Rho by smg-GDS, a rather nonspecific exchange factor with unknown function, was analyzed in great detail using mant fluorescence. In this particular instance, anisotropy of the fluorescence was used as an additional signal to monitor the release of mGDP from its complex with Rho on addition of GDS, and a transient increase of anisotropy could be observed which monitors ternary-complex Rho•mGDP•GDS formation ***(39)***. In conclusion, through the use of fluorescence spectroscopy, great progress has been made in understanding mechanistic principles of the GTPase switch, and we are confident that such studies will continue to throw light on the fundamental properties of these molecules.

4. Notes

1. The mant fluorescence, located in the "active site" of the GTPase ***(40)*** is the most versatile and sensitive reporter group to indicate local conformational changes and protein-protein and protein-ligand interactions. However, it should be noted that the mant nucleotide derivatives do not work in all cases. For example, the interaction between Rho and RhoGAP, although similar in the type of enzymatic catalysis as the Ras-RasGAP interaction ***(41)***, does not show a sufficiently large fluorescence change on interaction. This reaction has thus been investigated using other techniques, such as real-time monitoring the absorbance change produced by coupling P_i release to an enzymatic reaction ***(42)***.
2. A fluorescence change is also not observed in the interaction of Ran and RanGAP, making quantitative mechanistic studies difficult ***(43)***.
3. In cases in which the intrinsic fluorophores and the introduced mant reporter groups produce a convenient signal—for example, Rho-RhoGDI interaction—which show a weak change in fluorescence ***(44)***, other options are available. Thus, in one study RhoGDI has been labeled with a coumarin derivative on the single cysteine ***(45)***, whereas others have used fluorescence energy transfer between the mGDP bound to Cdc42 and fluorescein maleiimide coupled to that cysteine. Alternatively energy transfer between Cdc42•mGDP in the membrane and hexadecyl(amino)-fluorescein that was randomly inserted in the membrane bilayer was used to monitor the extraction by RhoGDI ***(46)***. Using similar

methods, the interaction of Arf with the plasma membrane upon GDP-GTP exchange has been investigated in great detail *(47)*.

References

1. Higashijima, T., Ferguson, K. M., Sternweis, P. C., Ross, E. M., Smigel, M. D., and Gilman, A. G. (1987) The effect of activating ligands on the intrinsic fluorescence of guanine nucleotide-binding regulatory proteins. *J. Biol. Chem.* **262,** 752–756.
2. Lambright, D. G., Noel, J. P., Hamm, H. E., and Sigler, P. B. (1994) Structural determinants for activation of the alpha-subunit of a heterotrimeric G protein. *Nature* **369,** 621–628.
3. Coleman, D. E., Berghuis, A. M., Lee, E., Linder, M. E., Gilman, A. G., and Sprang, S. R. (1994) Structures of active conformations of $G_{i\alpha 1}$ and the mechanism of GTP hydrolysis. *Science* **265,** 1405–1412.
4. Faurobert, E., Otto-Bruc, A., Chardin, P., and Chabre, M. (1993) Tryptophan W207 in transducin Talpha is the fluorescence sensor of the G protein activation switch and is involved in the effector binding. *EMBO J.* **12,** 4191–4198.
5. Rensland, H., John, J., Linke, R., Simon, I., Schlichting, I., Wittinghofer, A., et al. (1995) Substrate and product structural requirements for binding of nucleotides to h-ras p21—the mechanism of discrimination between guanosine and adenosine nucleotides. *Biochemistry* **34,** 593–599.
6. Ahmadian, M. R., Zor, T., Vogt, D., Kabsch, W., Selinger, Z., Wittinghofer, A., et al. (1999) Guanosine triphosphatase stimulation of oncogenic Ras mutants. *Proc. Natl. Acad. Sci. USA* **96,** 7065–7070.
7. Yamasaki, K., Shirouzu, M., Muto, Y., Fujita-Yoshigaki, J., Koide, H., Ito, Y., et al. (1994) Site-directed mutagenesis, fluorescence, and two-dimensional NMR studies on microenvironments of effector region aromatic residues of human c-Ha-Ras protein. *Biochemistry* **33,** 65–73.
8. Neal, S. E., Eccleston, J. F., Hall, A., and Webb, M. R. (1988) Kinetic analysis of the hydrolysis of GTP by p21N-ras. *J. Biol. Chem.* **263,** 19,718–19,722.
9. John, J., Sohmen, R., Feuerstein, J., Linke, R., Wittinghofer, A., and Goody, R. S. (1990) Kinetics of interaction of nucleotides with nucleotide-free H-ras p21. *Biochemistry* **29,** 6058–6065.
10. Klebe, C., Nishimoto, T., and Wittinghofer, F. (1993) Functional expression in Escherichia coli of the mitotic regulator proteins p24ran and p45rcc1 and fluorescence measurements of their interaction. *Biochemistry* **32,** 11,923–11,928.
11. Alexandrov, K., Scheidig, A. J., and Goody, R. S. (2001) Fluorescence methods for monitoring interactions of Rab proteins with nucleotides, Rab escort protein, and geranylgeranyltransferase. *Methods Enzymol.* **329,** 14–31.
12. Hiratsuka, T. (1983) New ribose-modified fluorescent analogs of adenine and guanine nucleotides available as substrates for various enzymes. *Biochim. Biophys. Acta* **742,** 496–508.

13. Rensland, H., Lautwein, A., Wittinghofer, A., and Goody, R. S. (1991) Is there a rate-limiting step before GTP cleavage by H-ras p21? *Biochemistry* **30,** 11,181–11,185.
14. Klebe, C., Prinz, H., Wittinghofer, A., and Goody, R. S. (1995) The kinetic mechanism of Ran—nucleotide exchange catalyzed by RCC1. *Biochemistry* **34,** 12,543–12,552.
15. Lenzen, C., Cool, R. H., and Wittinghofer, A. (1995) Analysis of intrinsic and CDC25-stimulated guanine nucleotide exchange of p21ras-nucleotide complexes by fluorescence measurements. *Methods Enzymol.* **255,** 95–109.
16. Prakash, B., Praefcke, G. J. K., Renault, L., Wittinghofer, A., and Herrmann, C. (2000) Structure of human guanylate-binding protein 1 representing a unique class of GTP-binding proteins. *Nature* **403,** 567–571.
17. Praefcke, G. J. K., Geyer, M., Schwemmle, M., Kalbitzer, H. R., and Herrmann, C. (1999) Nucleotide-binding characteristics of human guanylate-binding protein 1 (hGBP1) and identification of the third GTP-binding motif. *J. Mol. Biol.* **292,** 321–332.
18. Tucker, J., Sczakiel, G., Feuerstein, J., John, J., Goody, R. S., and Wittinghofer, A. (1986) Expression of p21 proteins in Escherichia coli and stereochemistry of the nucleotide-binding site. *EMBO J.* **5,** 1351–1358.
19. Ahmadian, M. R., Wiesmuller, L., Lautwein, A., Bischoff, F. R., and Wittinghofer, A. (1996) Structural differences in the minimal catalytic domains of the GTPase-activating proteins p120GAP and neurofibromin. *J. Biol. Chem.* **271,** 16,409–16,415.
20. Goody, R. S., Frech, M., and Wittinghofer, A. (1991) Affinity of guanine nucleotide binding proteins for their ligands: facts and artefacts. *Trends Biochem. Sci.* **16,** 327–328.
21. Zhang, B. L., Zhang, Y. Q., Wang, Z. X., and Zheng, Y. (2000) The role of Mg2+ cofactor in the guanine nucleotide exchange and GTP hydrolysis reactions of Rho family GTP-binding proteins. *J. Biol. Chem.* **275,** 25,299–25,307.
22. Richter, M. F., Schwemmle, M., Herrmann, C., Wittinghofer, A., and Staeheli, P. (1995) Interferon-induced MxA protein. GTP binding and GTP hydrolysis properties. *J. Biol. Chem.* **270,** 13,512–13,517.
23. Sydor, J. R., Engelhard, M., Wittinghofer, A., Goody, R. S., and Herrmann, C. (1998) Transient kinetic studies on the interaction of Ras and the Ras-binding domain of c-Raf-1 reveal rapid equilibration of the complex. *Biochemistry* **37,** 14,292–14,299.
24. Joseph, L., Lapid, S., and Sukhatme, V. (1987) The major ras induced protein in NIH3T3 cells is cathepsin L. *Nucleic Acids Res.* **15,** 3186.
25. van Horck, F. P. G., Ahmadian, M. R., Haeusler, L. C., Moolenaar, W. H., and Kranenburg, O. (2001) Characterization of p190RhoGEF: a RhoA-specific guanine nucleotide exchange factor that interacts with microtubules. *J. Biol. Chem.* in press.

26. Kuhlmann, J., Macara, I., and Wittinghofer, A. (1997) Dynamic and equilibrium studies on the interaction of Ran with its effector, RanBP1. *Biochemistry* **36,** 12,027–12,035.
27. Rudolph, M. G., Bayer, P., Abo, A., Kuhlmann, J., Vetter, I. R., and Wittinghofer, A. (1998) The Cdc42/Rac interactive binding region motif of the Wiskott Aldrich syndrome protein (WASP) is necessary but not sufficient for tight binding to Cdc42 and structure formation. *J. Biol. Chem.* **273,** 18,067–18,076.
28. Herrmann, C., Martin, G. A., and Wittinghofer, A. (1995) Quantitative analysis of the complex between p21ras and the Ras-binding domain of the human Raf-1 protein kinase. *J. Biol. Chem.* **270,** 2901–2905.
29. Herrmann, C., Horn, G., Spaargaren, M., and Wittinghofer, A. (1996) Differential interaction of the ras family GTP-binding proteins H-Ras, Rap1A, and R-Ras with the putative effector molecules Raf kinase and Ral-guanine nucleotide exchange factor. *J. Biol. Chem.* **271,** 6794–6800.
30. Herrmann, C. and Nassar, N. (1996) Ras and its effectors. *Prog. Biophys. Mol. Biol.* **66,** 1–41.
31. Moore, K. J. M., Webb, M. R., and Eccleston, J. F. (1993) Mechanism of GTP hydrolysis by $p21^{N\text{-}ras}$ catalyzed by GAP: studies with a fluorescent GTP analogue. *Biochemistry* **32,** 7451–7459.
32. Ahmadian, M. R., Hoffmann, U., Goody, R. S., and Wittinghofer, A. (1997) Individual rate constants for the interaction of Ras proteins with GTPase-activating proteins determined by fluorescence spectroscopy. *Biochemistry* **36,** 4535–4541.
33. Ahmadian, M. R., Stege, P., Scheffzek, K., and Wittinghofer, A. (1997) Confirmation of the arginine-finger hypothesis for the GAP-stimulated GTP-hydrolysis reaction of Ras. *Nat. Struct. Biol.* **4,** 686–689.
34. Eccleston, J. F., Moore, K. J. M., Brownbridge, G. G., Webb, M. R., and Lowe, P. N. (1991) Fluorescence approaches to the study of the p21ras GTPase mechanism. *Biochem. Soc. Trans.* **19,** 432–437.
35. Wu, S. K., Luan, P., Matteson, J., Zeng, K., Nishimura, N., and Balch, W. E. (1998) Molecular role for the rab binding platform of guanine nucleotide dissociation inhibitor in endoplasmic reticulum to golgi transport. *J. Biol. Chem.* **273,** 26,931–26,938.
36. Alexandrov, K., Simon, I., Iakovenko, A., Holz, B., Goody, R. S., and Scheidig, A. J. (1998) Moderate discrimination of rep-1 between rab7-gdp and rab7-gtp arises from a difference of an order of magnitude in dissociation rates. *FEBS Lett.* **425,** 460–464.
37. Alexandrov, K., Simon, I., Yurchenko, V., Iakovenko, A., Rostkova, E., Scheidig, A. J., et al. (1999) Characterization of the ternary complex between Rab7, REP-1 and Rab geranylgeranyl transferase. *Eur. J. Biochem.* **265,** 160–170.
38. Anant, J. S., Desnoyers, L., Machius, M., Demeler, B., Hansen, J. C., Westover, K. D., et al. (1998). Mechanism of rab geranylgeranylation—formation of the catalytic ternary complex. *Biochemistry* **37,** 12,559–12,568.
39. Hutchinson, J. P. and Eccleston, J. F. (2000) Mechanism of nucleotide release from Rho by the GDP dissociation stimulator protein. *Biochemistry* **39,** 11,348–11,359.

40. Scheidig, A. J., Franken, S. M., Corrie, J. E., Reid, G. P., Wittinghofer, A., Pai, E. F., et al. (1995) X-ray crystal structure analysis of the catalytic domain of the oncogene product p21H-ras complexed with caged GTP and mant dGppNHp. *J. Mol. Biol.* **253,** 132–150.
41. Scheffzek, K. and Wittinghofer, A. (1998) Structures of Ras and its complexes with Raf and GAP, in *G Proteins, Cytoskeleton and Cancer* (Maruta, H. and Kohama, K., eds.), R. G. Landes Company, Austin, TX, pp. 185–210.
42. Zhang, B. L., Wang, Z. X., and Zheng, Y. (1997) Characterization of the interactions between the small GTPase Cdc42 and its GTPase-activating proteins and putative effectors—comparison of kinetic properties of Cdc42 binding to the Cdc42-interactive domains. *J. Biol. Chem.* **272,** 21,999–22,007.
43. Hillig, R. C., Renault, L., Vetter, I. R., Drell, T., Wittinghofer, A., and Becker, J. (1999) The crystal structure of rna1p: a new fold for a GTPase-activating protein. *Molecular Cell.* **3,** 781–791.
44. Nomanbhoy, T. K. and Cerione, R. A. (1996) Characterization of the interaction between RhoGDI and Cdc42Hs using fluorescence spectroscopy. *J. Biol. Chem.* **271,** 10,004–10,009.
45. Newcombe, A. R., Stockley, R. W., Hunter, J. L., and Webb, M. R. (1999). The interaction between Rac1 and its guanine nucleotide dissociation inhibitor (GDI), monitored by a single fluorescent coumarin attached to GDI. *Biochemistry* **38,** 6879–6886.
46. Nomanbhoy, T. K., Erikson, J. W., and Cerione, R. A. (1999). Kinetics of Cdc42 membrane extraction by Rho-GDI monitored by real-time fluorescence resonance energy transfer. *Biochemistry* **38,** 1744–1750.
47. Antonny, B., Berauddufour, S., Chardin, P., and Chabre, M. (1997). N-terminal hydrophobic residues of the G-protein ADP-ribosylation factor-1 insert into membrane phospholipids upon GDP to GTP exchange. *Biochemistry* **36,** 4675–4684.

II

Specific Methods

Ras Family GTPases

5

Ras and Rac as Activators of Reactive Oxygen Species (ROS)

Herbert Archer and Dafna Bar-Sagi

1. Introduction

The investigation into the relationship between reactive oxygen species (ROS) and the signaling pathways of the Ras family of small GTPases has experienced a recent surge in interest. Known to cause DNA damage, and implicated in disease processes as wide-ranging as atherosclerosis and kidney disease, ROS have appeared in unlikely capacities within the Ras signaling pathway. Eschewing their evil ways, ROS have been shown to be essential for mitogenesis in *Ras*-transformed cells, and have been identified as mediators of a Rac-dependent anti-apoptotic signal in cells microinjected with activated *Ras* ***(1,2)***. The beneficence of ROS on *Ras*-transformed cells carries with it the macroscopic potential for selection and growth promotion of cancerous cells. Several investigations have suggested such a role for ROS, specifically superoxide ($O_2^{\bullet}$), in a Ras-independent manner. Populations of SV40-transformed cells have been observed to undergo phenotypic reversion when transfected with *MnSOD*, an enzyme that converts superoxide to hydrogen peroxide. Moreover, several investigations have demonstrated decreased levels of MnSOD in populations of cancerous cells ***(3,4)***. Together, these studies suggest a requirement for superoxide in the maintenance of the transformed state.

The elucidation of the exact role of ROS in the Ras signaling pathway and in other pathways involved in growth control depends on established methods of ROS detection. This chapter describes two reliable methods of monitoring ROS production at the single-cell level. Both methods take advantage of the inherent reactivity of radical oxygen. The nitro blue tetrazolium (NBT) assay derives its functionality from the ability of the ROS superoxide ($O_2^{\bullet}$) produced

From: *Methods in Molecular Biology, vol. 189: GTPase Protocols: The Ras Superfamily*
Edited by: E. J. Manser and T. Leung © Humana Press Inc., Totowa, NJ

by the NAD(P)H oxidase system to reduce NBT to a visible substance—blue formazon. In the presence of intracellular superoxide, NBT added to a cell will be reduced and visualized as blue precipitate in the location of its reduction. An investigator can expose cells to NBT, and continue to use those cells in other cytochemical staining protocols. The specificity of the reaction can be assessed by adding a scavenger of superoxides, such as superoxide dismutase (SOD). In its presence, reduction of NBT by superoxides of the NADH oxidase system will be blocked, thereby indicating whether superoxide is the true effector in the pathway of interest ***(5)***.

The second assay presented involves the visualization of ROS at the electron microscopic level. In this protocol, the detectable ROS is hydrogen peroxide (H_2O_2). The assay design is based on the ability of intracellular peroxidase to oxidize the reactant, diaminobenzidine (DAB), in the presence of hydrogen peroxide ***(6)***. Success of the assay requires either the phagocytosis or pinocytosis of DAB by the target-cell line. Typically, this involves the use of polymorphonuclear granular lymphocytes as the test-cell type. However, in studying the Ras pathway, the investigator can take advantage of the observation that both Ras and Rac GTPases induce fluid-phase pinocytosis when microinjected into quiescent cells ***(7,8)***. When pinocytosis is induced in the injected cells, DAB is engulfed and delivered to the newly formed pinosomes. If hydrogen peroxide is produced, the peroxidase found in the pinosomes will oxidize DAB to an insoluble polymer that can be visualized using electron microscopy. The detectable element of the assay, therefore, is the ROS hydrogen peroxide.

Together, the NBT reduction assay and the DAB oxidation assay provide two methods for investigation of the effect of ectopically expressed small GTPases (by either injection or transfection) on intracellular ROS production.

2. Materials

2.1. Cell Culture

1. Dulbecco's Modified Eagle's Medium (DMEM): (Gibco-BRL or Cellgro). Filter-sterilize prior to use.
2. Fetal calf serum (FCS): Stored at –20°C in 25-mL aliquots (Gibco-BRL).
3. Penicillin/streptomycin: Added to DMEM to a final concentration of 100 U/mL penicillin, 100 μg/mL streptomycin.
4. Phosphate-buffered saline (PBS): 8 g NaCl, 0.2 g potassium chloride (KCl), 1.44 g Na_2HPO_4, and 0.24 g KH_2PO_4 in 1 L dH_2O; pH 7.2.
5. 3.7% formaldehyde: Diluted in PBS from 37% stock (Sigma).

2.2. Calcium-Phosphate Transfection

1. Plasmid DNA—prepared via polyethylene glycol (PEG) precipitation, cesium chloride, or Qiagen preparation.

2. Solution 1: 40 m*M* HEPES/270 m*M* NaCl.
3. Solution 2: 35 m*M* NaH_2PO_4/35 m*M* Na_2HPO_4.
4. Solution 3: 1 m*M* Tris-HCl, 0.1 m*M* EDTA pH 7.0.
5. Solution 4: 2 *M* $CaCl_2$.

2.3. Microinjection

1. Plasmid DNA: prepared via polyethylene glycol (PEG) precipitation or cesium chloride.
2. DNA microinjection buffer: 50 m*M* HEPES (pH 7.2), 100 m*M* KCl, and 5 m*M* $NaPO_4$. The buffer is kept as a 2X stock solution, and brought to its final concentration in dH_2O when the plasmid DNA is added at the time of injection.
3. Protein microinjection buffer: 20 m*M* Tris-acetate (pH 7.4), 20 m*M* NaCl, 1 m*M* $MgCl_2$, 0.1 m*M* EDTA, 5 m*M* 2-mercaptoethanol.

2.4. Nitroblue Tetrazolium (NBT) Assay

1. Nitroblue tetrazolium at 0.5% (w/v) final concentration in DMEM (Sigma).
2. Superoxide dismutase (SOD) at 300 U/mL final concentration in DMEM (Sigma).

2.5. DAB Oxidation Assay

1. Plasmid DNA: prepared for microinjection (*see* **Subheading 2.3., item 1**).
2. Microinjection buffer: prepared as in **Subheading 2.3., item 1**).
3. REF-52 cells and DMEM supplemented with 10% FCS for cell culture. REF-52 cells survive better at 37°C and 3.5% CO_2.
4. 0.1 *M* Tris-maleate, 7% sucrose (pH 7.5).
5. 0.5 mg/mL 3,3′-diaminobenzidine tetrahydrochloride dihydrate (Aldrich Chemical Co., Inc.). Prepare in Hanks' Balanced Salt Solution (BSS) + 1 mg/mL glucose.
6. Preincubation mix: 0.1 *M* Tris-maleate, 7% sucrose (pH 7.5) plus 1 m*M* 3amino-1,2,4-triazole (AT) for inhibition of DAB oxidation. This is used in the negative control.
7. Incubation mix: 0.1 *M* Tris-maleate, 7% sucrose (pH 7.5), 1 m*M* $CeCl_3$, 0.71 m*M* NADH (+ 10 m*M* AT for negative control). Filter-sterilize with 0.22 millipore filter.
8. Fixation mix: 2% glutaraldehyde in 0.1 *M* cacodylate (pH 7.3), 5% sucrose.
9. Wash mix: 0.1 *M* cacodylate, 5% sucrose (pH 6.0).
10. Overnight wash mix: 0.1 *M* cacodylate, 5% sucrose (pH 7.4).

3. Methods

3.1. Introduction of DNA

3.1.1. Calcium-Phosphate Transfection

The following protocol is used for 100-mm cell-culture plates. The reagent volumes can be reduced by half for 60-mm cell-culture plates. The solutions referred to are from **Subheading 2.2.**

1. Grow cells to 70–80% confluence in their normal growth media.
2. Using two test tubes for each plate, add 470 μL of Solution 1 and 11 mL of Solution 2 to test tube 1.
3. Add 435 μL Solution 3 and the desired quantity of DNA to test tube 2. Slowly add 66 μL Solution 4 to test tube 2.
4. Slowly add the contents of test tube 2 to test tube 1 while bubbling the contents of test tube 1 with a glass pipet.
5. Add the final mixture to the cell plates and incubate at 37°C, 3.5% CO_2 for 12–16 h.
6. Wash cells 2× with PBS, and add back normal growth media.
7. Allow expression of the protein of interest for the optimal amount of time as determined by titration (*see* **Note 3**).

3.1.2. Microinjection

1. Grow cells on gridded glass cover slips to 70–80% confluence.
2. Serum-starve cells for 24 h (only if growth factors are known activators of the pathway of interest).
3. Bring the sample DNA (in suitable expression vectors) to the final desired concentration in 2X microinjection buffer and dH_2O. In the investigation of the Ras family of small GTPases, a final concentration of 50 ng/uL is usually sufficient (*see* **Note 2**). Use an empty vector as the negative control. When injecting proteins, use the protein-injection buffer as listed in **Subheading 2.3., item 1**, and inject proteins at a final concentration between 2 and 10 mg/mL.
4. Load a microinjection needle with 2 μL of injection solution containing DNA (or protein), and inject cells according to the guidelines for the individual microinjector.
5. Incubate injected cover slips in serum-free DMEM at 37°C for the appropriate amount of time to achieve optimal protein expression (*see* **Note 3**).

3.2. Quantitation of Superoxide Production Using the NBT Assay

1. Ectopically express the proteins of interest using either calcium-phosphate transfection or microinjection as described in **Subheadings 3.1.1.** and **3.1.2.**
2. Incubate in serum-free DMEM at 37°C for the appropriate amount of time to achieve optimal protein expression (*see* **Note 3**). Aspirate off the serum-free DMEM.
3. Replace media with 0.5% NBT in DMEM. Incubate at 37°C for 1 h. If inhibition of superoxide production is desired, first add 300 U/mL SOD in DMEM and incubate for 1 h at 37°C. Follow up with a 1-h incubation at 37°C in DMEM containing 0.5% NBT and 300 U/mL SOD (*see* **Notes 4** and **5**).
4. Wash cells 3× in PBS and fix in 3.7% formaldehyde for 15 min.
5. Visualize NBT under bright-field microscopy. If desired, co-staining for ectopically expressed protein (e.g., Ras or Rac), will allow correlation of ROS generation with transgene expression (*see* **Note 6**).

3.3. DAB Oxidation Assay

1. Ectopically express the proteins of interest using either calcium-phosphate transfection or microinjection as described above in **Subheadings 3.1.1.** and **3.1.2.**
2. Incubate cells at 37°C in DMEM for the optimal period of expression of the gene of interest (*see* **Note 3**).
3. Aspirate off media and wash briefly at room temperature in 0.1 *M* Tris maleate/7% sucrose pH 7.5.
4. Pre-incubate cells for 10 min at 37°C in 0.1 *M* Tris-maleate, 7% sucrose pH 7.5 (add 1 m*M* AT to the negative controls).
5. Incubate cells for 30 min at 37°C in 0.1 *M* Tris-maleate, 7% sucrose pH 7.5, 1 m*M* $CeCl_3$, 0.71 m*M* NADH. To test for DAB oxidation, include DAB in the incubation mix at a concentration of 0.5 mg/mL. Include 10 m*M* AT in the negative control's incubation mix.
6. Wash briefly 3× at room temperature in 0.1 *M* Tris-maleate, 7% sucrose pH 7.5.
7. Fix cells for 30 min at room temperature in 2% glutaraldehyde in 0.1 *M* cacodylate, 5% sucrose pH 7.3.
8. Wash cells for 1 hour at 4°C in 0.1 *M* cacodylate, 5% sucrose pH 6.0.
9. Wash cells overnight at 4°C in 0.1 *M* cacodylate, 5% sucrose pH 7.4.
10. Dehydrate cells through graded ethanol, and embed in Epon.
11. Cut thin sections and stain with uranyl acetate and lead citrate.
12. Visualize using a transmission electron microscope (*see* **Note 7**).

4. Notes

1. Visualize reduced NBT under bright-field microscopy. Blue formazon will be visible in the tetramethylrhodamine isothiocyanate (TRITC) channel. It will not fluoresce, but the dense precipitate may obscure any co-stain labeled with TRITC antibody. Use fluorescein-5-isothiocyanate (FITC) labeled antibodies for co-staining experiments.
2. If working with a well-expressed, nontoxic protein, a plasmid concentration of 20 ng/μL may be suitable for microinjection, and 25–50 ng/μL may be suitable for calcium-phosphate transfection. Prior to performing the NBT assay, one may consider titrating the concentration of DNA, and staining for expression at each concentration. Also, if expressing more than one plasmid, there may be problems with promoter competition that must be worked out with the proper controls.
3. The incubation time of 10 h post-DNA introduction was specifically developed for Ras and Rac expression plasmids containing the cytomegalovirus (CMV) promoter. This incubation time is intended to allow for optimal expression of the desired protein. Therefore, if your protein is well-expressed before 10 h, this incubation time may be appropriately shortened. It is also critical that the incubation is short enough to avoid any deleterious effects of prolonged expression of ectopic protein. It would be problematic if the population of cells has entered apoptosis before the NBT assay. The best way to resolve such issues

is to run a time-course experiment, staining for expressed protein at each time-point. Incubate for the shortest time-point that consistently yields substantial expression, and simply observe that the phenotype of the cells does not change significantly over this period.

4. When working with immortalized cell lines, there is a tendency to have background reduction of NBT. Be sure to always include a negative, empty vector control. It is also possible to shorten the incubation time with NBT to reduce background, but this is up to the investigator's discretion.
5. While it is possible to shorten the incubation time with NBT to reduce background, it is not advisable to lengthen the incubation time much past 1 h. After 1 h of incubation, the background reduction of NBT increases greatly, confounding the scoring technique of counting NBT-positive cells.
6. Scoring cells for superoxide production using the NBT-reduction assay consists of counting those cells in which the blue formazon precipitate is found. At least 100 cells should be counted. Co-staining for any ectopically expressed protein increases the accuracy of this scoring technique. If desired, staining of samples using indirect immunofluorescence should be performed after fixation as outlined in **Subheading 3.2., step 4**. In this manner, ROS production can be correlated with transgene expression.
7. In the DAB oxidation assay, consideration must be given to locating the protein-expressing cells. In the author's use of the assay, Ras-induced pinocytic vesicles were used to identify those cells expressing Ras.
8. Serum-starvation is a critical step in any assay involving the small GTPases. Many of the common effectors of the small GTPases are also activated by the elements of serum. Failure to serum-starve will result in background activation of many of these small GTPase effectors. Serum-starvation is best achieved by removal of serum-containing media for 12–36 h.

References

1. Irani, K., Xia, Y., Zweier, J. L., Sollott, S. J., Der, C. J., Fearon, E. R., et al. (1997) Mitogenic signaling mediated by oxidants in ras-transformed fibroblasts. *Science* **274,** 1649–1652.
2. Joneson, T. and Bar-Sagi, D. (1999) Suppression of ras-induced apoptosis by the Rac GTPase. *Mol. Cell. Biol.* **19,** 5892–5901.
3. Yan, T., Oberley, L. W., Zhong, W., and St. Clair, D. K. (1996) Manganese-containing superoxide dismutase overexpression causes phenotypic reversion in SV40-transformed human lung fibroblasts. *Cancer Res.* **56,** 2864–2871.
4. Oberley, L. W., McCormick M. L., Sierra-Rivera, E., and St. Clair, D. K. (1989) Manganese superoxide dismutase in normal and transformed human embryonic lung fibroblasts. *Free Radic. Biol. Med.* **6,** 379–384.
5. Akashi, M., Hachiya, M., Paquette, R. L., Osawa, Y., Shimizu, S., and Suzuki, G. (1995) Irradiation increases manganese superoxide dismutase mRNA levels in human fibroblasts. *J. Biol. Chem.* **270,** 15,864–15,869.

6. Briggs, R. T., Karnovsky, M. L., and Karnovsky, M. J. (1975) Cytochemical demonstration of hydrogen peroxide in polymorphonuclear leukocyte phagosomes. *J. Cell Biol.* **64,** 254–260.
7. Bar-Sagi, D. and Feramisco, J. R. (1986) Induction of membrane ruffling and fluid-phase pinocytosis in quiescent fibroblasts by ras proteins. *Science* **233,** 1061–1068.
8. Ridley, A. J., Paterson, H. R., Johnston, C. L., Diekmann, D., and Hall, A. (1992) The small GTP-binding protein rac regulates growth factor-induced membrane ruffling. *Cell* **70,** 401–410.

6

Interfering with Ras Signaling Using Membrane-Permeable Peptides or Drugs

Hiroshi Maruta, Hong He, and Thao Nheu

1. Introduction

1.1. Signaling Therapy of Cancers

During the last two decades since the first oncogene product called v-Src was identified as a constitutively activated mutant of a normal mitogenic gene (proto-oncogene) encoding a Tyr kinase called c-Src in early 1980s, it was firmly established that the malignant transformation of normal cells is caused by a combination of the following genetic events: overexpression of a proto-oncogene or constitutive activation of its gene product (mitogenic signal transducer), and deletion of a tumor-suppressor gene (also called anti-oncogene or anti-mitogenic gene) or dysfunction of its gene product (anti-mitogenic signal transducer).

A typical example of the latter is *p53*, a nuclear transcription factor, whose deletion or dysfunction is responsible for the development of more than 50% of all human cancers—notably, more than 70% of colon cancers. Typical examples of the former are RAS family GTPases or G proteins (Ha-RAS, Ki-RAS, and N-RAS), whose mutations contribute to the development of more than 30% of all human cancers—notably, 50% of colon cancers and 90% of pancreatic cancers *(1)*.

Thus, in principle, reversion of malignant transformation to the normal state could be achieved through so-called "signaling therapy of cancers,"—signaling cancer cells back to normal by either restoring the "missing" tumor-suppressor genes such as *p53* and retinoblastoma (RB) using a gene therapy technology, or blocking the function of oncogenic signal transducers such as RAS and Src

From: *Methods in Molecular Biology, vol. 189: GTPase Protocols: The Ras Superfamily*
Edited by: E. J. Manser and T. Leung © Humana Press Inc., Totowa, NJ

or their downstream effectors through peptido- or chemotherapy using their specific antagonists (peptides or chemical compounds).

1.2. Blocking Oncogenic RAS Signaling

1.2.1. Downstream of RAS/PI-3 Kinase

Extensive studies on the molecular mechanism underlying RAS-induced transformation, conducted during the last two decades, have revealed that oncogenic RAS mutants such as v-Ha-RAS—which are locked constitutively in the active guanosine triphosphate (GTP)-bound form—cause malignant transformation by activating several distinct effectors such as Raf and PI-3 kinase ***(2)***. PI-3 kinase is a complex of the catalytic subunit (*p110*) and the regulatory subunit (*p85*) that activates several downstream effectors through its end product (PIP3). At least two distinct members of Rho family GTPases—Rac and CDC42—are activated by this RAS/PI-3 kinase pathway, and both are required for RAS transformation ***(3–6)***. However, thus far no chemical compound or bacterial toxin has been found that selectively inhibits either Rac or CDC42.

Furthermore, the so-called dominant-negative mutant (N17) of either Rac or CDC42 is not a specific inhibitor for either Rac or CDC42, simply because it can sequester so-called nonspecific GDP-dissociation stimulators (GDSs) such as DBL, Vav, and Tiam-1 that activate Rac or CDC42 as well as Rho. Thus, our recent effort has been focused on the development of anti-RAS cancer molecules that selectively block the pathways downstream of either Rac or CDC42.

CDC42 only in the active GTP-bound form (but not in the inactive guanosine diphosphate (GDP)-bound form) activates several distinct effectors such as PAK family kinases (PAK1, PAK2, PAK3 and PAK4), ACK family kinases (ACK1 and ACK2), and WASP and N-WASP, and mainly induces the formation of microspikes and filopodia ***(7–10)***. Likewise, Rac only in the GTP-bound form (but not in the GDP-bound form) activates several distinct effectors such as PAKs, POR1, and SRA-1, and mainly induces membrane ruffling or disrupt actin stress fibers ***(7,11,12)***. Thus, it is conceivable that the minimal GTPase-binding domain of each effectors can selectively sequester only the GTP-bound form of CDC42 or Rac.

This chapter introduces a powerful new technology which is useful for determining a potential pharmacological function of relatively small peptides such as GTPase-binding fragments and SH3-binding Pro-rich motifs. The CDC42-binding fragment of 42 amino acids derived from ACK1 (residues 504-545, ACK42) blocks the interactions of CDC42 with all its effectors ***(6)***, whereas the CDC42/Rac-binding fragment of 40 amino acids derived from

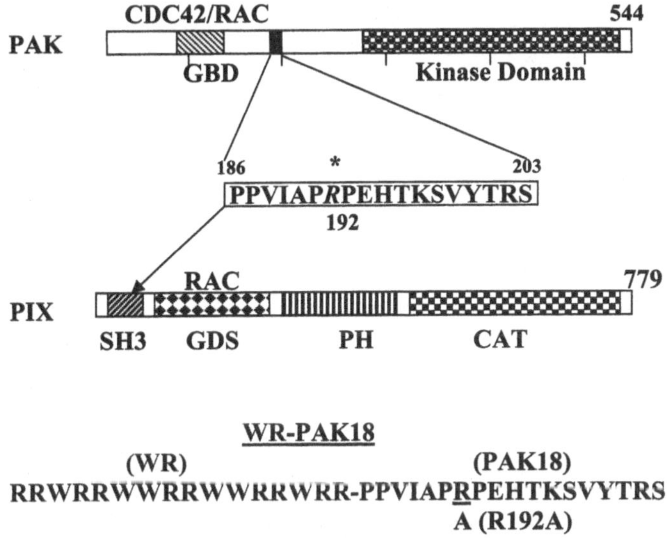

Fig. 1. PAK–PIX interaction.

PAK1 (residues 69-108, PAK40) blocks the interactions of CDC42/Rac with all their effectors. Furthermore, PAKs require not only CDC42 or Rac, but also a third protein called PIX for the full activation ***(13)***. As shown in **Fig. 1**, the N-terminal SH3 domain of PIX binds the Pro-rich motif of PAKs (residues 186-203, PAK18), and the PAK 18 selectively blocks the PIX-PAK interaction which is essential for PAK activation ***(14–16)***. To deliver these three peptides (ACK42, PAK40 and PAK18) into the target cells to demonstrate their pharmacological activity, particularly the suppression of RAS-induced malignant transformation, we have conjugated these peptides to a cell-permeable peptide vector of 16 amino acids called WR ***(17)*** through a recombinant technology or chemical synthesis (*see* **Fig. 1**).

1.2.2. Penetratin (Peptide Vector) Technology

In the past, the effect of each small peptide on the growth or other biological activities of target cells has been examined by transfecting the cells with a DNA encoding each peptide or microinjecting each peptide into individual cells. However, it is difficult to control the expression level of each peptide by a simple transfection without any inducible promoter, and to deliver each peptide into one million cells by micro-injection technology at a time or in a relatively short time for a biochemical analysis. To overcome such a difficulty, a series of both hydrophobic and positively charged peptide vectors of 16 amino acids called penetratins have been developed during last several years that can deliver efficiently any small peptides of less than 100 amino acids into the cytoplasm or nucleus of target cells *(17)*.

The first penetratin that was shown to translocate across both plasma membrane and nuclear envelope was derived from a Drosophila transcription factor called Antennapedia, which contains a DNA-binding homeodomain of 60 amino acids *(18)*. A subsequent mutational analysis of this homeodomain has revealed that the helix 3 of 16 amino acids (residues 43–58) in this homeodomain called penetratin-1 (or PEN) is sufficient for the translocation across the biological membranes *(19)*. Interestingly, a peptide of the reverse sequence called NEP (58–43 instead of 43–58) is also capable of penetrating across the membranes *(17)*. Through a further mutational analysis, a symmetric peptide consisting of only 10 Arg residues and 6 Trp residues, thus called WR, was found to be the most simplified version of penetratin *(20)*. Below 10 μ*M*, WR or other penetratins alone demonstrate no significant cytotoxicity *(17)*.

Penetratin- 1 (PEN):	RQ I K I WFQNRRMKWKK
Penetratin-2 (NEP):	KKWKMRRNQFW I K I QR
WR:	RRWRRWWRRWWRRWRR

1.2.3. Drugs That Block RAS-Induced Autocrine Pathways

Although these cell-permeable peptides, such as WR-PAK18, WR-ACK42 and WR-PAK40, are useful for blocking oncogenic RAS signaling in vitro (at cell-culture levels) as described shortly in detail, they may not be so effective in vivo (at experimental animal or clinical levels), as they are susceptible to a possible proteolytic degradation during their prolonged administration. Thus, "more metabolically stable" cell-permeable chemical compounds have been developed to block PAK activation by interfering with the RAS-induced autocrine pathways that are essential for RAS transformation, both in vitro and in vivo *(21–23)*.

Oncogenic RAS are known to activate through the Raf-MEK-MAP kinase (ERKs) cascade several distinct genes encoding EGF family ligands such as TGF-alpha, HB-EGF, amphi-regulin and heregulin *(24–27)*. These ligands then activate ErbB family receptors (Tyr kinases) including ErbB1, ErbB2, ErbB3, and ErbB4. We have recently found that RAS requires two independent autocrine pathways for both PAK activation and RAS transformation *(22)*. One pathway activates ErbB1, and the other activates ErbB2 *(22)*. Furthermore, both pathways require Src family kinase(s) for both PAK activation and RAS transformation *(21,22)*. In fact, either the ErbB1-specific inhibitor AG 1478, the ErbB2-specific inhibitors AG 825 and AG 879, or the Src family kinase-specific inhibitors PP1 and PP2 *(28,29)* strongly inhibit both PAK activation and RAS transformation in vitro and in vivo *(21,22)*. **Fig. 2** provides a summary

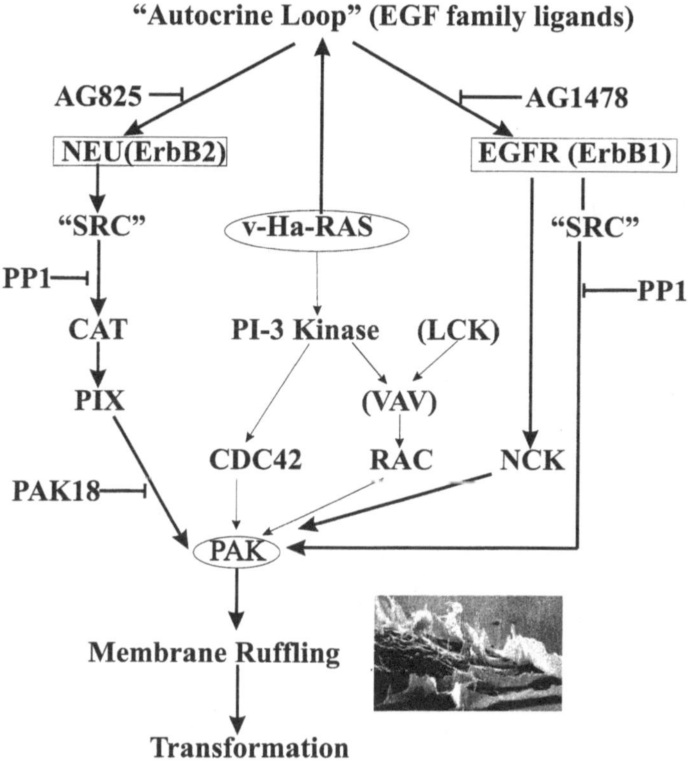

Fig. 2. Autocrine pathways essential for PAK activation.

of several distinct pathways, all of which are essential for both RAS-induced activation of PAKs and malignant transformation in mammalian fibroblasts such as NIH-3T3 cells.

2. Materials

1. The following peptides were chemically synthesized by Chiron Technologies (Clayton, Vic) according to the manufacturer's standard method. Peptides were purified to 95% by high-pressure liquid chromatography (HPLC).

penetratin	**WR**, RRWRRWWRRWWRRWRR
PAK18 alone	PPVIAPRPEHTKSVYTRS
WR-PAK18	**WR** -PPVIAPRPEHTKSVYTRS
WR-PAK18 (R192A)	**WR** -PPVIAP<u>A</u>PEHTKSVYTRS.

2. Using the vector pGEX-2TH (30), the following thrombin-cleavable glutathione-*S*-transferase (GST) fusion proteins were produced in *Escherichia coli* (*E. coli*) by IPTG induction at 28°C for 7 h and then affinity-purified on GSH-agarose

beads as described previously *(6)*: GST-HA-WR-ACK42 (residues 504–545 of human ACK-1), and GST-WR-PAK40 (residues 69–108 of human PAK1).

3. The following cell-permeable chemical compounds were generously provided by Dr. Alexander Levitzki of Hebrew University of Jerusalem, Israel: AG1478, AG 825, AG 879, and PP1. These chemical compounds were first dissolved in ethanol for sterilization, and these stock solutions (1–10 m*M*) were further diluted with PBS or the culture medium for in vitro or in vivo tests.

3. Methods

3.1. Preparation of Cell-permeable ACK42 or PAK40 Derivatives from GST Fusion Proteins

The cell-permeable derivative of ACK42 or PAK40 was separated from GST by digesting each GST-fusion protein with thrombin (50 U per mg of GST-fusion protein) at 37°C for 1 h in the presence of 2 m*M* $CaCl_2$ as described previously *(6)*. For the sterilization, each peptide solution in PBS was passed through a millipore membrane (0.5 μ in diameter).

3.2. Assay for Pharmacological Effects of Each Peptide on Cell Growth or PAK Activation

1.5 m*M* stock solution of each peptide in PBS was diluted to the final concentration 10 μ*M* or less in the standard culture medium (10% FCS plus DME medium for growth, and serum-free DME medium for PAK activation).

3.3. Assay for the Anchorage-Dependent Growth of Cells

1. 2×10^4 cells (normal or v-Ha-RAS-transformed NIH-3T3 fibroblasts) is seeded in 100 μL of the culture medium on a 96-well microtiter plate and cultured overnight.
2. The medium is removed and replaced with a fresh medium containing 10 μ*M* of various test peptides or the control WR alone.
3. Every 24 h the number of cells was counted after detaching the cells from the substratum by a brief trypsin-treatment, as previously described *(6)*.

3.4. Soft Agar Assay for Anchorage-Independent Growth

1. v-Ha-RAS transformed NIH-3T3 cells (10^3 cells/plate) were cultured in soft agar in the presence or absence of various concentrations of the ErbB1-inhibitor AG1478, the ErbB2-specific inhibitors AG 825 or AG 879 or the Src family-kinase-specific inhibitors PP1 or PP2 under standard culture conditions *(6,30)*.
2. After 15 d culture, the colonies formed were stained with 0.005% crystal violet and counted.

3.5. Assay for PAK Activation

1. All cells (normally 10^6 cells per each assay) were serum-starved overnight.
2. The v-Ha-RAS transformed NIH-3T3 cells were then incubated in the fresh DME alone or the medium containing 10 μ*M* of various test peptides, the control WR alone, or various concentrations of chemical drugs.
3. Normal NIH-3T3 cells were also incubated with various growth factors such as EGF (100 ng/mL), heregulin (50 ng/mL) or PDGF (10 ng/mL) in the fresh DME in the presence or absence of WR-PAK18 (10 μ*M*) or various chemical drugs for various time periods.
4. The cells were washed 2x with cold PBS, and disrupted in the lysis buffer.
5. The cell lysates were cleared by centrifugation at 20,000*g* for 10 min at 4°C.
6. The protein concentrations in the supernatants were determined, and each extract (containing 800 μg of proteins) was subjected to the following in vitro PAK kinase assays.
7. The kinase assays were performed on anti-PAK immunoprecipitates from the above cell extracts as described previously ***(31)***.
8. Briefly, extracts were incubated with anti-PAK antibody (Santa Cruz Biotech, CA) and protein A beads for 2 h at 4°C.
9. Immunoprecipitates were washed twice with the lysis buffer followed by 2× wash with the kinase buffer (10 m*M* $MgCl_2$, 40 m*M* HEPES, pH 7.4).
10. The immunoprecipitates were mixed with 5 μg of myelin basic protein (MBP, Sigma) on ice for 5 min.
11. Kinase assays were initiated by adding 10 m*Ci* of [γ-^{32}P]ATP and 20 μ*M* ATP, followed by incubation at 25°C for 10 min.
12. Reactions were stopped by the addition of 2 × SDS sample buffer and heated to 95°C.
13. The samples were run on 13% SDS-PAGE.
14. After the gels were fixed and dried, the radioactive bands (phosphorylated MBP) were scanned in phosphorImager (Molecular Dynamics).
15. The density of each band was measured with Image Quantify program.

3.6. Suppression of v-Ha-RAS-Induced Sarcoma Development in Nude Mice

1. Female ARC/s nu/nu nude mice (about 6 wk old) were obtained from Animal Resources Center (West Australia).
2. v-Ha-RAS transformed NIH-3T3 cells (10^5 cells/mouse) in 0.1 mL of PBS were first grown subcutaneously in each nude mouse (6 mice/cage) as described previously ***(21,32)***.
3. After 3 d, either PP1 or AG 879 suspended in 30% dimethyl sulfoxide (DMSO) with PBS was administered i.p. for the RAS-induced sarcoma bearing mice at 10–20 mg/kg on d 3, 5, 7, 10, 12, 14, and 17.
4. The nontreated (control) mice received only 30% DMSO in PBS.

4. Notes

1. At 10 μ*M*, either WR-PAK18, WR-ACK42, or WR-PAK40 almost completely inhibits the growth of RAS transformants, whereas none of these has any significant effect on the growth of the parental normal cells ***(6,16,21–23)***. Under the same conditions, either the control WR alone—a mutant of WR-PAK18 called R192A which no longer binds the SH3 domain of PIX ***(14)***—or PAK18 alone which does not enter the cells, shows no effect on their growth ***(6,16,22)***. At the same concentration, WR-PAK18 almost completely inhibits RAS-induced activation of PAKs, whereas either WR alone, PAK18 alone, or the mutant R192A has no effect on the PAK activation.
2. Unfortunately, since the highly positively-charged WR-PAK18 (and any other WR derivatives) is almost completely sequestered by the highly negatively-charged agar (a sulfated galactan), it is technically impossible to perform the soft agar assay for the effect of WR-PAK18 on their anchorage-independent growth. Focus formation is therefore used to examine the effect of WR-PAK18 on their anchorage-independent growth. When WR-PAK18 at 10 μ*M* is added to RAS transformants which had already formed many foci, the majority of cells in the foci detached and rapidly died within 24 h, and only a few cells remained attached to the substratum to survive. However, WR-PAK18 at the same concentration has no significant effect on the parental normal fibroblasts over 24 h. These observations indicate that WR-PAK18 selectively suppresses the anchorage-independent growth of RAS transformants which is important for their malignancy. In other words, in the presence of this peptide, the cells need to adhere to the substratum for their survival.
3. The anti-RAS action of WR-ACK42 in RAS transformants depends on the affinity of ACK42 for its target CDC42/GTP complex, and if it is higher, it would be more potent. As summarized in **Table 1**, our screening for the higher affinity mutant(s) of ACK42 has revealed that: (1) replacement of Arg at position 34 in ACK42 (corresponding to residue 537 of ACK1) with Lys dramatically increases the affinity for the constitutively active mutant (Leu 61) of CDC42, and (2) both these basic residues at position 34 and the highly conserved His at position 17 of ACK42 (corresponding to residue 520 of ACKs) are absolutely essential for ACK family kinases (ACK1 and ACK2) to bind the CDC42/GTP complex ***(23)***. These findings strongly suggest that the Lys34 mutant of WR-ACK42 would be much more potent than the wild-type to suppress RAS transformation.
4. Either the Tyr-kinase inhibitor AG 1478 (0.2 μ*M*), AG 825 (0.8 μ*M*) or PPI (10 nm) alone inhibits the PAK activation in RAS transformants by 50%, although none of them shows any significant effect on the anchorage-dependent growth of both RAS transformants or normal cells ***(22)***. However, the combination of AG 1478 and AG 825 almost completely inhibits the PAK activation, and the anchorage-independent growth (colony formation in soft agar) of RAS transformants as well ***(22)***. Furthermore, PP1 alone strongly suppresses the anchorage-independent growth of RAS transformants, and PPI or AG 879 treatment alone (20 mg/kg,

Table 1
Affinities of ACK42 (504-545) Mutants for CDC42 (L61)

1 5 10 17 20 26 30 34 42	
GLSAQDISQPLONSFIHTGHGDSDPRHCWGFPDRIDELYLGN	
504	545
ACK42 Mutants	Kd (nm)
ACK42 (Wild-type)	23
LEU 2 to ASP or LYS	23
ILE 16 to ASP	82
ILE 16 to LYS	104
HIS 17 to TYR	NB
HIS 17 to ARG	NB
HIS 17 to LYS	NB
HIS 20 to TYR	70
ARG 26 to LEU	23
ARG 34 to LEU	NB
ARG 34 to LYS	1.5*

NB: No binding.
*15-fold increase in affinity.

i.p., every other day) is sufficient to keep at least 50% of nude mice free of RAS-induced sarcoma development *(21,22)*.

Acknowledgment

We are grateful to Drs. Ed Manser and Louis Lim for their generous gift of a pGEX-2TK plasmid expressing a GST-fusion protein of human alpha-PIX SH3 domain, Dr. Alex Levitzki for his generous gift of several distinct Tyr kinase inhibitors discussed in this article, and Drs. Shin Kobayashi and Michiyuki Matsuda for their generous gift of the plasmid 513 pGEX4T-3 expressing a GST-fusion protein of the residues 67–150 of human PAK1.

References

1. Bos, J. L. (1989) Ras oncogenes in human cancer: a review. *Cancer Res.* **49,** 4682–4689.
2. Downward, J. (1998) Oncogenic RAS signaling network, in *G Proteins, Cytoskeleton and Cancer* (Maruta, H. and Kohama, K., eds.), Landes Bioscience, Austin, TX, pp. 171–183.
3. Qiu, R., Che, J., Kirn, D., McCormick, F., and Symons, M. (1995) An essential role for Rac in RAS transformation. *Nature* **374,** 457–459.

4. Qiu, R., Abo, A., McCormick, F., and Symons, M. (1997) CDC42 regulates anchorage-independent growth and is necessary for RAS transformation. *Mol. Cell Biol.* **17,** 3449–3458.
5. Lilienthal, J., Moon, S. Y., Lesche, R., Mammillapalli, R., Li, D., et al (2000) Genetic deletion of the PTEN tumor suppressor gene promotes cell motility by activation of Rac and CDC42 GTPases. *Curr. Biol.* **10,** 401–404.
6. Nur-E-Kamal, M. S. A., Kamal, J., Quresh, M., and Maruta, H. (1999) The CDC42-specific inhibitor derived from ACK-1 blocks v-Ha-RAS-induced transformation. *Oncogene* **18,** 7787–7793.
7. Manser, E., Leung, T., Salihuddin, H., Tan, L, and Lim, L. (1993) A nonreceptor Tyr kinase that inhibits the GTPase activity of CDC42. *Nature* **363,** 364–367.
8. Manser, E., Leung, T., Salihuddin, H., Zhao, Z., and Lim, L. (1994) A brain Ser/Thr kinase activated by CDC42 and Rac. *Nature* **367,** 40–46.
9. Symons, M., Derry, J., Karlak, B., Jiang, S., Lemahieu, V., and McCormick, F. (1996) Wiskott-Aldrich syndrome protein, a novel effector for the GTPase CDC42, is implicated in actin polymerization. *Cell* **84,** 723–734.
10. Miki, H., Sasaki, T., Takai, Y., and Takenawa, T. (1998) Induction of filopodium formation by a WASP-related actin depolymerizing protein N-WASP. *Nature* **391,** 93–96.
11. Van Aelst, L., Joneson, T., and Bar-Sagi, D. (1996) Identification of a novel Rac-interacting protein involved in membrane ruffling. *EMBO J.* **15,** 3778–3786.
12. Kobayashi, K., Kuroda, S., Fukata, M., Nakamura T., Nagase, T., et al. (1998) P140Sra-1 (specifically Rac-associated protein) is a novel specific target for Rac GTPase. *J. Biol. Chem.* **273,** 291–295.
13. Daniels, R., Zenke, F., and Bokoch, G. (1999) PIX stimulates PAK kinase activity through exchange factor-dependent and independent mechanisms. *J. Biol. Chem.* **274,** 6047–6050.
14. Manser, E., Loo, T., Koh, C., Zhao, Z., Chen, X., et al. (1998) PAK kinases are directly coupled to the PIX family of nucleotide exchange factors. *Mol. Cell* **1,** 183–192.
15. Obermeier, A., Ahmed, S., Manser, E., Yen, S., Hall, C., and Lim, L.(1998) PAK promotes morphological changes by actin upstream of Rac. *EMBO J.* **17,** 4328–4339.
16. Maruta, H., He, H., Anjali, T., and Nur-E-Kamal, M. S. A. (1999) Cytoskeletal tumor suppressors that block oncogenic RAS signaling. *Ann. NY Acad. Sci.* **886,** 48–57.
17. Derossi, D., Chassaing, G., and Prochiantz, A. (1998) Trojan peptides: the penetratin system for intracellular delivery. *Trends Cell Biol.* **8,** 84–87.
18. Joliot, A., Pernelle, C., Deagostini-Bazin, H., and Pronchiantz, A. (1991) Antennapedia homeobox peptide regulates neural morphogenesis. *Proc. Nat. Acad. Sci. USA 88,* 1864–1868.
19. Derossi, D., Joliot, A., Chassaing, G., and Prochiantz, A. (1994) The third helix of the Antennapedia homeodomain translocates through biological membranes. *J. Biol. Chem.* **269**, 10,444–10,450.

20. Williams, E., Dunican, D., Green, P., Howell, F., and Derossi, D. (1997) Selective inhibition of growth factor-stimulated mitogenesis by a cell-permeable Grb-2-binding peptide. *J. Biol. Chem.* **272,** 22,349–22,354.
21. He, H., Hirokawa, Y., Levitzki, A., and Maruta, H. (2000) An anti-RAS cancer potential of PP1, an inhibitor specific for SRC family kinases: in vitro and in vivo studies. *Cancer J. Sci. Am.* **6,** 243–248.
22. He, H., Hirokawa, Y., Manser, E., et al. (2001) Signal therapy of RAS-induced cancers in combination of AG 879 and PP1, specific inhibitors for ErbB2 and Src family kinases, that block PAK activation. *Cancer J.* **7,** 191–202.
23. Maruta, H., He, H., Tikoo, A.,Vuong-Nheu, T., and Nur-E-Kamal, M. S. A. (1999) G proteins, phosphatidylinositides, actin-cytoskeleton in the control of cancer growth. *Microsc. Res. Tech.* **47,** 61–66.
24. Glick, A., Sporn, M., and Yuspa, S. (1991) Altered regulation of TGFβ-1 and TGFα in primary keratinocytes and papillomas expressing v-Ha-RAS. *Mol. Carcinog.* **4,** 210–219.
25. Higashiyama, S., Abraham, J., Miller, J., Fiddes, J., and Klagsbrun, M. (1991) A heparin-binding growth factor secreted by macrophage-like cells is related to EGF. *Science* **251,** 936–939.
26. Normanno, N., Selvan, M., Qi, C., Saeki, T., Johnson, G., et al. (1994) Amphiregulin as an autocrine growth factor for c-Ha-Ras /c-ErbB2-transformed human mammary epithelial cells. *Proc. Nat. Acad. Sci. USA* **91,** 2790–2794.
27. Mincione, G., Bianco, C., Kannan, S., Colletta, G., Ciardiello, F., et al. (1996) Enhanced expression of heregulin in ErbB2 and c-Ha-RAS transformed mouse and human mammary epithelial cells. *J. Cell Biochem.* **60,** 437–446.
28. Levitzki A. and Gazit A. (1995) Tyrosine kinase inhibition: an approach to drug development. *Science* **267,** 1782–1788.
29. Hanke, J., Gardner, J., Dow, R., Changelian, P., Brissette, W., et al. (1996) Discovery of a novel, potent, and Src family-selective Tyr kinase inhibitor. *J. Biol. Chem. 271,* 695–701.
30. Maruta, H., Holden, J., Sizeland, A., and D'Abaco, G. (1991) The residues of RAS and Rap proteins that determine their GAP activities. *J. Biol. Chem.* **266,** 11,661–11,668.
31. Tang, Y., Yu, J., and Field, J. (1999) Signals from the RAS, Rac and Rho GTPases converge on the PAK kinase in Rat-1 fibroblasts. *Mol. Cell Biol.* **19,** 1881–1891.
32. Tikoo, A., Shakri, R., Connolly, L., Hirokawa, Y., Shishido, T., et al. (2000) Treatment of RAS-induced cancers by the F-actin bundling drug MKT-077. *Cancer J.* **6,** 162–168.

7

A Ras-Based Module to Generate ^{32}P-Labeled Fusion Proteins for Blot Overlays

Zhoushen Zhao and Louis Lim

1. Introduction

A variety of techniques can be used to find protein partners, including immunoprecipitation, affinity chromatography, blot overlays, and yeast two-hybrid screening. Of these, the blot overlay protocol of sodium dodecyl sulfate (SDS)-acrylamide-gel-separated proteins provides perhaps the most direct assessment of target binding, since one can simultaneously screen multiple cell types or tissue extracts, and infer the size—and in some cases the relative affinity—of binding proteins. Combined use with other techniques such as protein fractionation and affinity chromatography, blot overlay allows us to trace target proteins and undertake their purification. Overlays have also been successfully used for screening of expression libraries (*see* **refs.** ***1,2***), yet the success of the yeast two-hybrid system ***(3)*** has partially eclipsed the use of this method for cDNA screening.

Overlay protocols differ primarily in the method of probe labeling and subsequent detection. Probes for blot overlay can be labeled using nonradioactive moeties, such as biotin for detection by streptavidin ***(3)*** or horseradish peroxidase (HRP)-conjugated antibodies against the probe ***(4)***. Drawbacks of these protocols include lack of sensitivity and background problems caused by factors such as endogenous biotinylated proteins or antibodies. Radioisotopic labeling remains the method of choice, and for several decades, the incorporation of radiolabel into "probes" has relied on [^{125}I] iodination of proteins. However, this isotope is difficult to handle, and has been largely replaced by [^{32}P] ***(5,6)*** or [^{35}S]methionine labeling ***(7)***. Here we describe the use of a new expression vector to produce recombinant proteins for blot overlays, and

From: *Methods in Molecular Biology, vol. 189: GTPase Protocols: The Ras Superfamily*
Edited by: E. J. Manser and T. Leung © Humana Press Inc., Totowa, NJ

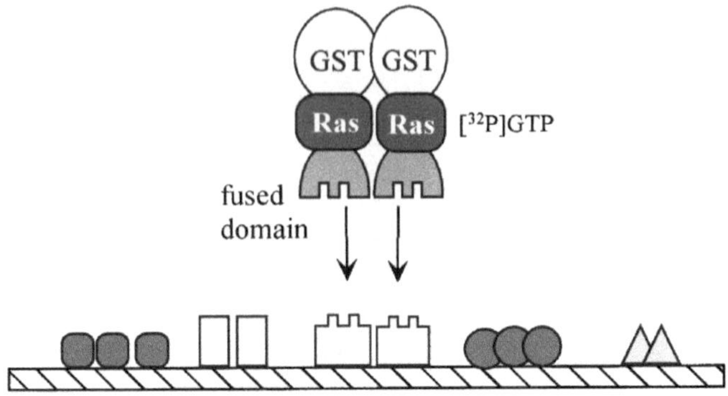

Fig. 1. A schematic diagram of the Ras-GTP module used in blot overlay. GST/Ras fused to a protein domain of choice is labeled with [γ^{32}P]GTP and used as a probe to detect various protein targets that are arrayed on the membrane. These usually represent SDS-polyacrylamide fractioned proteins from total tissue or cell extracts, but can also represent other recombinant proteins or peptide arrays. In its dimeric form the GST domain plays a role in stabilizing the association with targets probably by reducing the "off" rate (*see* **Note 7**).

illustrate its use in detecting SH3 targets. These modular domains comprise approx 55 amino acids. The presence of SH3 domains allows proteins to interact with proline-rich sequences in their binding partners with moderate affinity ***(8,9)***.

Ras binds guanosine diphosphate/guanosine triphosphate (GDP/GTP) binding with high affinity ***(10)*** with measured dissociation constants (K_d) in the picomolar range in the presence of Mg^{2+}. Removal of Mg^{2+} leads to a substantial decrease in affinity for nucleotide ***(11)***, and the rate of exchange of GDP for GTP increases substantially when Mg^{2+} <0.5 μ*M* ***(12)***. These properties of Ras small G proteins allow rapid radiolabeling with GTP by lowering the Mg^{2+} concentration. Subsequently, at concentrations of Mg^{2+} >1 m*M*, the nucleotide exhibits a very slow dissociation rate. Previously, overlays with the [γ^{32}P]GTP-labeled Rho proteins have identified both GTPase-activating protein (GAP) ***(13)*** and downstream target proteins, such as PAK ***(14)*** and ROK ***(15)***. However, [γ^{32}P]GTP-labeled Ras probes do not recognize their targets—such as Raf1, PI-3 kinase, or RasGAP—after gel electrophoresis in SDS (Zhao, Lim, unpublished data). This is probably more a result of stringent tertiary structural requirements for these Ras-associated proteins. Ras can thus be used as a tool to allow efficient labeling of fusion proteins (**Fig. 1**) without significant background binding. This chapter describes how the [γ^{32}P]GTP binding property of Ras allows noncovalent labeling of fusion proteins, and demonstrates overlay results with the Nck SH3 domain.

The protein-labeling method described in this chapter offers a number of advantages as compared to others that have been described *(2,7,16)*. This labeling method provides the following benefits:

1. speed: a 5-min "exchange" reaction is sufficient to label the probe.
2. simplicity: only standard purification methods and [γ^{32}P]GTP are required.
3. efficiency: >70% labeling efficiency is obtained (free label separation is not required).
4. reproducible: the recombinant protein for labeling is produced and stored indefinitely.
5. sensitive: high sensitivity is achieved through the use of the ^{32}P radioisotope.
6. reversible: the noncovalently attached label can be easily removed from the filter for subsequent re-probing.

Finally, the observation that glutathione S-transferase (GST)-mediated probe dimerization is responsible for stable probe association to the target explains how the technique is able to detect associations (such as SH3 binding) that occur with μ*M* affinity *(8,9)*.

2. Materials

2.1. Construction of pGEX-TK-Ras Expression Vector

The cDNA fragment encoding H-Ras residues 1-185 containing the codon G12V mutation (from pXJ40Flag-Ras^{G12V} or other source) was amplified by polymerase chain reaction (PCB) (Vent polymerase NEB) with forward primer (5′gaagatctatgacggaatataagctggtgg3′) and reverse primer (5′gggaattc ggggatcccttgcagctcatgcagccggg 3′). The *Bgl*II site in the forward primer and *Eco*RI and *Bam*HI site in the reverse primer are underlined. The PCR product was digested with *Bgl*II and *Eco*RI, and ligated to a *Bam*HI and *Eco*RI linearized pGEX-4TK vector (a derivative of pGEX-2TK, whose multiple cloning site is identical to that in pGEX-4T-3). The resulting vector pGEX-TK-Ras encodes a fusion protein which can be labeled with either [γ^{32}P]GTP or [γ^{32}P]ATP. Thus, direct comparison of the two labeling protocols (Ras GTP binding vs PKA phosphorylation) is possible.

2.2. Cloning of Nck SH3 Domain

The coding sequence corresponding to the second SH3 domain of Nck (residues 110–176) designated SH3[2] was amplified by PCR and subcloned into the pGEX-TK-Ras vector. The recombinant fusion protein GST/Ras/SH3[2] was purified from *Escherichia coli* (*E. coli*) as described in **Subheading 2.3., item 1**, and analyzed by SDS polyacrylamide gel electrophoresis (**Fig. 3**, *lane 1*).

2.3. Expression and Purification of Recombinant GST-Ras Fusions

1. GST purification buffer: phosphate-buffered saline (PBS) containing 50 m*M* Tris (pH 8.0), 0.5% Triton X-100, 0.5 m*M* $MgCl_2$.
2. The pGEXTK-Ras plasmids were transformed into *E. coli* BL21 (DE3)-competent cells (Novagen) and 4 mL of overnight culture were diluted in 400 mL of 1X Luria broth (LB)-ampicillin and grown to OD_{600} = 0.9 at 37°C. Expression of GST-Ras-SH3 fusion protein was induced by adding a final concentration of 0.5 m*M* isopropylthio-β-D-galactoside (IPTG, Gibco-BRL) and further incubation for 3–4 h at 25°C.
3. Cells were pelleted and resuspended in 20 mL of purification buffer + 1 µg/mL lysozyme, 0.5 m*M* phenylmethylsulfonyl fluoride (PMSF), and 5 m*M* dithiothreitol (DTT). Cells are lysed after 15 min incubation on ice by sonication for 3 × 1 min intervals at 4°C (HeatSystems XL2020).
4. The bacterial lysate is clarified by centrifugation at 100,000*g*, 4°C for 30 min with Beckman Ti50 rotor. The GST-fusion protein is purified by passing the bacterial lysate through a 400-µL glutathione Sepharose 4B column: the column is washed with 10 mL of PBS containing 0.1% Triton X-100 followed by elution in 2 column volumes (i.e., 800 µL) of purification buffer containing 5% glycerol and 10 m*M* reduced glutathione (Sigma).
5. The fusion protein contains an internal thrombin cleavage site to allow removal of the GST moiety (*see* **Fig. 2**). Run the purified protein on SDS polyacrylamide gel (9%) to confirm integrity of the product. The purified GST/Ras fusion proteins are stored in suitable aliquots at –80°C.

2.4. Preparation of Tissue and Cell Extracts for Analysis

1. Extraction buffer: PBS containing 50 m*M* Tris, pH 8.0, 5 m*M* DTT, "complete" protease inhibitor cocktail (Roche), 10 m*M* β-glycerolphosphate, 1 m*M* Na_3VO_4, 0.5% Triton X-100, and 5% glycerol.
2. Rat tissues are minced and freshly homogenized in extraction buffer (10 mL/g) using 20 strokes of a hand-held Dounce homogenizer. The extracts were clarified by spinning at 16,272*g* for 10 min (Sorvall GSA rotor) and 100,000*g* for 40 min (Beckman Ti50 rotor or equivalent).
3. The soluble protein is then diluted to 10 µg/µL with extraction buffer and stored at –80°C in aliquots.
4. To prepare soluble cultured-cell extracts, each 100-mm dish (~80% confluence) is harvested in 500 µL of extraction buffer using a rubber-cell scraper, and is completely lysed by a brief sonication. The lysate is spun for 10 min, (15,000*g* Eppendorf microcentrifuge) and the supernatant stored at –80°C.
5. Alternatively, 500 µL of 1X SDS *sample buffer* (1% SDS, 50 m*M* Tris.Cl pH 6.8, 10 m*M* DTT, 10% glycerol, 0.02% Bromophenol blue) can be used to harvest total cellular proteins. This requires sonication of genomic DNA for 20 s with a microprobe to reduce viscosity. Centrifugation for 15 min (15,000*g* Eppendorf microcentrifuge) removes insoluble debris.

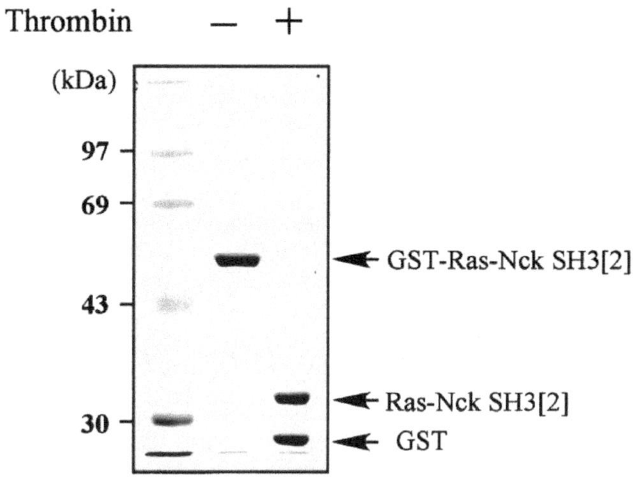

Fig. 2. Purified GST/Ras/NckSH3[2] protein used for labeling. Purified GST/Ras/SH3[2] (2 mg) was separated on a 10% SDS-PAGE and stained with Coomassie brilliant blue. The sample treated with thrombin (4 μg of purified GST/Ras/SH3[2] digested with 0.25 U of thrombin [Sigma] for 30 min at room temperature) to give complete cleavage. The various protein bands are indicated on the right.

6. The protein concentration of tissue and cell extracts (without SDS) is determined using a modified Bradford assay (Bio-Rad).

2.5. Electrophoretic Separation and Transfer of Protein Samples to PVDF Membrane

1. In each lane, 100 μg of total tissue or cell extract is separated on a SDS-polyacrylamide gel using the Bio-Rad Mini-PROTEAN system or equivalent (*see* **Note 1**).
2. Each polyvinylidene fluoride (PVDF) membrane (9 × 10 cm, NEN) is pre-wet in methanol and overlayed with a polyacrylamide gel for transfer using a "semidry" blotter (e.g., Bio-Rad). Efficient transfer of higher mol-wt proteins is achieved only by blotting overnight at 4°C (with 25 mA per gel—set voltage limit to 12V).
3. After transfer, proteins on the membrane can be visualized by staining (*see* **Note 2**).

3. Methods

3.1. Preparation of [γ^{32}P]GTP-labeled Probes

1. Nucleotide-exchange buffer (X2): 100 m*M* KCl, 50 m*M* N-2hydroxy-ethylpiperazine-N′-2-ethanesulfonic acid (HEPES, pH 7.3), 10 m*M* ethylene-diamine-tetraacetic acid (EDTA), 1 mg/mL bovine serum albumin (BSA), 0.1% Triton X-100. The buffer should be filtered through a 0.2-μm syringe filter (Sartorius), and can be stored at 4°C for up to 1 mo or at –20°C.

2. For each reaction, 10 μg of GST/Ras/SH3[2] fusion protein (first checked for integrity by SDS polyacrylamide) is labeled in 50 μL of exchange buffer containing 2 μL of [γ^{32}P]GTP (NEN ~6000 Ci/mmol). BSA and Triton X-100 in the reaction mix to stabilize the fusion protein and reduce nonspecific absorption to various surfaces.
3. The labeling reaction is performed at room temperature for 5 min, and the reaction mix is returned to ice and used within 15 min. Because GTP exchange requires low Mg^{2+} ***(11,12)*** and the overlay-binding buffer containing excessive amount of Mg^{2+}, the free radiolabeled GTP does not label other GTP-binding proteins on the filter under this condition (*see* **Note 3**).

3.2. Filter Overlay with ^{32}P-Labeled GST/Ras/SH3[2] Probes

1. Binding/wash buffer: PBS containing 10 m*M* HEPES, pH 7.3, 5 m*M* $MgCl_2$, and 0.05% Triton X-100 (store at 4°C for up to 1 wk).
2. The stained PVDF filter is destained with 100% MeOH briefly, washed in water, and blocked for 1 h at room temperature with 20 mL of PBS containing 10% BSA (Sigma) in a 50-mL falcon tube on a rolling mixer.
3. For each PVDF membrane, 4 mL of binding buffer together with either the [γ^{32}P]GTP-labeled probe is incubated on a roller for 3–4 h at 4°C.
4. The filter is washed 3× (5 min) in a container with 50 mL washing buffer, briefly blotted with Whatman 3MM paper, and placed to X-ray film at –70°C for 1–6 h. High backgrounds may be encountered. In this case, *see* **Note 4**.
5. For subsequent re-probing of the filter, the radiolabel can be removed under relatively mild conditions (*see* **Note 6**).

Figure 3 shows the overlay signal generated by the second SH3 domain (SH3[2]) of the adaptor Nck: a variety of binding proteins (left panel) are detected by the GST-Ras[γ^{32}P]GTP probe whose identities we have published previously ***(19)***. These Nck-binding proteins are: NIK (*p145*), Synaptojanin (*p135*), Dynamin (*p100*), Sam68, n-WASP, α-PAK (*p68*), and hnRNP-K (*p50*). For comparison, we show the signal obtained using protein kinase A (PKA) labeling of the fusion protein with [γ^{32}P]ATP. The Ras labeling system clearly produces stronger signals, which probably relate to the more efficient radiolabel incorporation, and incorporation of [γ^{32}P] into PKA rather than the probe. In both cases, the overlay signals actually depend on the presence of the GST moiety in the probe (*see* **Note 7**).

4. Notes

1. Glycerol (10%) is included in the polyacrylamide gel mix to increase the band sharpness and to slow down protein diffusion during electrophoretic transfer to PVDF.
2. PVDF membranes can be stained with Coomassie blue to determine the quality of transfer. In our hands, the final signal is unaffected by such staining. Mark

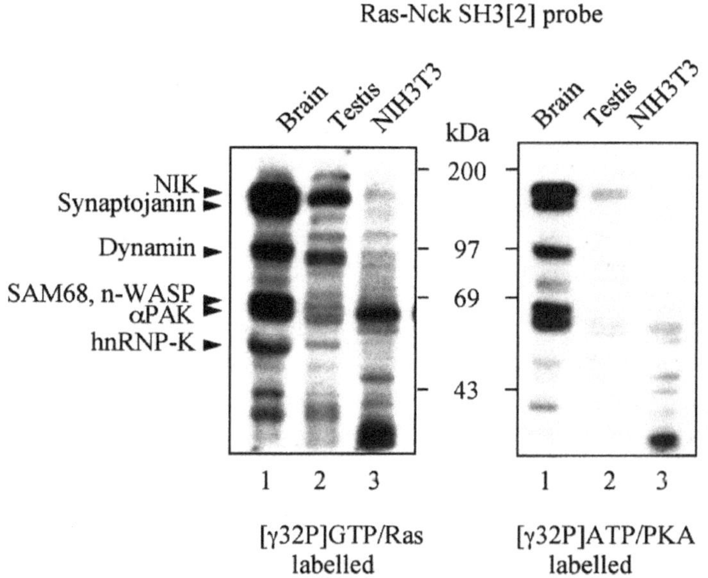

Fig. 3. A comparison of blot overlay performed with Ras [γ^{32}P]GTP labeled probe versus [g32P]ATP phosphorylated probe. Each lane contains 100 µg of rat brain, testis, or NIH-3T3 cell extract (as indicated) separated on a 7.5% SDS-polyacrylamide gel. Duplicate filters were probed with [γ^{32}P]GTP-labeled GST/Ras/SH3 probe *(left panel)* or the same protein labeled with [γ^{32}P]ATP and protein kinase A (Sigma). In each case, 10 µg of GST/Ras-Nck SH3[2] fusion protein was labeled either with 2 µL of [γ^{32}P]GTP (>6000 Ci/mmol) at room temperature for 5 min in a 50-µL reaction (*see* **Subheading 3.**) or with 5 µL of [γ^{32}P]ATP (>6000 Ci/mmol) and 25 U of PKA catalytic subunit (Sigma, P-2645) at 32°C for 30 min in a 50-µL reaction. After washing, the PVDF filters were exposed at –70°C for 30 min.

the position of the lower corners of the gel and the top edge of the separating gel with an oil-based pen. Immerse the filter in 0.1% Coomassie blue dissolved in 40% MeOH/10% acetic acid for 3 min., and destain in the same buffer. For convenience, a record of the blot can be scanned or photographed at this stage for comparison with autoradiographic signals.

3. Ras proteins are stabilized by the presence of Mg.GTP or Mg.GDP. It is therefore important to have $MgCl_2$ in any buffer in which the GTPase remains for any length of time.
4. If background signals are found to be a problem after developing the autoradiographic film, the PVDF filter can be rewashed at 4°C (15 min 50-mL buffer) blotted and re-exposed. This often improves the quality of the image.
5. More [γ^{32}P]GTP can be used to in the labeling reaction to compensate for specific activity loss with time. As the protein concentration recommended is

>10-fold excess over that of the [γ^{32}P]GTP added, up to 10 μL radiolabel can be added per reaction.

6. In order to strip the PVDF filters for subsequent reprobing, they are incubated in phosphate-buffered saline (PBS) containing 0.1% SDS and 10 m*M* EDTA. The amount of radiolabel remaining on the filter can be monitored with a hand-held Geiger counter.
7. Gel-filtration assays show that GST exists as dimers in solution ***(20)***, and this is seen in the crystal structure to be mediated by a complementary interface ***(21)***. Thus, GST-fusion probes used in overlay assays are present exclusively in a dimeric form. The fact that one never detects an interaction between labeled GST and the "monomeric" protein separated by SDS gel electrophoresis probably reflects the stability of the GST dimer in solution. If the GST/Ras/SH3 probe is cleaved with thrombin to remove the GST portion it is found that the labeled Ras/SH3 "probe" gives very little signal, indicating that the sensitivity of the technique derives from the ability of the probe to dimerize.

Acknowledgments

This work is supported by the Glaxo Singapore Research Fund.

References

1. Cicchetti, P., Mayer, B. J., Thiel, G., and Baltimore, D. (1992) Identification of a protein that binds to the SH3 region of Abl and is similar to Bcr and GAP-rho. *Science* **257,** 803–806.
2. Ren, R., Mayer, B. J., Cicchetti, P., and Baltimore, D. (1993) Identification of a ten-amino acid proline-rich SH3 binding site. *Science* **259,** 1157–1161.
3. Fields, S. and Song, O. (1989) A novel genetic system to detect protein-protein interactions. *Nature* **340,** 245–246.
4. Fouassier, L., Yun, C. C., Fitz, J. G., and Doctor, R. B. (2000) Evidence for ezrin-radixin-moesin-binding phosphoprotein 50 (EBP50). Self-association through PDZ-PDZ Interactions. *J. Biol. Chem.* **275,** 25,039–25,045.
5. Kaelin, W. G., Jr., Krek, W., Sellers, W. R., DeCaprio, J. A., Ajchenbaum, F., Fuchs, C. S., et al. (1992) Expression cloning of a cDNA encoding a retinoblastoma-binding protein with E2F-like properties. *Cell* **70,** 351–364.
6. Ron, D. and Dressler, H. (1992) pGSTag-a versatile bacterial expression plasmid for enzymatic labeling of recombinant proteins. *Biotechniques* **13,** 866–869.
7. Feller, S. M., Knudsen, B., Wong, T. W., and Hanafusa, H. (1995) Detection of SH3-binding proteins in total cell lysates with glutathione S-transferase-SH3 fusion proteins: SH3 blot assay. *Methods Enzymol.* **255,** 369–378.
8. Pawson, T. (1995) Protein modules and signalling networks. *Nature* **373,** 573–580.
9. Cohen, G. B., Ren, R., and Baltimore, D. (1995) Modular binding domains in signal transduction proteins. *Cell* **80,** 237–248.
10. John, J., Sohmen, R., Feuerstein, J., Linke, R., Wittinghofer, A., and Goody, R. S. (1990) Kinetics of interaction of nucleotides with nucleotide-free H-ras p21. *Biochemistry* **29,** 6058–6065.

11. Hara, M., Tamaoki, T., and Nakano, H. (1988) Guanine nucleotide binding properties of purified v-Ki-ras p21 protein produced in Escherichia coli. *Oncogene Res.* **2,** 325–333.
12. Hall, A. and Self, A. J. (1986) The effect of Mg^{2+} on the guanine nucleotide exchange rate of p21N-ras. *J. Biol. Chem.* **261,** 10,963–10,965.
13. Manser, E., Leung, T., Monfries, C., Teo, M., Hall, C., and Lim, L. (1992) Diversity and versatility of GTPase activating proteins for the p21rho subfamily of ras G proteins detected by a novel overlay assay. *J. Biol. Chem.* **267,** 16,025–16,028.
14. Manser, E., Leung, T., Salihuddin, H., Zhao, Z. S., and Lim, L. (1994) A brain serine/threonine protein kinase activated by Cdc42 and Rac1. *Nature* **367,** 40–46.
15. Leung, T., Chen, X. Q., Manser, E., and Lim, L. (1996) The p160 RhoA-binding kinase ROK alpha is a member of a kinase family and is involved in the reorganization of the cytoskeleton. *Mol. Cell. Biol.* **16,** 5313–5327.
16. Tanaka, S., Morishita, T., Hashimoto, Y., Hattori, S., Nakamura, S., Shibuya, M., et al. (1994) C3G, a guanine nucleotide-releasing protein expressed ubiquitously, binds to the Src homology 3 domains of CRK and GRB2/ASH proteins. *Proc. Natl. Acad. Sci. USA* **91,** 3443–3447.
17. Sastry, L., Lin, W., Wong, W. T., Di Fiore, P. P., Scoppa, C. A., and King, C. R. (1995) Quantitative analysis of Grb2-Sos1 interaction: the N-terminal SH3 domain of Grb2 mediates affinity. *Oncogene* **11,** 1107–1112.
18. Manser, E., Huang, H. Y., Loo, T. H., Chen, X. Q., Dong, J. M., Leung, T., et al. (1997) Expression of constitutively active alpha-PAK reveals effects of the kinase on actin and focal complexes. *Mol. Cell. Biol.* **17,** 1129–1143.
19. Zhao, Z. S., Manser, E., and Lim, L. (2000) Interaction between PAK and Nck: a template for Nck targets and role of PAK autophosphorylation. *Mol. Cell. Biol.* **20,** 3906–3917.
20. Park, D. (1997) Cloning, sequencing, and overexpression of SH2/SH3 adaptor protein Nck from mouse thymus. *Mol. Cells* **7,** 231–236.
21. Lim, K., Ho, J. X., Keeling, K., Gilliland, G. L., Ji, X., Ruker, F., et al. (1994) Three-dimensional structure of Schistosoma japonicum glutathione S-transferase fused with a six-amino acid conserved neutralizing epitope of gp41 from HIV. *Protein Sci.* **3,** 2233–2244.

II

Specific Methods

Rho Family GTPases

8

Determination of the Activity of Rho-Like GTPases in Cells

Angeliki Malliri, Jean P. ten Klooster, Christina Olivio, and John G. Collard

1. Introduction

Rho-like GTPases, including RhoA, Rac1 and Cdc42, have been implicated in the control of a wide range of biological processes such as the organization of cytoskeletal structures, adhesion, motility, transcriptional activation, and cell-cycle progression *(1–3)*. The Rho-like GTPases comprise a subfamily of the Ras superfamily of small guanosine triphosphate (GTP)-binding proteins. Like all members of the Ras superfamily, the Rho-GTPases function as molecular switches, cycling between an inactive guanosine diphosphate (GDP)-bound state and an active GTP-bound state. Among all Ras superfamily members, specific guanine-nucleotide-exchange factors (GEFs) activate Rho-like GTPases by promoting the transition from the GDP-bound to the GTP-bound state, while GTPase-activating proteins (GAPs), which increase the intrinsic rate of hydrolysis of bound GTP, inactivate Rho-like proteins. Rho-like GTPases are further regulated by guanine-nucleotide dissociation inhibitors (GDIs), which can inhibit both the exchange of GTP and the hydrolysis of bound GTP. Rho GDIs, as well as members of the ERM family (ezrin, radixin, and moesin), also appear to play a crucial role in the translocation of Rho-like GTPases between membranes and the cytoplasm—a process also associated with their activation. During recent years, numerous downstream effector proteins that bind Rho, Rac, and Cdc42 in their GTP-bound form have only been identified in an attempt to reveal the underlying molecular mechanisms by which the Rho-like GTPases mediate their various effects. Rac and Cdc42 target proteins include the p*21* Rac/Cdc 42-activated Kinases (Pak) family of serine/threonine kinases

From: *Methods in Molecular Biology, vol. 189: GTPase Protocols: The Ras Superfamily*
Edited by: E. J. Manser and T. Leung © Humana Press Inc., Totowa, NJ

(4–6). All PAK proteins identified to date share a similar motif stretching over 18 amino acids, which mediates the interaction with Rac or Cdc42. This motif is referred to as the Cdc42/Rac interacting binding (CRIB) domain. Rho target proteins include the serine/threonine kinases ROKα/RHO kinase and p160ROCK, as well as *p140*mDia, Rhophilin, and Rhotekin *(1)*.

Since Rho-like GTPases play a crucial role in a wide range of biological processes, the ability to determine their activation status in cells in various contexts is essential. In previous studies, the activation of a specific Rho-like GTPase in a selected cell type could only be inferred by comparing the resulting cytoskeletal changes with those induced by introduction of activated or dominant-negative mutants of the GTPase in the same cell type. This approach has a major disadvantage—it does not allow detection of active Rho-like GTPases in certain circumstances (such as transcriptional activation), or in cell types (e.g., lymphocytes) for which phenotypic changes are not apparent. Therefore, the development of biochemical methods to directly determine the activity of Rho-like GTPases is of critical importance.

The Rac activity assay, as described in Sander et. al. (*see* **Fig. 1**; **ref.** *7*), is based on the Rap1 and Ras activity assays described previously by Franke et al. *(8)* and Rooij and Bos *(9)*. Common to all these assays is a chimeric protein consisting of the glutathione-binding moiety of glutathione-S-transferase (GST) fused to part of an effector molecule which binds to the GTPase in its GTP-bound form. The complex of fusion protein and GTPase is then isolated from a cell lysate by immobilization ("pull-down") of the GST moiety on a sepharose substrate to which glutathione has been adsorbed. Finally, following elution from the glutathione Sepharose, the captured GTP-bound GTPase is detected by Western blotting. For the Rac and Cdc42 activity assays, the CRIB domain of the kinase PAK was fused to GST (GST-PAK-CD) *(7)*, and the Rho-binding domain of the Rho effector molecule Rhotekin fused to GST (GST-C21) was used in the Rho activity assay *(10,11)*.

The ability of these pull-down assays to distinguish between the activities of the various Rho-like GTPases is clearly demonstrated in **Fig. 2**. The assays have allowed the measurement of the activation of endogenous Rho-like proteins in response to various stimuli, such as growth-factor treatment *(11–13*; **Fig. 3**) or engagement of integrins *(7,10,14)*, and have been used to demonstrate cross-communication between different Rho-like GTPases (*11,12*; **Fig. 3**). The pull-down assays were also used to examine the specificity of novel putative GEFs or known GEFs in various cell types (*11,13*; **Figs. 2** and **3**). Previously, this could only be determined in vitro through radioactive GTP-loading assays, or indirectly in vivo by observing cytoskeletal changes following overexpression of the GEF under examination (*see* **ref.** *15*). In principle, it should be possible to adapt the pull-down assay to determine the activity of any other

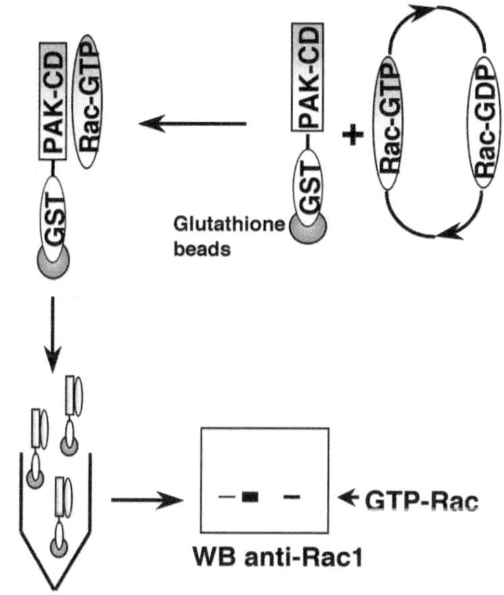

Fusion protein	Binding	Detection
GST-PAK-CD:	Rac1	_-Rac1
	Cdc42	_-Cdc42
GST-C21:	RhoA	_-RhoA

Fig. 1. Rac/Cdc42/Rho activity assays. Outline of the experimental scheme used to isolate Rho-like GTPases in their active GTP-bound state (here shown for Rac-GTP). Cell lysate is mixed with GST-fusion protein attached to glutathione-bearing Sepharose beads. Centrifugation of the sepharose beads is followed by washing and elution of the captured active-Rac in sample-loading buffer. Protein is then resolved by SDS-PAGE, transferred to membrane, and probed with a Rac1-specific antibody before detection by ECL. The fusion protein GST-PAK-CD is used to detect active Rac1 and Cdc42, whereas GST-C21 is used to measure the amount of active RhoA.

Rho family member, by generating a fusion protein that includes the binding domain of a specific effector molecule. This process also requires that a specific antibody is available to detect the Rho-like protein by Western blotting.

Recently, we have also begun to use Cdc42/Rac/Rho activity assays for the detection of GTP-bound Rho-like molecules in tissue samples. Measurement of the activity of these molecules in solid tumors should now be possible. Mutant Rho-like molecules have not yet been detected in tumors, and it is uncertain whether the activity of Rho-like molecules is required or altered during tumorigenesis in vivo. This is an important issue to resolve, considering

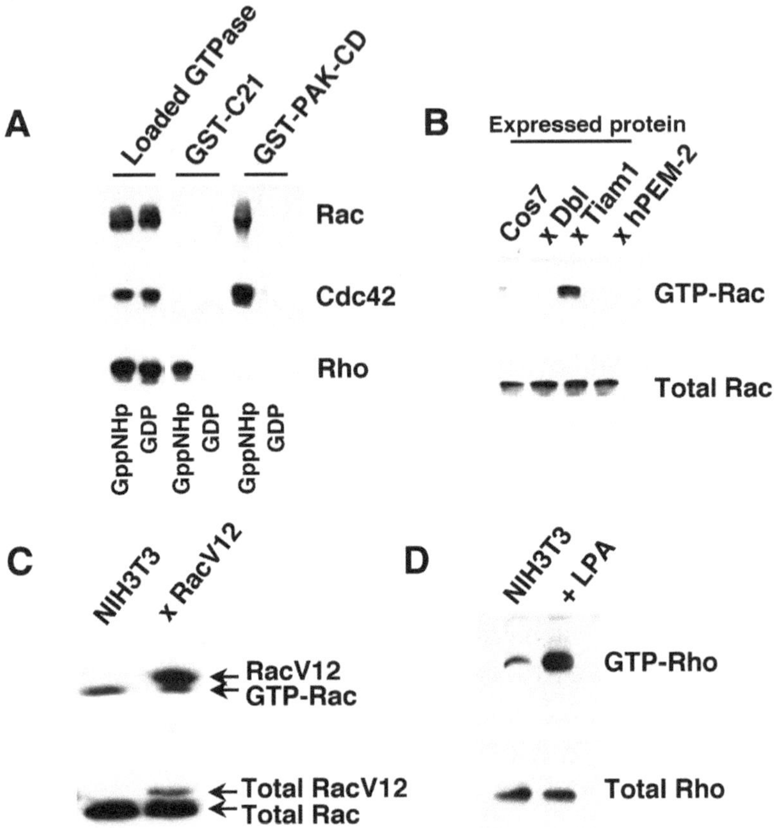

Fig. 2. Specificity of the Rac/Cdc42/Rho activity assays. (**A**) Recombinant GTPases shown were loaded in vitro with either Gpp(NH)p-a nonhydrolyzable GTP analog-or GDP and purified with either GST-PAK CD or GST-C21. Bound GTPases were then analyzed by Western blotting. The fusion proteins bind the GTPases in their GTP-bound form only *(13)*. (**B**) Cos7 cells were transfected with either empty vector, or with a vector expressing one of the following GEFs: Dbl, Tiam1, or hPEM-2. Cells were lysed after 24 h, and activated Rac was detected by pull-down with GST-PAK-CD and western blotting with an anti-Rac1 antibody. Only Tiam1 activates Rac *(13)*. (**C**) and (**D**) show positive controls used for Rac and Rho assays, respectively. (**C**) For Rac, NIH-3T3 cells were transfected with the constitutively active *myc*-tagged RacV12 mutant. 24 h later, active Rac protein was detected by pull-down with GST-PAK-CD and Western blotting with an anti-Rac1 antibody. A band for both the mutant RacV12 protein and endogenous Rac1 protein (migrates faster) is visible. Note the difference in relative amounts between V12Rac and endogenous Rac1 in total cell lysates and pull-downs. (**D**) For RhoA, serum-starved (24 h) NIH-3T3 cells were stimulated with 5 μ*M* LPA, a known activator of RhoA. After 15 min, active RhoA protein was detected by pull-down with GST-C21 and Western blotting with an anti-RhoA antibody.

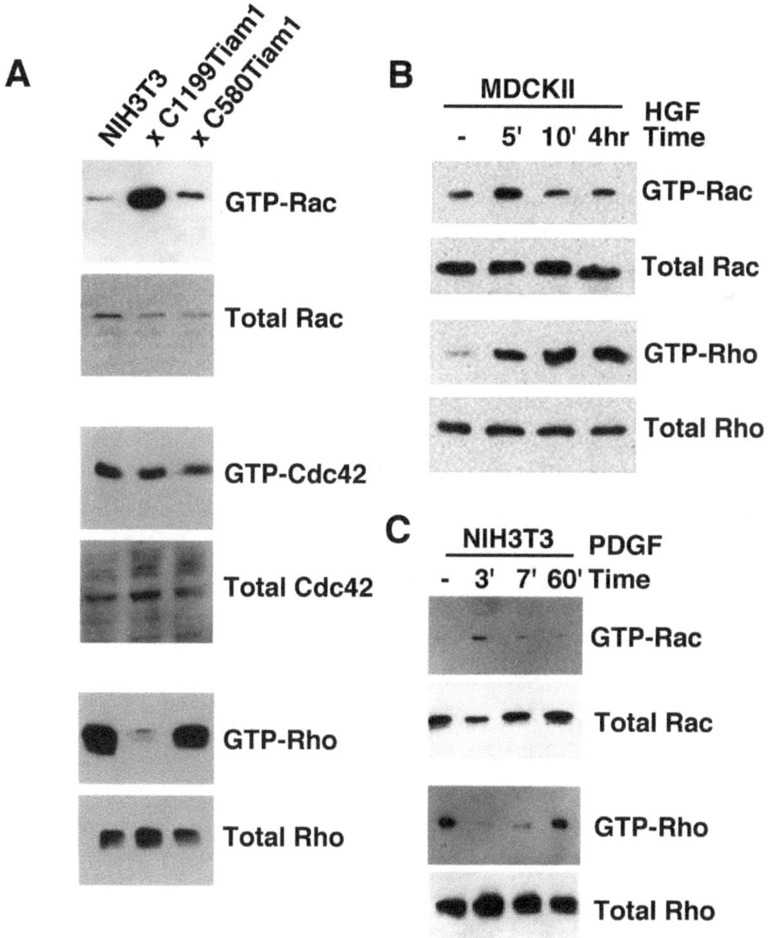

Fig. 3. Examples of applications of the Rac/Cdc42/Rho activity assays. (**A**) Downregulation of RhoA by activation of Rac1. NIH-3T3 cells were transduced with empty virus or virus expressing either a truncated active mutant form of Tiam1 (C1199Tiam1) or an inactive mutant of Tiam1 (C580Tiam1) *(19)*. In C1199Tiam1-expressing cells, Rac1 is strongly activated *(upper panels)* compared to cells expressing empty virus only or C580Tiam1, Cdc42 activity remains unaltered *(middle panels)*, while RhoA activity is suppressed (lower panels). (**B**) Hepatocyte growth factor/scatter factor treatment of serum-starved (for 24 h) MDCKII cells induces transient activation of Rac1, optimal by 5 min *(upper panels)* and a prolonged activation of RhoA *(lower panels)*. (**C**) PDGF treatment of serum-starved (for 24 h) NIH-3T3 cells transiently activates Rac1 *(upper panels)*, which is accompanied by downregulation of RhoA activity *(lower panels)*.

the key regulatory roles of Rho-like molecules in several of the signaling pathways that become deregulated in cancer cells ***(3,12,16,17)***. If the activities of Rho-like proteins are altered in tumors, then the assays may be of clinical interest as diagnostic tools, and it may be of further interest to develop methods to influence the function of Rho-like proteins in tumor cells.

One limitation of the present pull-down assays is that they measure the activity of bulk Rho-like GTPases within a population of cells. During a process as complex as directed cell migration, in which cell adhesions are continuously being broken and reformed and actin polymerization and depolymerization occur simultaneously and repeatedly, the activity of Rho-like molecules within an individual cell is also likely to differ spatially and temporally. Further, cells within a nonclonal population may respond differently to a given stimulus. Thus, the extent and magnitude of changes detected in an activity assay may not reveal the true subtlety with which the activities of Rho-like GTPases are coordinately regulated within individual cells. A major advancement in this field would be the development of probes that allow the activation of Rho-like molecules to be visualized within individual living cells in real time.

2. Materials

2.1. Expression and Purification of GST-Fusion Proteins

GST-fusion proteins were expressed from pGEX vectors (Promega) in *Escherichia coli* (*E. coli*), cultured in Luria-Bertani medium in the presence of ampicillin (50 µg/mL) (Roche). Isopropylthiogalactoside (IPTG) (Roche) was added to induce protein expression. For purification of fusion proteins, *see* **Subheading 3.**

1. Bacterial-lysis buffer: 50 m*M* Tris-HCl, pH 7.5, 1 m*M* EDTA, 100 m*M* NaCl, 5% glycerol, 0.1% Triton X-100. Add 1 m*M* dithiothreitol (DTT), 1 µg/mL leupeptin, 0.7 µg/mL pepstatin, and 2 µg/mL aprotinin immediately before use (all obtained from Roche).
2. Glutathione Sepharose was obtained from Amersham Pharmacia Biotech.

2.2. Protein Extraction from Cells/Tissues and Electrophoresis

1. Proteins were extracted from cells in culture or from homogenized tissues in the following lysis buffer (cell-lysis buffer): 50 m*M* Tris-HCl, pH 7.4, 2 m*M* $MgCl_2$, 1% NP-40, 10% glycerol, 100 m*M* NaCl, 1 m*M* benzamidine, 1 m*M* DTT, 1 µg/mL leupeptin, 1 µg/mL pepstatin, and 1 µg/mL aprotinin (benzamidine, DTT, leupeptin, pepstatin, and aprotinin were added to the lysis buffer just before use).
2. 2× Laemmli sample buffer: 62.5 m*M* Tris pH 6.8, 2% sodium dodecyl sulfate (SDS), 10% glycerol, 0.1% Bromophenol, 5% β-mercaptoethanol.

3. For electrophoresis of Cdc42/Rac/Rho-assay protein samples and protein from total cell lysates, 12% polyacrylamide-resolving gels were prepared (using premixed 40% acrylamide/*bis* solution [37.5 : 1 from Bio-Rad] in 375 m*M* Tris-HCl pH 8.8, 0.1% SDS) and overlaid with a 4% stacking polyacrylamide gel (using the above acrylamide/*bis* solution in 125 m*M* Tris-HCl pH 6.8, 0.1% SDS).
4. Western transfer buffer: 25 m*M* Tris (base), 192 m*M* glycine, 20% methanol.

2.3. Antibodies and Western Blot Solutions

1. Blocking buffer: 5% skimmed milk in 10 m*M* Tris-HCl pH 7.5, 150 m*M* NaCl, 0.1% Tween-20 (TBST).
2. Wash buffer: TBST.
3. Antibodies used: mouse monoclonal anti-Rac1 antibody was obtained from Transduction Laboratories, and mouse monoclonal anti-RhoA and rabbit polyclonal anti-Cdc42 P1 were obtained from Santa Cruz Biotechnology, Inc. Horseradish peroxidase (HRP)-conjugated secondary antibodies were purchased from Amersham Pharmacia Biotech. Enhanced chemiluminescence (ECL) detection kit was also purchased from Amersham Pharmacia Biotech.

3. Methods

3.1. Cloning and Production of GST-PAK-CD and GST-C21-Fusion Proteins

1. A fragment encoding amino acids 56-141 from human PAK1B (the CRIB domain, CD) was generated by standard PCR and inserted between the *Bam*HI and *Eco*RI sites of pGEX2TK to yield GST-PAK-CD *(7)*. A vector expressing GST-C21 containing the NH2-terminal 90 amino acids representing the Rho-binding domain from the Rho effector protein Rhotekin was similarly constructed ***(13,18)***.
2. *E. coli* BL21 cells transformed with the GST-fusion constructs are grown for 6 h in 25 mL LB/ampicillin at 37°C. Expression of recombinant protein is induced by addition of IPTG to 0.2 m*M* and further incubation in a total volume of 500 mL Luria broth (LB)/ampicillin at 25°C overnight (*see* **Note 1**). Cells are harvested by centrifugation (5 min at 2,000 *g*), resuspended in 10 mL bacterial lysis buffer, and then sonicated (2 × 30 s, 50% cycle, mark 6). Cell lysates are aliquoted (1 mL) and centrifuged at 4°C for 20 min at 21,000 *g*, and the supernatant is transferred to a fresh microcentrifuge tube and frozen at –80°C. The amount of fusion protein expressed is then determined. An aliquot of the supernatant is incubated with glutathione-coupled Sepharose 4B beads for 1 h at 4°C (in a ratio of 200 µL of 50% bead slurry per 1 mL of supernatant). Protein bound to the beads is washed 3× in cell lysis buffer and eluted in Laemmli sample buffer. The amount of bound fusion protein is estimated by comparing to bovine serum albumin (BSA) standards resolved in parallel on a 12% reducing polyacrylamide gel, and afterwards stained with Coomassie blue.

3. Prior to performing the GST-pull-down assay to determine the activity of specific Rho-like GTPases in cell lysate or tissue lysate samples, appropriate numbers of bacterial supernatant aliquots are defrosted on ice and incubated with glutathione-coupled Sepharose 4B beads as described in **Subheading 3.1., step 2** (*see* **Note 2**). Sufficient beads are then added to each sample to be analyzed to yield 20 μg of fusion protein per sample.

3.2. GST-Pull-Down Assays for Rho-Like GTPase Activity

1. When possible, an equivalent number of cells ($5–10 \times 10^6$) is analyzed for each different sample. Cell-culture dishes are placed on ice, and cells are washed with ice-cold phosphate-buffered saline (PBS) (containing 1 m*M* $MgCl_2$ and 0.5 m*M* $CaCl_2$) and incubated for 5 min on ice in cell-lysis buffer (500 μL–1 mL). A cell scraper is used to harvest cell lysates, which are then centrifuged for 5 min at 21,000 *g* at 4°C. Aliquots are taken from the supernatant at this stage to allow subsequent comparison of total Rho-like GTPase amounts by Western blotting (*see* **Notes 3** and **4**). The remainder of the cell lysate supernatant is incubated with bacterially produced GST-fusion protein bound to glutathione-coupled Sepharose beads (*see* **Subheading 3.1., step 2**) at 4°C for 30 min for GST-PAK-CD and 1 h for GST-C21 (*see* **Note 5**). The beads and proteins bound to the fusion protein are washed 3× in an excess of cold cell-lysis buffer, eluted in Laemmli sample buffer, boiled for 5 min, and then analyzed for bound Rho-like molecules by SDS-PAGE and Western blotting using the appropriate primary antibody (*see* **Notes 6** and **7**). Detection is by ECL following incubation with a HRP-conjugated secondary antibody. Signal is detected by exposure of Western blots to X-ray film (*see* **Note 8**).
2. It is also possible to perform these same pull-down assays on tissue samples. In this case, tissues of interest are removed surgically from the organism and immediately frozen in liquid nitrogen. Tissues are then homogenized to a fine powder by grinding in a pestle while still frozen, and lysed directly by addition of cell-lysis buffer to the residual powder. Lysates are cleared by centrifugation as in **step 1** above. To standardize the assay, it is necessary to first quantitate the protein concentration in a fraction of the sample lysate (the remainder of the lysate is aliquoted, snap-frozen, and stored at –80°C) (*see* **Note 9**). Once the protein concentration is known, then sample lysates can be thawed on ice, and volumes containing equivalent amounts of protein can be added to fusion-protein-bound GST-beads and further processed as described for cell lysates.

4. Notes

1. We perform GST-fusion protein induction with IPTG overnight at 25°C instead of 37°C for 2 h. Thus, most of the GST-fusion proteins remain in their soluble form.
2. Couple the GST-fusion proteins with the glutathione-Sepharose beads just before use, or at least on the same day it will be used. It is not advisable to store coupled beads.

3. It is important to include total cell lysates among the samples for Western blotting, even if apparently equivalent numbers of cells or amounts of protein are being used for the pull-down assays. This will control for inconsistencies in sample handling or measure possible effects of expressing certain genes in various cell lines on the expression levels of Rho-like GTPases. If cell lines are believed to proliferate at different rates, it is advisable to quantitate protein levels following lysis in cell-lysis buffer (*see* **Note 9**).
4. Although it is preferable to perform the Rho-like GTPase activity assays immediately after lysing cells, if this proves difficult or if protein concentration needs to be determined first, protein lysates can be snap-frozen and then defrosted on ice before being assayed.
5. Rho-like molecules have high intrinsic GTPase activity. Therefore, it is necessary to lyse samples in the minimum time and to add GST-PAK-CD or GST-C21 as soon as possible. Do not rotate protein lysates with the fusion-protein coupled beads for more than 30–60 min. During the pull-down assays, keep samples cold at all times.
6. It is advisable to include a positive control whenever performing Rho-like GTPase activity assays. *Myc*-tagged RacV12 and Cdc42V12 expressing cells are good positive controls for the Rac and Cdc42 assays, respectively. The constitutively active mutant proteins can be distinguished from endogenous active proteins by their reduced electrophoretic mobility on SDS-polyacrylamide gels (*see* **Fig. 2**). Although overexpression of RhoV14 induces an activated Rho phenotype, it hardly binds GST-C21 for unknown reasons (*see also* ***18***). Therefore, we prefer to use a lysate from serum-starved fibroblasts treated with Lysophosphatidic acid (LPA) as a positive control in the Rho-activity assays (*see* **Fig. 2**).
7. PVDF membranes can be stained with Coomassie blue to determine the quality of the protein transfer following blotting. This does not appear to affect the subsequent binding of the anti-Cdc42, -Rac, or -Rho antibodies. For Coomassie staining, we immerse membranes in 0.1% Coomassie blue dissolved in 50% methanol/10% acetic acid for approx 3 min and then destain in the same buffer. We then completely destain the membranes by washing in methanol, and place blots in TBST before blocking them.
8. A suitable exposure time must be determined empirically during X-ray film processing, so that the density of the protein band detected varies linearly with the amount of protein present on the blot. The densities of the protein bands detected can then be quantitated by scanning densitometry.
9. The cell-lysis buffer does not interfere with the Bicinchoninic acid-based protein assay (purchased from Omnibabo), which can be used for the determination of protein concentration.

Acknowledgment

The work of J.G. Collard is supported by grants from the Dutch Cancer Society. A. Malliri is supported by a Marie Curie fellowship from the European

Community. We thank Leo S. Price and Rob A. van der Kammen for stimulating discussions.

References

1. van Aelst, L. and D'Souza-Schorey, C. (1997) Rho GTPases and signaling networks. *Genes Dev.* **11,** 2295–2322.
2. Hall, A. (1998) Rho GTPases and the actin cytoskeleton. *Science* **279,** 509–514.
3. Zohn, I. M., Campbell, S. L., Khosravi-Far, R., Rossman, K. L., and Der, C. J. (1998) Rho family proteins and Ras transformation: the RHOad less traveled gets congested. *Oncogene* **17,** 1415–1438.
4. Manser, E., Leung, T., Salihuddin, H., Zhao, Z. S., and Lim, L. (1994) A brain serine threonine protein kinase activated by Cdc42 and Rac1. *Nature* **367,** 40–46.
5. Manser, E. and Lim, L. (1999) Roles of PAK family kinases. *Prog. Mol. Subcell. Biol.* **22,** 115–133.
6. Knaus, U. G. and Bokoch, G. M. (1998) The p21Rac/Cdc42-activated kinases (PAKs). *Int. J. Biochem. Cell Biol.* **30,** 857–862.
7. Sander, E. E., van Delft, S., ten Klooster, J. P., Reid, T., van der Kammen, R. A., Michiels, F., et al. (1998) Matrix-dependent Tiam1/Rac signaling in epithelial cells promotes either cell-cell adhesion or cell migration and is regulated by phosphatidylinositol 3-kinase. *J. Cell Biol.* **143,** 1385–1398.
8. Franke, B., Akkerman, J. W. N., and Bos, J. L. (1997) Rapid Ca^{2+}-mediated activation of Rap1 in human platelets. *EMBO J.* **16,** 252–259.
9. de Rooij, J. and Bos, J. L. (1997) Minimal Ras-binding domain of Raf1 can be used as an activation-specific probe for Ras. *Oncogene* **14,** 623–625.
10. Ren, X. D., Kiosses, W. B., and Schwartz, M. A. (1999) Regulation of the small GTP-binding protein Rho by cell adhesion and the cytoskeleton. *EMBO J.* **18,** 578–585.
11. Sander, E. E., ten Klooster, J. P., van Delft, S., van der Kammen, R. A., and Collard, J. G. (1999) Rac downregulates Rho activity: reciprocal balance between both GTPases determines cellular morphology and migratory behavior. *J. Cell Biol.* **147,** 1009–1022.
12. Zondag, G. C., Evers, E. E., ten Klooster, J. P., Janssen, L., van der Kammen, R. A., and Collard, J. G. (2000) Oncogenic Ras downregulates Rac activity, which leads to increased Rho activity and epithelial-mesenchymal transition. *J. Cell Biol.* **149,** 775–782.
13. Reid, T., Bathoorn, A., Ahmadian, M. R., and Collard, J. G. (1999) Identification and characterization of hPEM-2, a guanine nucleotide exchange factor specific for Cdc42. *J. Biol. Chem.* **274,** 33,587–33,593.
14. del Pozo, M. A., Price, L. S., Alderson, N. B., Ren, X. D., and Schwartz, M. A. (2000) Adhesion to the extracellular matrix regulates the coupling of the small GTPase Rac to its effector PAK. *EMBO J.* **19,** 2008–2014.
15. Michiels, F., Habets, G. G. M., Stam, J. C., van der Kammen, R. A., and Collard, J. G. (1995) A role for Rac in Tiam1-induced membrane ruffling and invasion. *Nature* **375,** 338–340.

16. Qiu, R. G., Chen, J., Kirn, D., McCormick, F., and Symons, M. (1995) An essential role for Rac in Ras transformation. *Nature* **374,** 457–459.
17. Mira, J. P., Benard, V., Groffen, J., Sanders, L. C., and Knaus, U. G. (2000) Endogenous, hyperactive Rac3 controls proliferation of breast cancer cells by a p21-activated kinase-dependent pathway. *Proc. Natl. Acad. Sci. USA* **97,** 185–189.
18. Reid, T., Furayashiki, T., Ishizaki, T., Watanabe, G., Watanabe, N., Fujisawa, K., et al. (1996) Rhotekin, a new putative target for Rho bearing homology to a serine/threonine kinase, PKN, and rhophilin in the Rho-binding domain. *J. Biol. Chem.* **271,** 13,556–13,560.
19. Michiels, F., Stam, J. C., Hordijk, P. L., van der Kammen, R. A., Ruulsvanstalle, L., Feltkamp, C. A., et al. (1997) Regulated membrane localization of Tiam1, mediated by the NH2-terminal pleckstrin homology domain, is required for Rac-dependent membrane ruffling and c-jun NH2-terminal kinase activation. *J. Cell Biol.* **137,** 387–398.

9

Introduction of Dominant Inhibitory Proteins Directed Against ROK and MRCK Kinases

Ivan Tan, Louis Lim, and Thomas Leung

1. Introduction

Gene-specific disruption and protein-specific inhibition techniques are powerful tools for the study of protein functions in vivo. A number of methods have been applied to investigate the loss of protein function. These include generation of knockout animals or cell lines, application of mRNA or protein ablation techniques, use of neutralizing antibodies or chemically synthesized inhibitory compounds (such as anti-sense oligonucleotides and enzyme inhibitors), and the important application of dominant-inhibitory mutants. The aim is to inhibit either the expression of genes of interest or the normal function of gene products. Here, we describe methods by which dominant interfering proteins for two related kinases can be expressed in mammalian cells, and show that the biological effects of inhibiting their respective signaling pathways are markedly different. Although dominant inhibitory GTPases mutants have provided a great deal of information about the specific pathways regulated by such GTPases, because of the plethora of effector proteins (in many cases), new reagents are required that can take into account each of these effectors in turn.

Our laboratory has used two different cell systems to investigate the roles of Rho GTPase-associated kinases. In order to investigate the effect of Rho and Cdc42 signaling on the actin cytoskeleton, plasmid expression vectors are introduced into HeLa cells by microinjection or liposome-mediated transfection. These cells show clear morphological changes associated with Cdc42 activation, such as cell rounding, loss of actin stress fibers, and the production of filopodia. RhoA activation leads to enhanced actin stress fiber and focal-

From: *Methods in Molecular Biology, vol. 189: GTPase Protocols: The Ras Superfamily*
Edited by: E. J. Manser and T. Leung © Humana Press Inc., Totowa, NJ

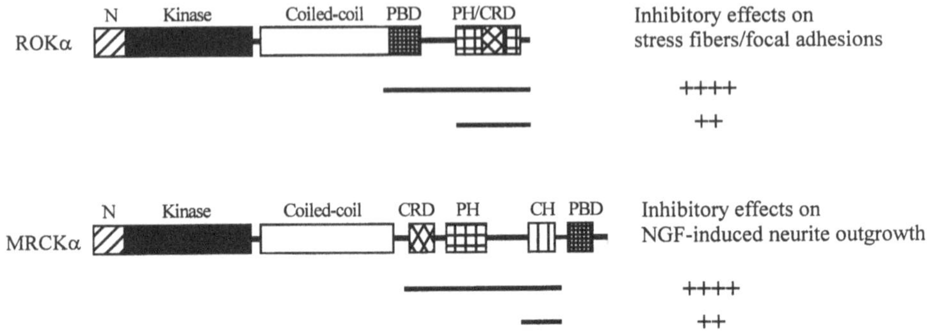

Fig. 1. Primary structural features of MRCKα and ROKα. A number of domains have been detected in these proteins. These include the N terminus (N), kinase domain, coiled-coil domain, cysteine-rich domain (CRD), pleckstrin homology domain (PH), citron homology domain (CH), and *p21*-binding domain (PBD). The smaller C-terminal constructs that are discussed in the text are illustrated by bars, and alongside their effectiveness as dominant inhibitors of kinase function is indicated.

adhesion-complex formation. The role of Rho GTPases can also be explored using PC12 cells, which differentiate and produce neurites in response to nerve growth factor (NGF). This process depends on inhibition of RhoA and activation of Cdc42. Therefore, one can study the role of downstream kinases in this process by scoring for process formation after exposure to NGF.

2. Effects of Dominant-Negative Mutants of ROKα and MRCKα

2.1. Inhibiting ROK-Kinase Signaling

Rho-kinase (ROK/ROCK) and myotonic dystrophy kinase-related Cdc42-binding kinase (MRCK) are related downstream effectors of RhoA and Cdc42, respectively ***(1–3)***. ROK has been implicated in the assembly of actin-based stress fibers and the formation of focal adhesions regulated by RhoA. Both ROK and MRCK are relatively large molecules (160 kDa for ROK and 180 kDa for MRCK), with some similarity of structural arrangements. These include a conserved N-terminal serine/threonine kinase domain, an extended coiled-coil domain, a *p21*-GTPase-binding domain (PBD), a pleckstrin homology domain (PH), and a cysteine-rich zinc-finger domain (**Fig. 1**).

The use of ROK dominant-inhibitory mutants has been instrumental in elucidating their in vivo functions. In the first instance, expression of ROKα^{K112A} (kinase-dead mutant), N-terminally truncated ROKα$^{78\text{-}1379}$ (kinase-domain truncation) and ROKα$^{971\text{-}1379}$ constructs (*see* **Fig. 1**) were all effective in inducing the loss of stress fibers and focal adhesions in HeLa cells, consistent with a loss of ROK function downstream of RhoA (**Fig. 2A**; *see* **ref. *1***).

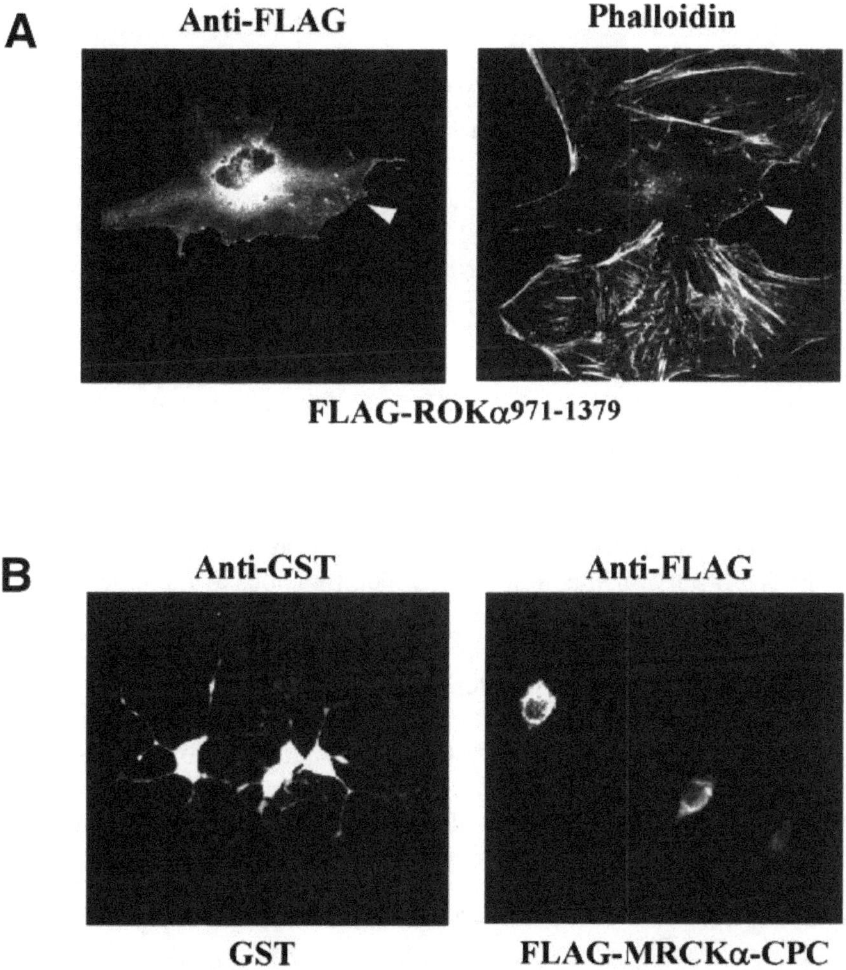

Fig. 2. Effects of dominant-negative ROKα and MRCKα. (**A**) Microinjected with the pXJ-FLAG-ROKα$^{971-1379}$ construct (illustrated in **Fig. 1** by the larger of the two bars) causes a loss of actin stress fibers as stained by phalloidin staining *(right)*. Arrow shows the injected Hela cell. (**B**) PC12 cells were separately transfected with the pXJ-FLAG-MRCKα-CPC construct encoding the CRD-PH-CH region of MRCKα (longer bar in **Fig. 1**) followed by treatment with nerve growth factor (NGF) for 48 h. PC12 cells expressing the vector control responded to NGF by producing neurites (left), whereas cells expressing pXJ-FLAG MRCKα-CPC failed to produce substantial neurites (right).

Additionally RB/PH (TT) domain (Rho-kinase$^{941\text{-}1388}$, also RhoA-binding defective) expression was found to inhibit the formation of stress fibers and focal adhesions induced by both RhoA^{V14} (dominant-positive mutant) and ROKα-CAT (constitutively active kinase domain) ***(4)***. Markedly different effects are seen in the case of ROKα^{K112A} expression in PC12 cells, leading to nerve growth factor (NGF)-independent neurite outgrowth, reminiscent of the inhibition of RhoA activity in PC12 cells ***(5–8)***. Taken together, these results suggest that the C terminus of ROKα/Rho-kinase functions as the dominant-inhibitory mutant, and they confirm the involvement of ROK in actin/focal adhesion organization and regulation of neural cells differentiation downstream of RhoA.

The Rho-binding/PH region of ROK and the distal coiled-coil domain of MRCKα constitute auto-inhibitory domains of the respective kinases ***(4,9,10)***, providing direct inhibition of kinase. In addition, because the ROKα C terminus plays a role in membrane translocation of ROK, overexpression of the C-terminal region potentially interferes with endogenous ROK protein by blocking its correct location. The distal coiled-coil domain, which functions as a kinase inhibitor in vitro ***(10)*** may also be useful for cell studies. Because both ROK and MRCK ***(10)*** exist in multimeric high mol-wt complexes through their extended coiled-coil domains and the conserved N-terminal sequence, it is likely that full-length kinase-dead mutants form unproductive multimers with endogenous counterparts, similar to receptor tyrosine kinases. Taken together, it is possible that the effects of dominant inhibitory proteins occur through multiple mechanisms. Overexpression of the dominant-negative mutants of ROKα seems capable of disrupting the functions of the β isoforms that are expressed in most cultured cells.

2.2. Inhibiting MRCK-Kinase Signaling

MRCK has been shown to be involved in Cdc42-mediated periphery actin formation and neurite-outgrowth in HeLa and PC12 cells, respectively. As for ROK, expression of the mutant MRCKα^{K106A,H1579A,H1581A} (kinase-dead and Cdc42-binding-defective mutant) construct effectively blocked the ability of Cdc42 to drive actin microspikes and focal complexes formation in HeLa cells ***(3)***. Furthermore, its inhibitory effect on NGF-induced neurite outgrowth in PC12 cells suggested a role for MRCK in the differentiation of neural cells in which Rac and Cdc42 are believed to be involved ***(5,8)***. Further mapping analyses of the basis of the inhibitory effect implicated to MRCKα-CPC (C-terminal fragment including the CRD, PH and a novel citron homology (CH) domain) or even the CH domain alone being sufficient for this inhibition (**Fig. 2B**). Consistent with this observation, MRCKα-CPC was equally effective in inhibiting the formation of filopodia and the accumulation of C-terminal threonine-phosphorylated ERM (ezrin, radixin, and moesin, putative substrates of MRCK) proteins at filopodia induced by Cdc42^{V12} *(9)*. Given the fact that the expression of constitutively active kinase domains of both ROKα and MRCKα cause similar retraction phenotypes on NGF-induced neurites (probably because of the similarity of cellular kinase targets), the opposite effects of the dominant-inhibitory constructs in PC12 cells clearly reflects their distinct in vivo functions *(5)*.

Although the mechanism of action for these dominant-negative constructs is not established, a number of experiments point to the titration of critical partner proteins. Mutations that abolish *p21*-binding do not affect the inhibition, indicating the mechanism does not occur through the "upstream" partner. The identified functional-inhibitory domains present in the C termini of ROK and MRCK are potential protein-protein and protein-lipid interaction modules (**Fig. 1**). Thus C-terminal dominant-inhibitory constructs may compete with the endogenous ROK or MRCK for putative binding partners. This is exemplified by the inhibitory effects of MRCKα-CPC and MRCKα-CH on PC12-cell differentiation. In general, despite limited knowledge of the details of the inhibitory mechanism, it is clear that that ROKα and MRCKα dominant-inhibitory construct can produce specific effects in the cellular context.

3. Methods

Here we describe the two most widely used methods (plasmid transfection and microinjection) in our laboratory for the introduction of ROKα and MRCKα dominant-inhibitory constructs into HeLa or PC12 cells. The plasmids contain promoter sequences from cytomegalovirus (CMV) upstream of a Kozak translation initiation sequence and short peptide "tag" sequence that allows detection of any stable protein product. Such vectors allow expression in a variety of cell types, and are available from many commercial sources.

3.1. Liposome-Based Delivery of Plasmids for HeLa Cells

1. HeLa cells are cultured in Modified Eagle's medium containing 10% fetal bovine serum to 70–80% confluence on cover slips placed in 35-mm tissue-culture dishes. The humidified incubator is maintained at 37°C with 5% CO_2.
2. Prepare DNA-liposome complexes by gently mixing 5 µL of Lipofectamine reagent (Gibco-BRL) with 1 µg of DNA plasmid (*see* **Note 1**) in 200 µL serum-free medium in a 5-mL sterile tube.
3. Leave for 45 min at room temperature to allow complex formation.
4. Dilute the DNA-liposome mix with 800 µL serum-free medium and transfer the content to the cells onto the cover slip (cells prewashed with serum-free medium).
5. Leave for 3 h (longer incubation with transfection mix causes cell death) in the 37°C incubator, and then replace the transfection mix with fresh serum-containing medium.

6. After overnight incubation or longer (16–24 h), fix with paraformaldehyde (**Subheading 3.4.**).

3.2. Liposome-Based Delivery of Plasmids for PC12 Cells

1. Grow PC12 cells in Dulbecco's Modified Eagle's Medium (DMEM) containing 10% horse serum and 5% fetal bovine serum (FBS) overnight to reach 70–80% confluency on laminin-coated cover slips.
2. Laminin-coated coverslips are prepared by overlaying sterile glass cover slips with 20 μg/mL purified laminin (Upstate Biotechnology) in phosphate-buffered saline (PBS) at room temperature for 3 h or longer.
3. Wash with PBS and allow to air-dry in the laminar flow hood.
4. The rest of the transfection procedure for PC12 cells is essentially similar to the one for HeLa cells, as described in **Subheading 3.1., steps 2–5.**
5. Depending on the experiment, nerve growth factor (NGF, at 50 ng/mL in medium with 1% serum) can be added to induce PC12-cell differentiation after the 3-h incubation with transfection mix.
6. Cells are maintained in NGF for 48 h to allow differentiation and production of neurites of substantial length for scoring. In some experiments (*see* **Note 2**) it is preferable to examine the cells following overnight incubation in serum-containing medium.
7. Cells are fixed and stained as described in **Subheading 3.4.**

3.3. Microinjection of Plasmid DNA

Descriptions of both protein and plasmid-cell microinjection techniques can also be found elsewhere ***(11,12)***. Here we provide the basic elements of the procedure for microinjection of plasmid DNA constructs, allowing expression of proteins within ~30 min of the injection.

1. Dishes containing PC12 or HeLa cells for microinjection are prepared as described in the transfection protocol (*see* **Subheadings 3.1., step 1** and **3.2., steps 1–2**). A cell density of less than 50% confluency is desirable. The cover slips can be marked with a diamond pen (cross) to allow easier location of injected cells.
2. Keep cells in an incubator, close to the injection apparatus if possible to avoid excessive changes in temperature.
3. Plasmid DNA samples must be clarified by centrifugation at 14,000 *g* for 10 min, then diluted to 50 ng/mL with microinjection buffer (50 m*M* Tris-HCl, pH7.5, 50 m*M* (KCl) and 5 m*M* $MgCl_2$).
4. Load the DNA sample (10 μL) into a glass capillary needle (Femtotip from Eppendorf) using a microloader pipet (Eppendorf).
5. Monitoring the field of cells in question on a Zeiss Axiovert 35 microscope with heated stage, the needle should be aligned with the nucleus of each chosen cell (**Note 3**) and injected either manually or in semi-automatic mode at 300 hPa (air pressure), using an Eppendorf Microinjector model 5242.

6. Return injected cells to the incubator for 1–4 h (*see* **Note 4**) to allow expression and monitor for changes in morphology by phase-contrast microscopy before fixing and processing for immunocytochemistry (*see* **Subheading 3.4.**).
7. In some cases it is desirable to monitor live cells for plasmid expression. In this case, the cells can be co-injected with an expression plasmid for green fluorescent protein (GFP), which can be detected as early as 1 h after injection.

3.4. Immunofluorescent Staining of Cells

1. Following a specific incubation period after transfection or microinjection, wash cells with phosphate-buffered saline (PBS), then fix in 4% paraformaldehyde in PBS for 20 min at room temperature.
2. After removing the fixing solution, permeabilize cells on cover slips with 0.2% Triton X-100 in PBS for 10 min, and then block with 10% goat serum in PBS for 10 min.
3. Incubate cells with primary antibodies at 10 μg/mL (mouse anti-HA or anti FLAG monoclonal antibodies [MAbs]) in 0.5% Triton X-100 in PBS for 2 h at 37°C. To conserve reagents, cover slips may be laid face down onto parafilm squares onto which is spotted 100 μL of antibody solution.
4. Wash twice (10 min each) with 0.1% Triton X-100 in PBS.
5. Incubate cover slips (*see* **step 3**) with 10 μg/mL fluorescein isothiocyanate (FITC)- or tetramethylrhodamine isothiocyanate (TRITC)-conjugated secondary antibodies against mouse immunoglobulin IgG (Jackson laboratories) for 1 h at room temperature in a light-tight box.
6. Wash with 0.1% Triton X-100 in PBS twice (5 min each) and mount the cover slips onto slides.
7. TRITC-phalloidin from Sigma (*see* **Note 5**) can be used to visualize the effects of the dominant-inhibitory mutants on actin-filament organization. This is added at **step 5** in place of one of the secondary antibodies. The effects on focal adhesions can be monitored by using antibodies directed against vinculin or paxillin (Transduction Laboratories).
8. For evaluating the effects of MRCKα dominant-negative mutant on nerve growth factor (NGF)-induced neurite outgrowth, PC12 cells are treated with NGF (50 ng/mL) for 2 d before fixing. A total number of 50 cells expressing the tagged-construct are analyzed, and the proportion of cells with neurites of longer than twice the body width scored.

4. Notes

1. Plasmid DNA can be purified by a variety of methods. The most common is bacterial alkaline lysis followed by either ion-exchange purification, or cesium chloride centrifugation in the presence of ethidium bromide. In our hands, the latter tends to yield preparations that are more suitable for liposome-mediated transfection.
2. Although NGF-induced neurites take more than 48 h to achieve sufficient length to be scored with any accuracy, similar structures are produced much more

rapidly if inhibitors of RhoA or ROK are being assayed. In this case, overnight (16 h) expression may be sufficient.

3. Cells characterized by a more rounded appearance may be in the process of dividing, and are often less firmly attached to the substratum. Such cells are usually avoided because they are probably at inappropriate stages of the cell cycle for analysis.
4. Since many laboratories do not have an enclosed cell microinjection apparatus with regulated CO_2 environmental chamber, it is important to keep the cells out of the incubator for as short a time as possible (typically 5–20 min). Beyond this time, irreversible changes to the cells often occur.
5. TRITC-phalloidin is highly light-sensitive, and should be aliquoted and frozen in a light-tight box. Freeze-thaw such aliquots a maximum of 2× and during the incubation of cover slips with solution containing TRITC-phalloidin, wrap the dish in aluminum foil.

Acknowledgment

This work is funded by the Glaxo Singapore Research Fund.

References

1. Leung, T., Chen, X.-Q., Manser, E., and Lim, L. (1996) The p160 RhoA-binding kinase ROKα is a member of a kinase family and is involved in the reorganization of the cytoskeleton. *Mol. Cell. Biol.* **16,** 5313–5327.
2. Matsui, T., Amano, M., Yamamoto, T., Chihara, K., Nakafuku, M., Ito, M., et al. (1996) Rho-associated kinase, a novel serine threonine kinase, as a putative target for the small GTP-binding protein Rho. *EMBO J.* **15,** 2208–2216.
3. Leung, T., Chen, X.-Q., Tan, I., Manser, E., and Lim, L. (1998) Myotonic distrophy kinase-related Cdc42-binding kinase acts as a Cdc42 effector in promoting cytoskeletal reorganization. *Mol. Cell. Biol.* **18,** 130–140.
4. Amano, M., Chihara, K., Nakamura, N., Kaneko, T., Matsuura, Y., and Kaibuchi, K. (1999) The COOH terminus of Rho-kinase negatively regulates Rho-kinase activity. *J. Biol. Chem.* **274,** 32,418–32,424.
5. Chen, X.-Q., Tan, I., Leung, T., and Lim, L. (1999) The myotonic dystrophy kinase-related Cdc42-binding kinase is involved in the regulation of neurite outgrowth in PC12 cells. *J. Biol. Chem.* **274,** 19,901–19,905.
6. Jalink, K., van Corven, R. J., Hengeveld, T., Morii, N., Narumiya, S., and Moolenaar, W. H. (1994) *J. Cell Biol.* **126,** 801–810.
7. Tigyi, G., Fischer, D., Sebok, S., Yang, C., Dyer, D. L., and Miledi, R. (1996) *J. Neurochem.* **66,** 537–548.
8. Kozma, R., Sarner, S., Ahmed, S., and Lim, L. (1997) Rho family GTPases and neuronal growth cone remodeling: relationship between increased complexity induced by Cdc42Hs, Rac1, and acetylcholine and collapse induced by RhoA and lysophosphatidic acid. *Mol. Cell. Biol.* **17,** 1201–1211.

9. Nakamura, N., Oshiro, N., Fukata, Y., Amano, M., Fukata, M., Kuroda, S., et al. (2000) Phosphorylation of ERM proteins at filopodia induced by Cdc42. *Genes Cells* 571–581.
10. Tan, I., Seow, K. T., Lim, L., and Leung, T. (2001) Intermolecular and intramolecular interactions regulate catalytic activity of myotonic dystrophy kinase-related Cdc42-binding kinase α. *Mol. Cell. Biol.* **21,** 2767–2778.
11. Ridley, A. J. (1998) Mammalian Cell Microinjection Assay to Study the Function of Rac and Rho. *Methods Mol. Biol.* **84,** 153–160.
12. Leung, T., Manser, E., and Lim, L. (1999) Cytoskeletal regulation by Rho-p21 associated kinases: analysis of ROK and PAK function by plasmid microinjection. Microinjection in: *Methods and Tools in Biosciences and Medicine* (Lacal, J. C., Perona, R., Ferancisco, J., eds.) Birkhauser Verlag, Boston, MA, pp. 94–105.

10

Effects of Rho Family GTPases on Cell–Cell Adhesion

Masaki Fukata, Masato Nakagawa, Shinya Kuroda, and Kozo Kaibuchi

1. Introduction

Cell–cell adhesion is dynamically rearranged in various cellular processes, including cell scattering, wound healing, developmental morphogenesis, and tumor metastasis. Cadherins are Ca^{2+}-dependent homophilic cell–cell adhesion molecules *(1)* and bind β-catenin or plakoglobin (also known as γ-catenin), which in turn is linked to the actin cytoskeleton via α-catenin. This linkage between cadherins and the actin cytoskeleton is required for the development of a strong adhesive state *(2)*. However, the mechanism underlying the dynamic rearrangement in cell–cell adhesion remains unclear.

Recently, several investigators have proposed that the Rho family GTPases are required for the regulation of cadherin-mediated cell–cell adhesion *(3)*. Microinjection or transfection of constitutively active or dominant negative mutants of Rho family GTPases affects the immunofluorescent intensity of E-cadherin at cell–cell contact sites in some epithelial cells. However, the immunofluorescent intensity of E-cadherin at cell–cell contact sites does not always correlate with the strength of the E-cadherin-mediated cell–cell adhesion. In fact, we have confirmed that microinjection of either dominant-negative Cdc42 ($Cdc42^{Asn17}$) or dominant-negative Rac1 ($Rac1^{Asn17}$) into L cells expressing E-cadherin (EL cells) does not affect the localization of E-cadherin or the immunofluorescent intensity of E-cadherin at cell–cell contact sites. On the other hand, the E-cadherin activity markedly decreases in EL cells expressing $Cdc42^{Asn17}$ or $Rac1^{Asn17}$ on cell dissociation assay, a quantitative

From: *Methods in Molecular Biology, vol. 189: GTPase Protocols: The Ras Superfamily*
Edited by: E. J. Manser and T. Leung © Humana Press Inc., Totowa, NJ

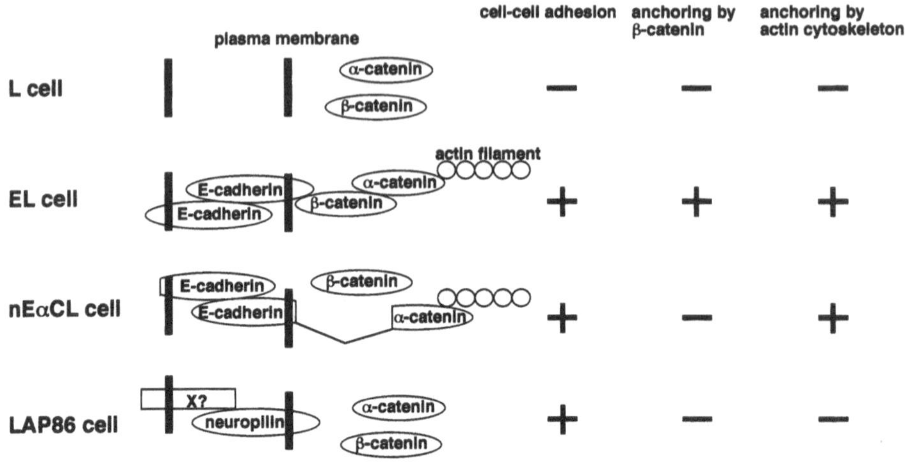

Fig. 1. Characteristics of L, EL, nEαCL, and LAP86 cells. In parental L cells, because cadherin is not expressed there is no cadherin-dependent cell-cell adhesion. EL cells are L fibroblast cells stably expressing wild-type E-cadherin, and adhere to each other in an E-cadherin-dependent manner. nEαCL cells are L cells stably expressing the E-cadherin-α-catenin chimeric protein. Although nEαCL cells also adhere to each other in a mutant E-cadherin-dependent manner, this adhesion does not require β-catenin, and the cadherin-catenin complexes are not remodeled. LAP86 cells express neuropilin-1, which is a calcium-independent cell-adhesion molecule *(7)*, and do not require an actin-based cytoskeleton (which EL and nEαCL cells require) or catenins for adhesion. Neuropilin-1 mediates cell adhesion by heterotypic interaction with an unknown protease-sensitive molecule (indicated as **X**?).

assay for E-cadherin activity *(4)*. Furthermore, since the Rho family regulates the actin-based cytoskeleton, an analysis of the immunofluorescent intensity of E-cadherin cannot determine whether the Rho family regulates the cadherin-mediated cell–cell adhesion by acting directly on the cadherin-catenin complexes or by acting indirectly on the actin cytoskeleton or other cytoskeletal components. These issues have been addressed through the use of the cell dissociation assay *(4–6)*, using three types of cell lines derived from mouse L fibroblasts (which do not express endogenous cadherins): EL cells, nEαCL cells *(5)*, and LAP86 cells *(7)* (**Fig. 1** and *see* **Subheading 2.1.**).

Here we describe a method that was developed to analyze the effects of Rho family GTPases and their effectors on the cadherin-mediated cell–cell adhesion. This method includes three steps:

1. Transfection of target genes (e.g., Rho family GTPases or their effectors) together with a marker gene (e.g., Aic2, *see* **Subheading 3.1.**) into EL, nEαCL, and LAP86 cells.

2. Concentration and isolation of the transfected cells by panning *(4,6,8,9)*. (*See* **Subheading 3.2.**)
3. The cell dissociation assay *(4–6)*. (*See* **Subheading 3.3.**)

The use of this method has several advantages over methods such as immunofluorescent analysis of E-cadherin.

1. By using EL cells, E-cadherin-mediated cell–cell adhesive activity can be directly and quantitatively measured excluding other cell adhesion molecules.
2. The proteins of interest in E-cadherin-mediated cell–cell adhesion can be classified as acting directly on cadherin–catenin complexes or indirectly on the actin-based cytoskeleton by using three types of cell lines: EL, nEαCL, and LAP86 cells.

2. Materials

2.1. Cell lines (L, EL, nEαCL, and LAP86 Cells) (Fig. 1)

EL cells and nEαCL cells were generated by Dr. A. Nagafuchi (Kyoto University, Kyoto, Japan), and cultured in Dulbecco's Modified Eagle's Medium (DMEM) (Gibco-BRL) supplemented with 10% fetal calf serum (FCS) (Cansera International Inc.) containing 0.1 mg/mL of G418 (Gibco-BRL) *(5)*. LAP86 cells were generated by Dr. H. Fujisawa (Nagoya University, Nagoya, Japan), and cultured in DMEM supplemented with 10% FCS containing 0.1 mg/mL of G418 *(7)*.

1. EL cells are L fibroblasts expressing E-cadherin, and adhere to each other in an E-cadherin-dependent manner.
2. nEαCL cells are L cells expressing an E-cadherin mutant whose cytoplasmic domain is deleted and replaced by the C-terminal domain of α-catenin. In nEαCL cells, the cadherin-catenin complexes are not remodeled.
3. LAP86 cells are L cells expressing neuropilin-1, which is a calcium-independent cell-adhesion molecule *(7)*, and does not require an actin-based cytoskeleton (which EL and nEαCL cells require) or catenins for adhesion.

2.2. Panning

Anti-Aic2 antibody *(8,9)* was kindly provided by Dr. S. Yonehara (Kyoto University, Kyoto, Japan).

2.3. Cell Dissociation Assay

1. HCMF solution: 137 m*M* NaCl, 5 m*M* potassium chloride (KCl), 0.33 m*M* Na_2HPO_4, 5.6 m*M* glucose, 10 m*M* HEPES (N-2-hydroxyethyl-piperazine-N′-2-ethanesulfic acid) at pH 7.4.
2. TC solution: HCMF containing 0.01% trypsin, and 1 m*M* $CaCl_2$.
3. TE solution: HCMF containing 0.01% trypsin, and 1 m*M* EGTA.

3. Methods

3.1. Transfection

Transfect the expression plasmids of interest together with pME18-tAic2A (*8,9*), which harbors cDNA of an interleukin (IL)-3β1 receptor lacking a cytoplasmic domain and is used as a marker, into EL cells, nEαCL cells, or LAP86 cells using lipofectamine according to the manufacturer's instructions (Gibco-BRL).

1. Seed EL cells, nEαCL cells, or LAP86 cells (3×10^6/10-cm dish) and incubate at 37°C for 16 h.
2. After washing the cells with 3 mL of Opti-MEM (Gibco-BRL), add 5 mL of DNA mixture (plasmid ratio; pME18-tAic2A:plasmids of interest = 1:5–10) and incubate the cells at 37°C for 6 h.
3. After changing the medium to growth medium, incubate the cells at 37°C for 24 h.

3.2. Concentration of Transfected Cells by Panning (Fig. 2)

Panning is a powerful method used to isolate and concentrate the transfected cells. Under these conditions, recombinant proteins are expressed among about approx 80–90% of the cells collected.

3.2.1. Preparation of Anti-Aic2 Antibody-Coated Dishes

1. Add 8 mL of 50 m*M* Tris-HCl, pH 9.5 to a noncoated 10-cm dish (e.g., BioBik), then add 40 μg of anti-Aic2 antibody, and immediately rock the dish well.
2. Incubate the dish at 37°C for 1 h without moving it.
3. Discard the solution containing anti-Aic2 antibody and wash the dish with 3 mL of PBS 2×.
4. Finally add 5 mL of DMEM for blocking to the dish, and incubate for 30 min at 37°C.

3.2.2. Panning

1. Wash the transfected cells with PBS twice.
2. Add 4 mL of PBS containing 5 mM EGTA pH 8.0 to the cells and incubate at room temperature for 20 min.
3. Add 2 mL of DMEM to the cells (final, 3.3 m*M* EGTA).
4. Scrape the cells from 3–4 dishes with silicon-rubber policemen (SUMILON) and suspend well, pipetting 4–5 times to obtain single cells. Then, transfer the cells to a dish precoated with anti-Aic2 antibody.
5. Incubate the cells at 37°C for 30 min. Do not rock the cells.
6. Wash the cells with 3 mL of DMEM 3×, and remove the unattached cells. Add 1 mL of DMEM to the cells and scrape them with the SUMILON. Suspend well, transfer the cells to a new 15-mL tube, and count the number of cells.

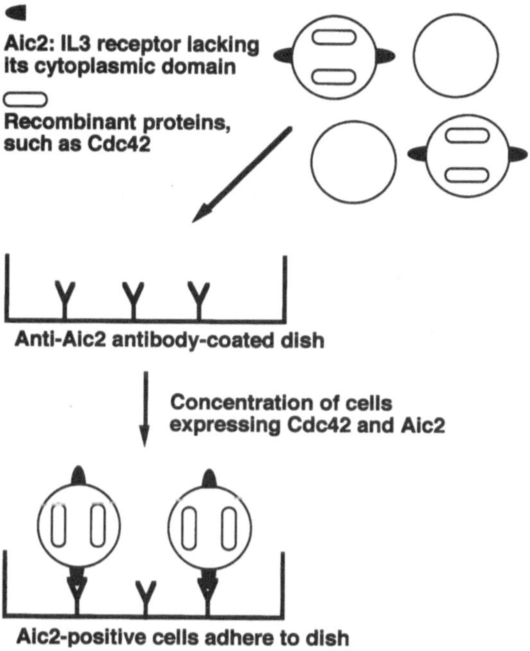

Fig. 2. Procedure used for panning. Proteins of interest are co-expressed together with Aic2, which is the IL3-receptor lacking the cytoplasmic domain. Aic2 is used as a marker. Aic2-expressing cells are immunologically collected using anti-Aic2 antibody.

7. Seed the immunologically collected cells (4×10^5 cells) onto a 48-well culture dish and incubate for 24 h.

3.3. Cell Dissociation Assay (Fig. 3 and Fig. 4)

3.3.1. TC or TE Treatment

1. After 24-h incubation, wash the immunologically collected cells with 1 mL of HCMF 3×, and then add 100 µL of TC (TC treatment) or TE (TE treatment) solution *(4–6)*. When LAP86 cells are used instead of EL cells or nEαCL cells, add 1 m*M* EDTA either in the presence (TEd treatment) or absence (Ed treatment) of trypsin *(4)*.
2. Incubate the cells for 20 min at 37°C.
3. Dissociate the cells by pipetting with a 200-µL tip (the edges cut with scissors) 10×.

3.3.2. Measurement of Cell Dissociation

The total particle numbers after TC or TE treatment are designated as N_{TC} and N_{TE}, respectively. When LAP86 cells are used, the total particle numbers

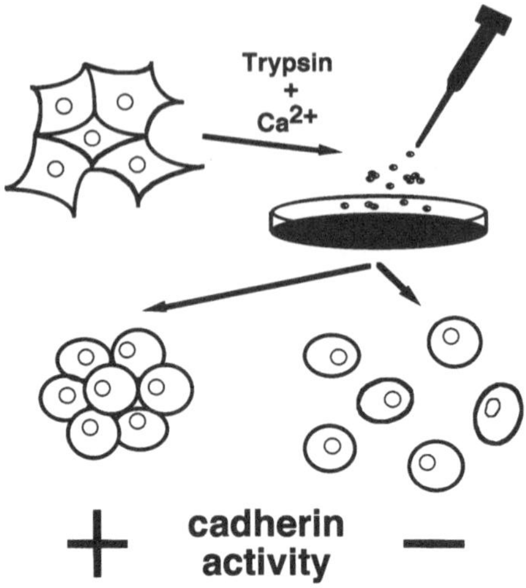

Fig. 3. Procedure for the cell dissociation assay. The cells, which are concentrated by panning, are treated with trypsin in the presence or absence of Ca^{2+}. After treatment, they are suspended by pipetting several times, and the dissociation of the cells is scored. In the presence of Ca^{2+}, cadherin is resistant to trypsin, and the cells form large aggregates. In contrast, in the absence of Ca^{2+}, cadherin is cleaved by trypsin, and the cells easily disperse after the pipetting.

after Ed or TEd treatment are designated as N_{Ed} and N_{TEd}, respectively. The extent of dissociation of cells is represented by the cell-dissociation index N_{TC}/N_{TE} or N_{Ed}/N_{TEd}. The lower the values of these indexes, the higher the activity of cell adhesion ***(4–6)***.

4. Notes

1. Since anti-Aic2 antibody, which is used for panning, must precipitate cells, its affinity to Aic2-antigen must be very high. Based on our experience, anti-Aic2 antibody can be stably stored at –80°C for up to 2 mo. Beyond this, the antibody suddenly loses activity. Never freeze and thaw.
2. Since the transfection efficiency is about 30%, and panning collects only 10% of transfected cells, the final isolated cells compose about 3% of the original population. If we start with 5–10 dishes (3×10^6/10-cm dish) for one assay, we can collect approx $5–10 \times 10^5$ cells by panning.
3. This method can be applied to the examination of the effects of the Rho family GTPases as well as their effectors. In fact, we found that overexpression of IQGAP1, an effector of Cdc42 and Rac1, decreases E-cadherin-mediated cell-cell adhesion through direct action on the cadherin-catenin complexes ***(6)***.

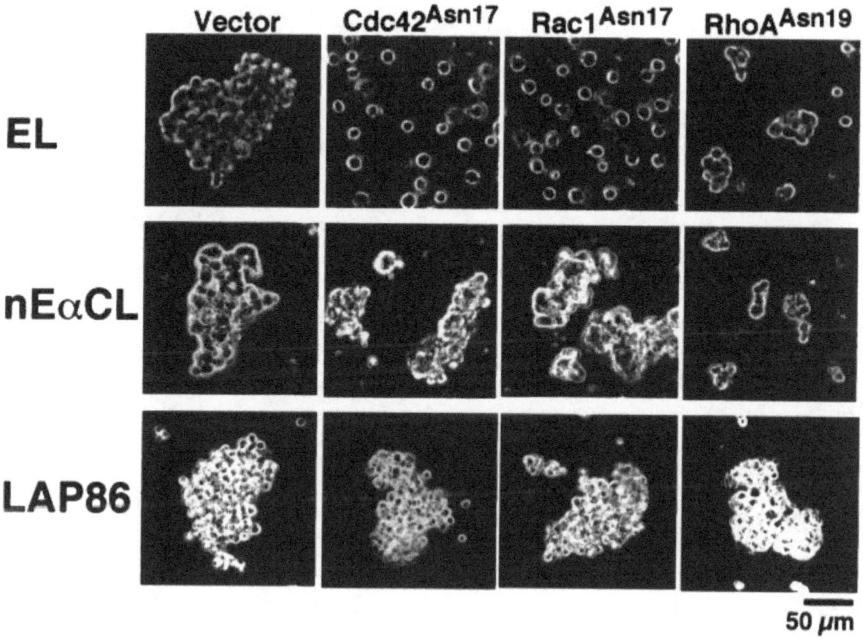

Fig. 4. Effects of Rho family GTPases on the cadherin-mediated cell-cell adhesion. Expression of Cdc42^{Asn17} or Rac1^{Asn17} in EL cells markedly reduces E-cadherin activity, whereas expression of dominant-negative RhoA (RhoA^{Asn19}) only slightly reduces it (under TC treatment). In contrast, expression of either Cdc42^{Asn17}, Rac1^{Asn17} or RhoA^{Asn19} in nEαCL cells slightly reduces the mutant E-cadherin activity (under TC treatment). In addition, the expression of Cdc42^{Asn17}, Rac1^{Asn17}, and RhoA^{Asn19} in LAP86 cells does not affect adhesion (under Ed treatment). Since β-catenin, the cytoplasmic domain of E-cadherin, and the N-terminal region of α-catenin are not required for the adhesion in nEαCL cells and the effect of Cdc42^{Asn17} or Rac1^{Asn17} in nEαCL cells is less than that in EL cells, Cdc42 and Rac1 appear to regulate E-cadherin-mediated cell adhesion, acting directly on the cadherin-catenin complexes. Since RhoA^{Asn19} has a similar effect on the E-cadherin activity in EL cells and in nEαCL cells (whose adhesion requires actin cytoskeleton) and no effect in LAP86 cells (whose adhesion does not require actin cytoskeleton), Rho presumably regulates E-cadherin activity by acting indirectly on the actin cytoskeleton *(4)*.

Acknowledgment

We thank Dr. S. Yonehara (Kyoto University, Kyoto, Japan) for providing anti-Aic2 antibody and the construct harboring Aic2; Drs. A. Nagafuchi and S. Tsukita (Kyoto University, Kyoto, Japan) for providing L, EL, and nEαCL cells; and Dr. H. Fujisawa (Nagoya University, Nagoya, Japan) for providing LAP86 cells.

References

1. Takeichi, M. (1995) Morphogenetic roles of classic cadherins. *Curr. Opin. Cell Biol.* **7,** 619–627.
2. Tsukita, S., Tsukita, S., Nagafuchi, A., and Yonemura, S. (1992) Molecular linkage between cadherins and actin filaments in cell-cell adherens junctions. *Curr. Opin. Cell Biol.* **4,** 834–839.
3. Kaibuchi, K., Kuroda, S., Fukata, M., and Nakagawa, M. (1999) Regulation of cadherin-mediated cell-cell adhesion by the rho family GTPases. *Curr. Opin. Cell Biol.* **11,** 591–596.
4. Fukata, M., Kuroda, S., Nakagawa, M., Kawajiri, A., Itoh, N., Shoji, I., et al. (1999) Cdc42 and Rac1 regulate the interaction of IQGAP1 with β-catenin. *J. Biol. Chem.* **274,** 26,044–26,050.
5. Nagafuchi, A., Ishihara, S., and Tsukita, S. (1994) The roles of catenins in the cadherin-mediated cell adhesion: functional analysis of E-cadherin-α-catenin fusion molecules. *J. Cell Biol.* **127,** 235–245.
6. Kuroda, S., Fukata, M., Nakagawa, M., Fujii, K., Nakamura, T., Ookubo, T., et al. (1998) Role of IQGAP1, a target of the small GTPases Cdc42 and Rac1, in regulation of E-cadherin-mediated cell-cell adhesion. *Science* **281,** 832–835.
7. Kawakami, A., Kitsukawa, T., Takagi, S., and Fujisawa, H. (1996) Developmentally regulated expression of a cell surface protein, neuropilin, in the mouse nervous system. *J. Neurobiol.* **29,** 1–17.
8. Yonehara, S., Ishii, A., Yonehara, M., Koyasu, S., Miyajima, A., Schreurs, J., et al. (1990) Identification of a cell surface 105 kd protein (Aic-2 antigen) which binds interleukin-3. *Int. Immunol.* **2,** 143–150.
9. Itoh, N., Yonehara, S., Schreurs, J., Gorman, D. M., Maruyama, K., Ishii, A., et al. (1990) Cloning of an interleukin-3 receptor gene: a member of a distinct receptor gene family. *Science* **247,** 324–327.

11

Cell Motility and Invasion Assays

Jackie Banyard and Marc Symons

1. Introduction

Cell motility and invasion play an essential role in a wide range of biological functions, including many stages of development, wound healing, and immune function. Deregulated motile behavior is believed to contribute to pathological processes such as metastasis, tumor angiogenesis, and atherosclerosis *(1–3)*.

Rho family members control a multitude of functions that have been implicated in the regulation of cell migration. Most notable among these are the organization and dynamics of the actin cytoskeleton, intercellular and cell-substrate adhesion, vesicle trafficking, and lipid metabolism *(4)*. Recently, Rac has been shown to stimulate the transcription of collagenase-1, a metalloproteinase involved in the remodeling of extracellular matrix *(5)*.

Therefore, it is not surprising that several Rho family proteins have been implicated in the control of cell motility and invasion. Rac, Cdc42, and Rho have been shown to play essential roles in growth factor- and cytokine-stimulated chemotaxis in fibroblasts, macrophages, and neutrophils *(6–9)*. There is also substantial evidence that activation of Rac, Cdc42, and Rho is necessary for the invasive behavior of carcinoma cells and fibroblasts *(7,10–12)*. There is still much uncertainty however regarding the signaling pathways activated by Rho proteins to regulate motile behavior.

The identification of the biological functions of Rho proteins has mostly relied on the specific inhibitory effects of dominant-negative mutants of these GTPases, and on the stimulation of signaling pathways by constitutively active versions. To introduce these mutants into cells, researchers have employed microinjection of plasmids or recombinant proteins *(8,13,14)*, transient transfections, or the establishment of stable cell lines *(7)*. More recently, cell-permeable peptides corresponding to the Cdc42/Rac-interacting-binding

From: *Methods in Molecular Biology, vol. 189: GTPase Protocols: The Ras Superfamily*
Edited by: E. J. Manser and T. Leung © Humana Press Inc., Totowa, NJ

(CRIB) motifs have also been successfully used to interfere with Cdc42- and Rac-mediated functions ***(15)***.

We have focused on the use of stable cell lines for our studies on the role of Rho GTPases in cell motility for several reasons. Foremost among these is the possibility of using the same lines to study the roles of these GTPases in other functions that are relevant to cell motility. In addition, the characterization of these and other parameters provides a large number of internal controls for the specificity of action of the dominant-negative and constitutively active GTPase mutants. Thus, we have established a panel of Rat1 fibroblasts expressing either dominant-negative or constitutively active versions of Rac1, Cdc42, and RhoA ***(16–18)***. The motility parameters that we have determined for this panel of cell lines include chemotaxis toward platelet-derived growth factor (PDGF) and soluble fibronectin, random motility stimulated by Lysophosphatidic acid (LPA), and PDGF-stimulated invasion into three-dimensional collagen gels ***(6,7)***. We have also used these lines to study the role of Rho GTPases in adhesion signaling, cell proliferation, and lipid metabolism ***(16–21)***.

The use of stable cell lines also avoids the pitfall of employing exceedingly high expression levels of mutant GTPases that may lead to nonspecific cross-communication and/or cell toxicity. Cell lines that express mutant GTPases from an inducible promoter have several additional advantages. Control of the level of inducer allows careful titration of GTPase expression levels. Inducible expression also largely eliminates the possibility of long-term adaptation of the cells to the presence of recombinant proteins.

2. Materials

2.1. Cell Culture

1. Rat1 fibroblasts are maintained in Dulbecco's Modified Eagle's Medium (DMEM) containing 10% fetal bovine serum (FBS), 2 m*M* glutamine, 4.5 g/L glucose, 100 U penicillin, and 2 µg/mL streptomycin.
2. Cdc42-expressing stable cell lines stably transfected with the pCMVneo*Myc* vector are maintained in the same media supplemented with 400 µg/mL G418. Rac- and Rho-expressing stable cell lines, containing a tetracycline-sensitive promotor, as detailed in **Subheading 2.1.1.**, are maintained in the same media supplemented with 400 µg/mL G418, 2.5 µg/mL puromycin, and 2 µg/mL tetracycline. Expression of Rac or Rho is induced by withdrawal of tetracycline 2 d before experimentation. Cell culture reagents were obtained from Life Technologies, MD.

2.1.1. Stable Lines

1. The tetracycline-regulatable expression system that we have used to derive stable lines expressing mutant Rho proteins was developed by Gossen and Bujard

(*22*,**23**). This system is based on two regulatory elements that are derived from the *Escherichia coli* (*E. coli*) tetracycline-resistance operon, the *tet* repressor protein, and the *tet* operator DNA sequence. The tet repressor protein has been fused to the activation domain of virion protein 16 of herpes simplex virus, to make the tetracycline-controlled transactivator (tTA). This transactivator in turn stimulates transcription from a tetracycline-response element (TRE) containing a minimal promoter sequence derived from the human cytomegalovirus (CMV) promoter, combined with seven repeats of the *tet* operator sequence.

2. The establishment and characterization of the stable cell lines was carried out in two steps, as previously described in detail ***(16,22)***. Briefly, in the first step a Rat1 fibroblast line that expresses the tTA transactivator was established. To test for expression and inducibility, approx 75 clones were isolated and transfected with a luciferase reporter that is controlled by a tTA-dependent promoter. Several clones were selected that showed 5–10-fold repression by tetracycline. In the second step, the tTA-expressing Rat1 line was transfected with plasmids containing the respective cDNAs for the epitope-tagged mutant GTPases controlled by the TRE. From the experience shared by many laboratories, it appears that the selection of a highly inducible tTA-expressing clone is the most tedious part of the cell-line construction. Often, a large number of clones (>100) must be inspected to yield a satisfactory cell line. However, several tTA-expressing cell lines representing a variety of different tissues are now commercially available (Clontech).
3. A number of other inducible expression systems have been described in the literature ***(24–26)*** and several of these are commercially available.

2.1.2. Transient Transfections

We have transiently overexpressed mutants of the Rho family GTPases in Rat1 fibroblasts using pEXV, pCMV, pcDNA3 (Invitrogen, CA) and pIRES2-EGFP (Clonetech, CA) vectors. Transfection was achieved using FuGENE 6 reagent (Roche, IN) that results in high transfection levels (approx 35%) and low toxicity in Rat1 fibroblasts. Transfection was performed on cells at 50–70% confluence according to the supplier's protocol: 2 µg DNA is used per 60-mm tissue-culture dish. Sorting of the transfected cell population is a valuable step, allowing interpretation of results without contamination by the untransfected population. Transfection using a green fluorescent protein (GFP) vector enables fluorescence-activated cell sorting (FACS) to be used. We have found that FACS sorting did not affect Rat1 fibroblast migratory response under the conditions used ***(7)***.

2.2. Boyden-Chamber Migration Assay

The 48-well modified Boyden chamber used in our studies is obtained from Neuroprobe Inc. (Cabin John, MD). The 8.0-µm pore PVP-F polycarbonate membranes are available from Nucleopore (Corning Separations, MA). Membranes are precoated with an extracellular matrix macromolecule before use, to enhance cell attachment. We routinely use gelatin, fibronectin, or collagen. Gelatin (Difco, MI) is prepared as a 1.5% solution, autoclaved, and stored at 4°C. The sterile gelatin solution is dissolved at 37°C before use. Fibronectin, collagen, and Matrigel are obtained from Becton Dickinson, MA. The assay solution used to dilute all test solutions and to resuspend cells is serum-free cell-culture media—i.e., DMEM supplemented with 0.1% bovine serum albumin (BSA), and filter-sterilized. Cells are stained using Gill's Hematoxylin #3 (Polysciences, PA). We use Permount mountant (Fisher Scientific, NJ), but other mountants may be used. We obtained PDGF-BB from R&D Systems, MN.

2.3. 3D Collagen Invasion Assay

1. Collagen used to form gels in this assay was extracted from rat tail tendons ***(27)***. Rat tails are cut from young control rats at sacrifice, wiped with 70% ethanol, and stored at –20°C until use. Tails are thawed in 70% ethanol overnight. Under sterile conditions, a pair of serrated edge clamp forceps are positioned 2 cm apart from each other, toward the tip of the tail. The forceps are twisted, which snaps the tail, allowing the tendons to be pulled out. The tendons are cut using sterile scissors into cold 3% acetic acid (99%+ purity). Tendon extraction is repeated for each tail, proceeding toward the body end of the tail, giving 3–4 tendon pulls per tail. Drop tendons from 10 tails into 500 mL acetic-acid solution. A sterile stir bar is added to each bottle and collagen is extracted by stirring overnight at 4°C.
2. The extract is centrifuged at 2,500*g* for 2.5 h at 4°C. The collagen supernatant is poured into sterile dialysis tubing, using a sterile funnel. The dialysis tubing is pretreated by boiling in 0.07 *M* EDTA/0.05 *M* sodium bicarbonate solution before being thoroughly washed with distilled water and autoclaved. Dialyze the collagen extract against cold distilled water at 4°C for 2 d, changing water twice per day. The dialysis tubing is sprayed with ethanol, then cut and poured into sterile centrifuge tubes and recentrifuged at 2,500*g* for 2.5 h at 4°C, and supernatant is collected. Concentration is determined against a standard curve of Vitrogen 100 (Cohesion) at 230 nm and diluted to 2 mg/mL using sterile, cold distilled water. 1% penicillin/streptomycin is added, and the stock collagen solution is stored at 4°C.
3. Obviously, this bulk extraction protocol—although cost-efficient for a laboratory that frequently uses the 3D collagen-gel invasion assay—is somewhat laborious. Many commercial sources of collagen are unsuitable for this assay, but we have successfully used collagen purchased from Vitrogen, and other investigators have reported using rat-tail collagen from Collaborative Biomedical Products ***(28)***. Collagen is mixed with 10X Minimal Essential Medium (MEM) solution and 7.5% sodium bicarbonate solution (Life Technologies, MD) to form gels.

3. Methods

3.1. Boyden-Chamber Migration Assay

The Boyden chamber is a transmembrane assay in which cell migration across a membrane is measured in response to a chemoattractant. This is a useful method for examining the effect of Rho GTPase activity on the migration of stable cell lines. The use of transiently transfected cells in this assay is unlikely to provide clear results, as any effect of the Rho GTPases will be masked by the normal response of the untransfected cell population. This problem may be circumvented by sorting of transiently transfected cells. We have, for example, sorted transiently expressed GFP plasmids using FACS *(7)*. Another alternative may be to use retrovirus or adenovirus vectors, which typically result in much higher efficiencies of cell infection.

We use a 48-well Microchemotaxis chamber, which is a modified version of the Boyden chamber that allows multiple conditions and replicates to be run in one chamber. The chamber apparatus consists of lower and upper wells that are separated by the insertion of an extracellular matrix-coated porous membrane. Chemotactic stimulus is placed in the lower wells, and cells are plated in the upper wells. Cells migrate through the membrane pores to the lower side. The chamber is disassembled and the membrane is fixed and stained. Nonmigrating cells on the upper side of the membrane are removed, and the migrated cells on the lower side of the membrane are counted.

1. Membranes are precoated with extracellular matrix. A number of different matrices are suitable depending upon cell type. We usually use fibronectin at 7.5 μg/mL in PBS coated at room temperature for 30–60 min. Alternatives we have used are:
   a. Gelatin—1.5% solution, *see* **Subheading 2.2.**, immerse membrane for 1–2 h at room temperature
   b. Collagen—100 μg/mL solution in phosphate-buffered saline (PBS). Immerse membrane overnight at room temperature.
   c. Matrigel—150 μg/mL in PBS. Immerse membrane for 1–2 h at 37°C. After coating, wash membranes twice in PBS and dry on Whatman paper before use. It is helpful to mark the orientation of the membrane by cutting the upper left-hand corner.
2. Chamber assembly: lower wells are filled with chemoattractant to form a slight positive meniscus. Exact volume varies between different chambers, but is typically 28 μL. The coated membrane is placed over the lower wells, and must not be moved once positioned, or the media will mix. The membrane is overlaid with a silicone gasket, and then the upper-well chamber is added, using even pressure while the chamber screws are fastened. The chamber is then equilibrated in a 37°C, 5% CO_2-humidified incubator. We find it helpful to place the chamber

in a plastic box containing dampened kimwipes to ensure proper humidity, which is left ajar for gas exchange.

3. Cells are trypsinized according to regular protocol for the cell type in question, then pelleted and resuspended in a set volume of assay media—serum-free-culture media supplemented with 0.1% bovine serum albumin (BSA). An aliquot is removed for cell counting, and the remainder is pelleted to wash. The cell pellet is resuspended at 3×10^5 cells/mL in assay media.
4. 50 μL of cell suspension is plated in each upper well. To avoid air bubbles being trapped at the membrane, pipet by placing the tip close to the membrane (do not touch membrane, as this may damage the matrix coating) at a 45° angle and raising the pipet tip as the cell suspension is expelled.
5. The chamber is then incubated at 37°C in a 5% CO_2-humidified incubator for 4 h.
6. Disassemble the chamber by removing the screws and lifting off the upper-well section. The gasket and membrane will be removed with the upper section, leaving the migrated cells on the uppermost side. The cut corner should now be on the upper right-hand side.
7. The membrane is removed using forceps, and is gently immersed in PBS to remove any cells not attached to the membrane. The membrane is then fixed in 10% buffered formalin for 10 min, rinsed in PBS, and stained in Gill's hematoxylin overnight.
8. The membranes are washed in several changes of tap water. Nonmigrating cells are then scraped off the membrane. This can be done using a rubber squeegee apparatus provided with the Microchemotaxis chamber, but we find it more controllable to do this manually, as follows. The membrane is cut vertically into two sections, and the corner of the second section is cut to ensure correct orientation. The first membrane section is then picked up at an edge using forceps, and the nonmigrating cells are scraped off by wiping the membrane across damp Whatman paper wrapped on the outside of a beaker. The scraping is repeated several times until all nonmigrating cells are removed. Keeping the membrane wet will prevent tearing. The membrane is then placed on a glass slide. While the membrane is still wet, an imprint of the wells can be seen. It is helpful to mark the positions of the wells using a red marker pen so that any wells in which few cells have migrated can be easily located. After the membrane has completely dried, mountant is added and either a large cover slip or a second glass slide is overlaid.
9. Analysis: Cell migration from the upper to the lower side of the membrane is assessed microscopically. A grid eyepiece reticule is used to define a counting area. Magnification of ×100 or ×200 may be used, depending on the level of cell migration. Control-cell migration should be greater than 10 cells per field so that any effect on basal migration is detectable. Cell number in the defined reticule area is counted in the center of each well, with 3–4 replicate wells per experimental condition. The mean and standard error are then calculated.

3.2. 3D Collagen-Gel Invasion Assay

Movement of cells into a matrix of type I collagen fibers is measured by the 3D collagen-gel invasion assay *(28a)*. In this assay, cells are plated as a monolayer on the gel surface. Cells are stimulated to invade the gel, either by the addition of growth factor to the culture media or by incorporation into the gel. We have used this assay to determine the effect of Rho GTPase expression on cell invasion in response to PDGF-BB in stable cell lines *(7)*. In addition, transient overexpression of V12-Rac increased Rat1 fibroblast invasion into collagen gels *(7)*, using cells plated in the invasion assay 24 h after transfection using FuGENE reagent, as described in **Subheading 2.1.2.** As discussed for the Boyden chamber, transient transfection may not be adequately sensitive to detect an inhibitory effect of Rho GTPase expression on the cell population, as the response of untransfected cells may dilute any effect. However, we have observed an inhibitory effect of N17-Rac expression on Rat1 fibroblast invasion in response to PDGF-BB *(7)*.

3.2.1. Assay Set-Up

1. Collagen gels are assembled by mixing 10.2 mL 2 mg/mL collagen solution with 1.2 mL 10X MEM and 0.6 mL 7.5% sodium bicarbonate solution. Mix while cold, then immediately pipet 1 mL into each well of a 12–well tissue-culture plate. Allow to gel undisturbed at room temperature for 15 min, then incubate at 37°C in a 5% CO_2-humidified incubator for 2 h before use.
2. 0.5 mL of serum free cell-culture media is then plated on the surface of the gel, with or without the addition of PDGF-BB. This is allowed to equilibrate for 1–2 h.
3. Cells are trypsinized, collected by centrifugation, and resuspended at 3×10^5 cells/mL in assay media: DMEM supplemented with 4% fetal bovine serum (FBS) and glutamine, so that the final concentration of each is 1%, considering the total 2-mL volume of the gel culture. Then, 1 mL of cell suspension is plated onto the surface of the gel, swirled to cover, then incubated at 37°C in a 5% CO_2-humidified incubator for 96 h.
4. Analysis: Cell invasion into the collagen matrix is microscopically assessed at ×200 magnification. The cell number is counted on the surface of the collagen matrix (approx 70–100 cells/field with a confluent-cell monolayer), and the number migrated into the matrix is counted in 10 random fields per well in duplicate gel cultures. A cell is defined as having invaded the gel when the cell surface is no longer focused. Data is expressed as percentage of cell invasion, calculated as follows:

$$[\text{cells in gel}/(\text{total cells: in gel + on gel})] \times 100$$

Consideration of the total number of cells allows any change in proliferation to be monitored, although we generally observe negligible proliferation of the contact-inhibited monolayer, as growth and saturation density are suppressed on collagen matrices *(29)*. We routinely count collagen-gel cultures using phase optics, but the ease of counting can be enhanced by fixing the gel cultures in 10% buffered formalin for 15 min, gently rinsing the gels 2× in water, then staining the cells with Toluidine blue for 15 min before rinsing off the stain with water. Fixed gel cultures can be stored at 4°C for several days before counting.

4. Notes

1. The Rho family GTPases play a critical role in cell migration and invasion. We have demonstrated this using two methods: the modified Boyden chamber, which measures cell migration across a membrane, and the 3D collagen-gel-invasion assay, which measures cell movement into a matrix. A number of other methods have been used to study the effects of the Rho family GTPases on cell motility, and a brief overview of some of these techniques follows.
   a. In-vitro wound-healing assay: The in-vitro "wound healing" assay analyzes cell migration away from the edge of a wounded monolayer into the denuded space. Migration of cells from the front of the wound edge is assessed microscopically. Cells on the wound edge can also be microinjected to determine the effect of protein expression on cell migration into the wound *(14)*. A possible caveat is that cell migration may be confounded by the fact that cells at the wound edge are essentially released from contact inhibition, since they have been grown to a quiescent monolayer. However, the assay set-up is very simple, and can readily be used as a first line of investigation.
   b. The Dunn chamber: Cell movement across a planar surface in a gradient of chemoattractant can be measured in the Dunn chamber (Weber Scientific International Ltd., Teddington, U.K.). This chamber is a modified Helber cell-counting slide. The slide consists of two concentric wells separated by a bridge marked with orientation points. Cells are plated onto a cover slip that is inverted onto the chamber and sealed to leave a 20-μm gap between the bridge and the cover slip. The outer well is drained and replaced with a chemotactic factor, forming a concentration gradient between the inner and outer wells. Cell movement in a demarcated area of the cover slip is tracked by time-lapse video microscopy over several hours. The directionality and speed of the cell-migration response is determined by analysis of the trajectories of individual cells. This assay has been used in combination with microinjection to investigate the role of Rho GTPases in macrophage chemotaxis toward CSF-1 *(8)*. This assay may be limited to motile cell types, for which cell translocation is sufficient to allow quantitative evaluation in the time frame in which the concentration gradient persists.
   c. Cell tracking of random motility: A number of motility parameters (including speed and directional persistence) can be obtained by computer-aided cell tracking of cells in combination with video microscopy. This approach has been used to study the role of PAK, an effector of Cdc42 and Rac, in directed cell migration *(30)*. Similarly, in the phagokinetic track assay, cells move

across a planar surface that has been coated with colloidal gold particles. The particles are phagocytosed, leaving tracks to indicate where the cells have moved, which can be analyzed in terms of both length and directionality. A detailed methodology for this assay has been described elsewhere ***(31)***.

d. Invasion assays: Transwell invasion assays operate on a principle similar to the Boyden chamber. Cells migrate across a porous membrane toward a chemoattractant in a tissue-culture well. However, the upper-chamber unit containing the membrane is coated with a layer of extracellular matrix (rather than a thin adhesive coating as in the Boyden-chamber assay), through which the cells must invade. The most commonly used matrix is Matrigel, although this invasion assay is also adaptable for use with other gel matrices, such as collagen gels. V12-Cdc42, Rac1, or L61-Rac overexpression increased epithelial-cell invasion across 2-mm collagen gels in the transwell system ***(10)***.

An adaptation of this assay makes it possible to simultaneously evaluate both cell motility and invasion properties ***(32)***. In this "inverse invasion" assay, cells are first attached to the underside of a Transwell™, and subsequently allowed to move through the pores into Matrigel. The number and position of cells is determined by confocal microscopy. Cells at or above 20 µm from the bottom of the filter are considered to have invaded the matrix, whereas cells that migrate through the pores, but are located below 20 µm from the bottom of the filter, are used to quantify chemotaxis.

Collagen-gel cultures can also be used to examine other invasive characteristics. For example, Madin-Darby canine kidney (MDCK)-epithelial cells seeded throughout collagen disperse and invade the collagen matrix. V12-Rac expression reduces this invasive phenotype, resulting in more compact colonies ***(33)***.

Cell-monolayer invasion assays represent an alternative axis through which to study effects on cell migration, looking at cell–cell interactions as compared to cell-matrix-based invasion assays, as discussed in **Subheadings 3.1.** and **3.2.** The invasive capacity of cells is determined by plating them onto a confluent-cell monolayer. Invasive cells infiltrate the monolayer and form foci. Overexpression of V12-Rac1 or V12-Cdc42 in T-lymphoma cells, or V14-Rho in hepatoma cells, increases invasion of a cell monolayer ***(12,34,35)***.

2. The 48-well modified Boyden chamber is a somewhat finicky assay that requires at least one practice-run setup. The chamber needs to be assembled fairly quickly, or the media in the bottom wells will begin to evaporate. If this happens, placing of the membrane on top often results in the trapping of air bubbles. Conversely, if the base wells are overfilled, then placing of the membrane will cause spillover between wells. If different stimuli are being used in the base wells, it is crucial that this contamination does not occur, and is especially critical for negative and positive controls. One precaution is to separate the different stimuli with empty wells.
3. Cells to be used in the Boyden-chamber assay should not have a fresh media change the day before the assay, in order to maximize growth-factor response.
4. We note that alternative Boyden-chamber protocols often remove nonmigrating cells from the upper side of the membrane before fixation and staining. However,

we find that the mechanical scraping easily removes nonmigrating cells even when fixed, and staining the membrane before scraping provides a very valuable source of information. We have found that treating the cells with various factors before assessing their migratory response can affect their adhesion to the membrane. A decrease in the number of cells that adhere reduces the final number of migrating cells. Using this protocol allows this change to be detected, avoiding the incorrect conclusion that cell migration is inhibited, when in fact this could be a secondary effect to that on cell adhesion.

5. Cell-migration analysis in the Boyden chamber can be semi-automated by using a florescent microscope and image-capture software. We have found that positive results can be achieved by staining the membrane with 4,6-diamindino-2-phenylindole (DAPI), as cell nuclei can easily be distinguished using image-analysis software. We have had accurate counts using Adobe Photoshop to capture microscopic fields, followed by the use of IPLab software to quantify cell number.

References

1. Banyard, J. and Zetter, B. R. (1998) The role of cell motility in prostate cancer. *Cancer Metastasis Rev.* **17,** 449–458.
2. Zetter, B. R. (1998) Angiogenesis and tumor metastasis *Annu. Rev. Med.* **49,** 407–424.
3. Heldin, C. H. and Westermark, B. (1999) Mechanism of action and in vivo role of platelet-derived growth factor. *Physiol. Rev.* **79,** 1283–1316.
4. Van Aelst, L. and D'Souza-Schorey, C. (1997) Rho GTPases and signaling networks. *Genes Dev.* **11,** 2295–2322.
5. Kheradmand, F., Werner, E., Tremble, P., Symons, M., and Werb, Z. (1998) Role of Rac1 and oxygen radicals in collagenase-1 expression induced by cell shape change. *Science* **280,** 898–902.
6. Anand-Apte, B., Zetter, B. R., Viswanathan, A., Qiu, R. G., Chen, J., Ruggieri, R., et al. (1997) Platelet-derived growth factor and fibronectin-stimulated migration are differentially regulated by the Rac and extracellular signal-regulated kinase pathways. *J. Biol. Chem.* **272,** 30,688–30,692.
7. Banyard, J., Anand-Apte, B., Symons, M., and Zetter, B. R. (2000) Motility and invasion are differentially modulated by Rho family GTPases. *Oncogene* **19,** 580–591.
8. Allen, W. E., Zicha, D., Ridley, A. J., and Jones, G. E. (1998) A role for Cdc42 in macrophage chemotaxis. *J. Cell Biol.* **141,** 1147–1157.
9. Roberts, A. W., Kim, C., Zhen, L., Lowe, J. B., Kapur, R., Petryniak, B., et al. (1999) Deficiency of the hematopoietic cell-specific Rho family GTPase Rac2 is characterized by abnormalities in neutrophil function and host defense. *Immunity* **10,** 183–196.
10. Keely, P. J., Westwick, J. K., Whitehead, I. P., Der, C. J., and Parise, L. V. (1997) Cdc42 and Rac1 induce integrin-mediated cell motility and invasiveness through PI(3)K. *Nature* **390,** 632–636.

11. Shaw, L. M., Rabinovitz, I., Wang, H. H., Toker, A., and Mercurio, A. M. (1997) Activation of phosphoinositide 3-OH kinase by the alpha6beta4 integrin promotes carcinoma invasion. *Cell* **91,** 949–960.
12. Itoh, K., Yoshioka, K., Akedo, H., Uehata, M., Ishizaki, T., and Narumiya, S. (1999) An essential part for Rho-associated kinase in the transcellular invasion of tumor cells. *Nat. Med.* **5,** 221–225.
13. Ridley, A. J. (1998) Mammalian cell microinjection assay to study the function of Rac and Rho. *Methods Mol. Biol.* **84,** 153–160.
14. Nobes, C. D. and Hall, A. (1999) Rho GTPases control polarity, protrusion, and adhesion during cell movement. *J. Cell Biol.* **144,** 1235–1244.
15. Vastrik, I., Eickholt, B. J., Walsh, F. S., Ridley, A., and Doherty, P. (1999) Sema3A-induced growth-cone collapse is mediated by Rac1 amino acids 17-32. *Curr. Biol.* **9,** 991–998.
16. Qiu, R. G., Chen, J., Kirn, D., McCormick, F., and Symons, M. (1995) An essential role for Rac in Ras transformation. *Nature* **374,** 457–459.
17. Qiu, R. G., Chen, J., McCormick, F., and Symons, M. (1995) A role for Rho in Ras transformation. *Proc. Natl. Acad. Sci. USA* **92,** 11,781–11,785.
18. Qiu, R. G., Abo, A., McCormick, F., and Symons, M. (1997) Cdc42 regulates anchorage-independent growth and is necessary for Ras transformation. *Mol. Cell Biol.* **17,** 3449–3458.
19. Clark, E. A., King, W. G., Brugge, J. S., Symons, M., and Hynes, R. O. (1998) Integrin-mediated signals regulated by members of the rho family of GTPases. *J. Cell Biol.* **142,** 573–586.
20. Peppelenbosch, M. P., Qiu, R. G., de Vries-Smits, A. M., Tertoolen, L. G., de Laat, S. W., McCormick, F., et al. (1995) Rac mediates growth factor-induced arachidonic acid release. *Cell* **81,** 849–856.
21. Hess, J. A., Ross, A. H., Qiu, R. G., Symons, M., and Exton, J. H. (1997) Role of Rho family proteins in phospholipase D activation by growth factors. *J. Biol. Chem.* **272,** 1615–1620.
22. Gossen, M. and Bujard, H. (1992) Tight control of gene expression in mammalian cells by tetracycline-responsive promoters. *Proc. Natl. Acad. Sci. USA* **89,** 5547–5551.
23. Gossen, M., Bonin, A. L., and Bujard, H. (1993) Control of gene activity in higher eukaryotic cells by prokaryotic regulatory elements. *Trends Biochem. Sci.* **18,** 471–475.
24. Baim, S. B., Labow, M. A., Levine, A. J., and Shenk, T. (1991) A chimeric mammalian transactivator based on the lac repressor that is regulated by temperature and isopropyl beta-D-thiogalactopyranoside. *Proc. Natl. Acad. Sci. USA* **88,** 5072–5076.
25. Braselmann, S., Graninger, P., and Busslinger, M. (1993) A selective transcriptional induction system for mammalian cells based on Gal4-estrogen receptor fusion proteins. *Proc. Natl. Acad. Sci. USA* **90,** 1657–1661.
26. No, D., Yao, T. P., and Evans, R. M. (1996) Ecdysone-inducible gene expression in mammalian cells and transgenic mice. *Proc. Natl. Acad. Sci. USA* **93,** 3346–3351.

27. Schor, S. L. and Court, J. (1979) Different mechanisms in the attachment of cells to native and denatured collagen. *J. Cell Sci.* **38,** 267–281.
28. Keely, P. J., Fong, A. M., Zutter, M. M., and Santoro, S. A. (1995) Alteration of collagen-dependent adhesion, motility, and morphogenesis by the expression of antisense alpha 2 integrin mRNA in mammary cells. *J. Cell Sci.* **108,** 595–607.
28a. Schor, S. L. (1980) Cell proliferation and migration on collagen substrata in vitro. *J. Cell Sci.* **41,** 159–175.
29. Kono, T., Tanii, T., Furukawa, M., Mizuno, N., Taniguchi, S., Ishii, M., et al. (1990) Correlation of contractility and proliferative potential with the extent of differentiation in mouse fibroblastic cell lines cultured in collagen lattices. *J. Dermatol.* **17,** 149–154.
30. Sells, M. A., Boyd, J. T., and Chernoff, J. (1999) p21-activated kinase 1 (Pak1) regulates cell motility in mammalian fibroblasts. *J. Cell Biol.* **145,** 837–849.
31. Takaishi, K., Sasaki, T., and Takai, Y. (1995) Cell motility assay and inhibition by Rho-GDP dissociation inhibitor. *Methods Enzymol.* **256,** 336–347.
32. Malliri, A., Symons, M., Hennigan, R. F., Hurlstone, A. F., Lamb, R. F., Wheeler, T., et al. (1998) The transcription factor AP-1 is required for EGF-induced activation of rho-like GTPases, cytoskeletal rearrangements, motility, and in vitro invasion of A431 cells. *J. Cell Biol.* **143,** 1087–1099.
33. Hordijk, P. L., ten Klooster, J. P., van der Kammen, R. A., Michiels, F., Oomen, L. C., and Collard, J. G. (1997) Inhibition of invasion of epithelial cells by Tiam1-Rac signaling. *Science* **278,** 1464–1466.
34. Michiels, F., Habets, G. G., Stam, J. C., van der Kammen, R. A., and Collard, J. G. (1995) A role for Rac in Tiam1-induced membrane ruffling and invasion. *Nature* **375,** 338–340.
35. Stam, J. C., Michiels, F., van der Kammen, R. A., Moolenaar, W. H., and Collard, J. G. (1998) Invasion of T-lymphoma cells: cooperation between Rho family GTPases and lysophospholipid receptor signaling. *EMBO J.* **17,** 4066–4074.

II

Specific Methods

Rab Family GTPases

12

Isolation of Regulatory Proteins for the Rab3 Subfamily GTPases

Hiroyuki Nakanishi, Takuya Sasaki, and Yoshimi Takai

1. Introduction

The Rab small GTPase family consists of over sixty members, and is implicated in intracellular vesicle trafficking, which includes exocytosis, endocytosis, and transcytosis *(1–5)*. All the Rab GTPases have unique C-terminal structures, which undergo posttranslational modifications with geranylgeranyl moieties in most cases. Rab GTPases cycle between the guanosine diphosphate (GDP)-bound inactive and guanosine triphosphate (GTP)-bound active forms and between the cytosol and the membranes, and these cyclical activation, inactivation, and translocation processes are regulated by at least three types of regulatory proteins: GDP dissociation inhibitors (GDIs), GDP/GTP exchange proteins (GEPs), and GTPase-activating proteins (GAPs). In the cytosol, a Rab GTPase is maintained in the GDP-bound inactive form by a GDI. The GDP-bound form is first released from the GDI by an unknown mechanism that is coupled to the delivery of a Rab GTPase to a specific membrane compartment *(6,7)*; the Rab GTPase is subsequently converted to the GTP-bound form by the action of a GEP. The GTP-bound form then interacts with a downstream effector(s). Thereafter, the GTP-bound form is converted to the GDP-bound form by the action of a GAP. The GDP-bound form produced on the membrane then complexes with the GDI and returns to the cytosol as a Rab GTPase-Rab GDI complex. Thus, a characteristic of Rab GTPases is that a cycle of membrane association/dissociation is superimposed onto their GDP/GTP cycle.

The Rab3 subfamily belongs to the Rab family and consists of four members: Rab3A, -3B, -3C, and -3D *(1)*. Rab3A and -3C are present in cells with Ca^{2+}-

From: *Methods in Molecular Biology, vol. 189: GTPase Protocols: The Ras Superfamily*
Edited by: E. J. Manser and T. Leung © Humana Press Inc., Totowa, NJ

dependent exocytosis, particularly in neurons, whereas Rab3B and -3D are expressed in epithelial cells and adipocytes, respectively ***(8,9)***. Mounting evidence indicates that Rab3A is involved in Ca^{2+}-dependent exocytosis, particularly neurotransmitter release ***(1)***. The Rab3A-knockout mouse analysis has revealed that Rab3A plays two roles: to efficiently dock synaptic vesicles to the presynaptic plasma membrane; and to regulate the efficiency of the fusion process ***(10,11)***. It has also been reported that Rab3A is involved in the formation of long-term potentiation in hippocampus ***(12)***.

Of all the Rab family members, the Rab3 subfamily has been studied most extensively in regard to the three types of regulatory proteins ***(5)***. Rab GDI has been originally isolated as a cytosolic protein, which interacts with GDP-bound Rab3A and thereby inhibits the dissociation of GDP from and the binding of GTP to Rab3A ***(13)***. Three isoforms of Rab GDI have thus far been isolated: Rab GDIα, -β, and -γ ***(5)***. The originally isolated isoform corresponds to Rab GDIα. Rab GDIα and -β have been well characterized. Rab GDIβ is ubiquitously expressed, whereas Rab GDIα is abundantly expressed in neuronal cells ***(14)***. Despite the different tissue distribution, both isoforms show similar biochemical properties ***(5)***: (1) Both proteins bind the lipid-modified, GDP-bound form, but no other forms such as the lipid-modified, GTP-bound form, the lipid-unmodified GDP-bound form, and the lipid-unmodified GTP-bound form; (2) Both proteins inhibit the basal and GEP-stimulated dissociation of GDP from the GDP-bound form; (3) Both proteins inhibit the binding of the GDP-bound form to the membrane and keep it in the cytosol; (4) When the GDP-bound form is produced from the GTP-bound form, both proteins bind the GDP-bound form and retrieve it from the membrane; and (5) Both proteins are active on all the Rab GTPases. In addition to these activities, Rab GDI plays an important role in the specific delivery of Rab GTPases to their target membranes ***(6,7)***.

Rab3 GEP is ubiquitously expressed, but is abundant in the brain ***(15,16)***. It is active on Rab3A, -3C, and -3D, but hardly on Rab3B. It is inactive on other Rab GTPases, including Rab2, -5A, -10, and -11. It prefers the lipid-modified form to the lipid-unmodified form as a substrate. Rab3 GAP is also ubiquitously expressed and enriched in the synaptic soluble fraction in the brain ***(16–18)***. Rab3 GAP consists of two subunits, catalytic (p130) and noncatalytic (p150) subunits ***(17,18)***. It is active on all the Rab3 subfamily GTPases, but the activity on Rab3B is slightly weaker than those on Rab3A, -3C, and -3D. Rab3 GAP is inactive on other Rab GTPases, including Rab2, -5A, and -11. The preference of Rab3 GAP for the lipid-modified form is similar to that of Rab3 GEP. Thus, it is likely that each member or each subfamily of the Rab family interacts with specific GEP and GAP.

This chapter describes the assays for Rab GDI, Rab3 GEP, and Rab3 GAP, and the purification procedures for these regulatory proteins for the Rab3 subfamily GTPases.

2. Materials

2.1. Preparation of Recombinant Lipid-Modified Rab3A

Recombinant lipid-modified Rab3A is expressed in *Spodoptera frugiperda* (Sf9) cells transfected with a baculovirus carrying the cDNA of Rab3A, and purified from the membrane fraction of the Sf9 cells ***(19)***. The purified sample of Rab3A is dissolved in a buffer containing 20 m*M* Tris-HCl, pH 8.0, 5 m*M* $MgCl_2$, 1 m*M* EDTA, 1 m*M* dithiothreitol (DTT), and 0.6% 3-[(3-cholamidopropyl)dimethylammonio]-1-propane sulfonic acid (CHAPS). The sample can be stored for at least 6 mo at –80°C.

2.2. Preparation of the Synaptic Soluble Fraction from Rat Brain

1. Cerebra are rapidly removed from 40 rats after decapitation.
2. Homogenization is performed with 12 up-and-down strokes of a Potter-Elvehjem Teflon-glass homogenizer in a homogenization buffer [0.32 *M* sucrose, 1 m*M* $NaHCO_3$, 1 m*M* $MgCl_2$, 0.5 m*M* $CaCl_2$, and 1 µ*M* (*p*-amidinophenyl) methanesulfonyl fluoride (APMSF)]: Cerebral tissues 10 g wet weight per 40 mL of the homogenization buffer. The homogenates are combined, diluted to 400 mL with the same buffer, and filtrated through four layers of gauze.
3. The homogenate is centrifuged at 1,400 *g* for 10 min.
4. The pellet is resuspended with three strokes of the homogenizer in 350 mL of the same buffer and centrifuged at 710 *g* for 10 min.
5. The resultant pellet, designated as P1 fraction, contains nuclei and cell debris. The supernatants from the two steps of centrifugation are combined and centrifuged at 13,800 *g* for 10 min.
6. The pellet is resuspended and rehomogenized in 350 mL of the same buffer, and then centrifuged again at 13,800 *g* for 10 min. The resultant pellet, designated as P2 fraction, contains myelin, mitochondria, and synaptosomes.
7. This pellet is resuspended with three strokes of the homogenizer in 100 mL of a buffer (0.32 *M* sucrose, 1 m*M* $NaHCO_3$, and 10 µg/mL leupeptin). The suspension is diluted with 450 mL of another buffer (6 m*M* Tris-HC, pH 8.0, and 10 µg/mL leupeptin), stirred for 45 min, and centrifuged at 32,800 *g* for 20 min.
8. The supernatant is further centrifuged at 78,000 *g* for 120 min. The final supernatant is pooled as the synaptic soluble fraction. This synaptic soluble fraction can be stored at –80°C for at least 3 mo.

2.3. Buffers for Purification of Rab GDIα, Rab3 GEP, and Rab3 GAP

1. Buffer A: 25 m*M* Tris-HCl, pH 7.5, 1 m*M* DTT, 1 m*M* $MgCl_2$, and 1 m*M* EGTA.

2. Buffer B: 25 m*M* Tris-HCl, pH 7.5, 1 m*M* DTT, and 0.5 m*M* EDTA.
3. Buffer C: 25 m*M* Tris-HCl, pH 7.5, 1 m*M* DTT, 0.5 m*M* EDTA, 320 m*M* sucrose, and 10 μ*M* APMSF.
4. Buffer D: 20 m*M* Tris-HCl, pH 7.5, and 1 m*M* DTT.
5. Buffer E: 20 m*M* potassium phosphate at pH 7.8, 1 m*M* DTT, 0.6% CHAPS, and 10% glycerol.
6. Buffer F: 20 m*M* *bis*-Tris-HCl, pH 5.5, 0.5 m*M* EDTA, 1 m*M* DTT, 0.6% CHAPS, and 10% glycerol.
7. Buffer G: 20 m*M* Tris-HCl, pH 7.5, 0.5 m*M* EDTA, 1 m*M* DTT, 0.6% CHAPS, 0.45% sodium cholate, 10% glycerol, and 0.15 *M* NaCl.
8. Buffer H: 20 m*M* Tris-HCl, pH 7.9, 140 m*M* NaCl, 3 m*M* potassium chloride (KCl), 1 m*M* $CaCl_2$, 0.5 m*M* $MgCl_2$, and 0.9 m*M* Na_2PO_4.
9. Buffer I: 20 m*M* Tris-HCl, pH 7.5, 1 m*M* DTT, and 0.6% CHAPS.
10. Buffer J: 20 m*M* Tris-HCl, pH 7.5, 0.5 m*M* EGTA, 0.5 m*M* EDTA, and 1 m*M* DTT.
11. Buffer K: 20 m*M* potassium phosphate at pH 7.5, 0.5 m*M* EDTA, and 1 m*M* DTT.
12. Buffer L: 50 m*M* sodium phosphate at pH 7.5 and 50 m*M* NaCl.

3. Methods

3.1. Rab GDI

3.1.1. Assays for Rab GDI

The activity of Rab GDI to regulate the GDP/GTP exchange reaction of Rab3A is assayed by measuring either the dissociation of [^{3}H]GDP from or the binding of [^{35}S]GTPγS to lipid-modified Rab3A (*see* **Note 1**). The initial velocities for the dissociation of [^{3}H]GDP from and the binding of [^{35}S]GTPγS to Rab3A depend on the free Mg^{2+} concentration in the assay mixture. In the experiments described in **Subheadings 3.1.1.1.** and **3.1.1.2.**, the free Mg^{2+} concentration is 0.5 μ*M*, because the initial velocities are so rapid at 0.5 μ*M* free Mg^{2+} concentration that the Rab GDI activity to inhibit the GDP/GTP exchange reaction is easily detected. Rab GDI is active on lipid-modified Rab3A, but is nearly inactive on lipid-unmodified Rab3A.

3.1.1.1. Dissociation Assay

1. [^{3}H]GDP-bound Rab3A is made by incubating Rab3A (2 pmol) for 20 min at 30°C with 1 μ*M* [^{3}H]GDP (7–9 × 10^3 cpm/pmol) in a reaction mixture (24 μL) containing 20 m*M* Tris-HCl, pH 7.5, 5 m*M* $MgCl_2$, 10 m*M* EDTA, 1 m*M* DTT, 1 m*M* L-α-dimyristoylphosphatidylcholine (DMPC), and 0.25% CHAPS.
2. After the first incubation, 1 μL of 375 m*M* $MgCl_2$ is added to yield a final concentration of 20 m*M* to prevent the dissociation of [^{3}H]GDP from Rab3A, and the mixture is immediately cooled on ice.

3. The second incubation is performed by adding a mixture (75 μL) containing 20 m*M* Tris-HCl, pH 7.5, 10 m*M* EDTA, 1 m*M* DTT, 200 μ*M* GTP, and various amounts of Rab GDI for 20 min at 30°C.
4. The reaction is stopped by adding 2 mL of an ice-cold solution containing 20 m*M* Tris-HCl, pH 7.5, 25 m*M* $MgCl_2$, and 100 m*M* NaCl to the reaction mixture, followed by rapid filtration on BA-85 nitrocellulose filters (pore size, 0.45 μm, Schleicher & Schuell [Dassel, Germany]) and washing with 2 mL of the same solution 4×. The radioactivity trapped on the filters is measured by liquid scintillation counting.

3.1.1.2. Binding Assay

1. GDP-Bound Rab3A (2 pmol) is incubated with various amounts of Rab GDI in a reaction mixture (100 μL) containing 20 m*M* Tris-HCl, pH 7.5, 10 m*M* EDTA, 5 m*M* $MgCl_2$, 1 m*M* DTT, and 0.75 m*M* DMPC, 0.06% CHAPS, and 1 μ*M* [^{35}S]GTPγS (6–8 × 10^3 cpm/pmol) for 20 min at 30°C. GDP-bound Rab3A used is the sample purified from Sf9 cells because it is purified as the GDP-bound form.
2. The reaction is stopped and the radioactivity is trapped on the filters is counted as described in **Subheading 3.1.1.1., step 4**.

3.1.2. Purification of Rab GDIα

3.1.2.1. Native Rab GDIα

All the purification procedures are carried out at at 0–4°C. Rab GDIα is detected by measuring the dissociation of [^{3}H]GDP from Rab3A (*see* **Subheading 3.1.1.1.**). In the Mono Q column chromatography described in **Subheading 3.1.2.1., step 4**, Rab GDIα can be easily detected on sodium dodecyl sulfate (SDS)-polyacrylamide gel electrophoresis (SDS-PAGE) followed by protein staining with Coomassie brilliant blue, because Rab GDIα appears as only a single protein band with a molecular mass of approx 54 kDa on SDS-PAGE.

1. Preparation of the cytosol fraction from bovine brain: Bovine cerebral tissue (450 g, wet wt) is homogenized for 1 min 2× at 1-min intervals in a Waring blender with 450 mL of buffer A containing 10 μ*M* APMSF. The homogenate is further homogenized in a Potter-Elvehjem Teflon-glass homogenizer and centrifuged at 100,000 *g* for 60 min. The supernatant (350 mL, 6.0 g of protein) is collected.
2. DEAE-Sephacel-column chromatography: The supernatant is applied to a DEAE-Sephacel column (Amersham-Pharmacia Biotech. Ltd., 7.5 × 18 cm) equilibrated with buffer A. After the column is washed with 8 L of buffer A and 400 mL of buffer A containing 0.3 *M* NaCl, Rab GDIα is eluted by 800 mL of buffer A containing 0.3 *M* NaCl.

3. Ammonium sulfate precipitation: Solid ammonium sulfate is added to this eluate (800 mL, 1.7 g of protein) to give a final concentration of 40% saturation. The sample is centrifuged at 20,000*g* for 20 min. Rab GDIα is mostly recovered in the supernatant. Solid ammonium sulfate is added to the supernatant to yield a final concentration of 60% saturation. The sample is centrifuged at 20,000*g* for 20 min. Rab GDIα is mostly precipitated. The 40–60% precipitate is dissolved in 26 mL of buffer B and dialyzed overnight against buffer B. The dialyzed sample is centrifuged at 100,000*g* for 60 min and the volume of the supernatant is adjusted to 32 mL by the addition of buffer B.
4. Mono Q HR10/10-column chromatography (*see* **Note 2**): One-fourth of the supernatant (8 mL, 147 mg of protein) is supplemented with 2.7 mL of buffer B containing 4% sodium cholate. The sample is adjusted to pH 8.0 by 1.5 *M* Tris and applied to a Mono Q HR10/10 column (Amersham-Pharmacia Biotech. Ltd.) equilibrated with buffer B containing 1% sodium cholate. The column is washed with 190 mL of the same buffer, and fractions of 4 mL each are collected. Most of Rab GDIα does not adsorb to the column, and is eluted as a single peak in Fractions 25–34. The active fractions in the pass fractions are collected and concentrated to 5 mL by an Amicon ultrafiltration cell equipped with a PM-10 filter. The concentrate (1.9 mg of protein) is dialyzed against buffer B and used as a purified sample.
5. The rest of the supernatant is subjected to the same Mono Q column chromatography in a similar manner. Approximately 8 mg of pure Rab GDIα is obtained from 450 g of bovine brain. It can be stored at -80°C for at least 6 mo.

3.1.2.2. Recombinant Glutathione S-Transferase Fusion Protein of Rab GDIα (GST-Rab GDIα)

1. Recombinant GST-Rab GDIα is expressed with the pGEX-2T construct containing the cDNA of Rab GDIα in *Escherichia coli* (*E. coli*).
2. The bacteria are harvested, suspended in 20 mL of phosphate-buffered saline (PBS), and washed with 20 mL of PBS. The pellet is frozen at –80°C.
3. The pellet is quickly thawed at 37°C and suspended in 20 mL of buffer C, and the suspension is sonicated on ice for 30 s 4× at 30-s intervals. The homogenate is centrifuged at 100,000*g* for 1 h.
4. Twenty mL of the crude supernatant is applied to glutathione Sepharose 4B beads (4 mL of bed volume) equilibrated with buffer B. After the column is washed with 20 mL of buffer B, GST-Rab GDIα is eluted with 5 mL of 50 m*M* Tris-HCl at pH 8.0 containing 5 m*M* reduced glutathione. This eluate is dialyzed against buffer B and used as a purified GST-Rab GDIα. It can be stored at –80°C for at least 6 mo.

3.2. Rab3 GEP

3.2.1. Assays for Rab3 GEP

The Rab3 GEP activity to stimulate the GDP/GTP exchange reaction of Rab3A is assayed by measuring either the dissociation of [^{3}H]GDP from or the binding of [^{35}S]GTPγS to lipid-modified Rab3A.

3.2.1.1. Dissociation Assay

1. [^{3}H]GDP-bound Rab3A is made by incubating Rab3A (3 pmol) at 30°C for 20 min with 3 μ*M* [^{3}H]GDP (7–9 × 10^3 cpm/pmol) in a reaction mixture (5 μL) containing 50 m*M* Tris-HCl, pH 8.0, 5 m*M* $MgCl_2$, 10 m*M* EDTA, 0.5 m*M* DTT, and 0.12% CHAPS.
2. The reaction is stopped by adding 2 μL of 100 m*M* $MgCl_2$ and 5 μL of a solution containing 50 m*M* Tris-HCl, pH 8.0, 5 m*M* $MgCl_2$, 0.5 m*M* EDTA, and 1 m*M* DTT, and the mixture is immediately cooled on ice.
3. The sample to be assayed is incubated at 30°C for 10 min with [^{3}H]GDP-bound Rab3A (3 pmol) in a reaction mixture (50 μL) containing 50 m*M* Tris-HCl, pH 8.0, 12 m*M* $MgCl_2$, 2 m*M* EDTA, 0.2 mg/mL bovine serum albumin (BSA), 12 μ*M* GTPγS, and 0.06% CHAPS.
4. The reaction is stopped and the radioactivity trapped on the filters is counted as described in **Subheading 3.1.1.1., step 4**.

3.2.1.2. Binding Assay

1. The sample to be assayed is incubated at 30°C for 10 min with GDP-bound Rab3A (3 pmol) in a reaction mixture (50 μL) containing 50 m*M* Tris-HCl, pH 8.0, 12 m*M* $MgCl_2$, 2 m*M* EDTA, 0.2 mg/mL BSA, 12 μ*M* [^{35}S]GTPγS (6–8 × 10^3 cpm/pmol), and 0.06% CHAPS.
2. The reaction is stopped and the radioactivity trapped on the filters is counted as described in **Subheading 3.1.1.1., step 4**.

3.2.2. Purification of Rab3 GEP

3.2.2.1. Native Rab3 GEP

All the purification procedures are carried out at 0–4°C. Rab3 GEP is detected by measuring the dissociation of [^{3}H]GDP from lipid-modified Rab3A (*see* **Subheading 3.2.1.1.**). In the Superdex 200-column chromatography described in **Subheading 3.2.2.1., step 6**, Rab3 GEP is also detected on SDS-PAGE (6.5% polyacrylamide gel) followed by protein staining with silver. Rab3 GEP shows a protein band with a molecular mass of about 200 kDa, which coincides well with the GEP activity. In the last column chromatography described in **Subheading 3.2.2.1., step 7**, Rab3 GEP appears as two peaks. The first and second peaks are named Rab3 GEPI and -II. Both Rab3 GEPI and -II show a protein band with a molecular mass of approx 200 kDa on SDS-PAGE followed by protein staining with silver. Rab3 GEPI and -II show similar properties, including the substrate specificity, the requirement of the lipid-modifications of Rab3A, and the ineffectiveness to Rab3A complexed with Rab GDI. When the peptide maps of Rab3 GEPI and -II are determined, they appear to be identical. The exact relationship between Rab3 GEPI and -II is unknown, but they are probably splicing variants.

1. The synaptic soluble fraction is prepared from 80 rat brains (*see* **Subheading 2.2.**).
2. Q-Sepharose-column chromatography: One-half of the synaptic soluble fraction (550 mL, 0.9 g of protein) is adjusted to 0.2 *M* NaCl and applied to a Q-Sepharose FF column (Amersham-Pharmacia Biotech. Ltd., 2.6 × 10 cm) equilibrated with buffer D containing 0.2 *M* NaCl. Elution is performed with 350 mL of buffer D containing 0.5 *M* NaCl at a flow rate of 5 mL/min. Fractions of 10 mL each are collected. Rab3 GEP appears in Fractions 5–19.
3. Phenyl-Sepharose-column chromatography: The active fractions of the Q-Sepharose column chromatography (150 mL, 159 mg of protein) are collected, and NaCl is added to give a final concentration of 2 *M*. The sample is applied to a phenyl-Sepharose column (Amersham-Pharmacia Biotech. Ltd., 2.6 × 10 cm) equilibrated with buffer D containing 2 *M* NaCl. Elution is performed with a 360-mL linear gradient of NaCl (2–0 *M*) in buffer D, followed by 180 mL of buffer D, at a flow rate of 3 mL/min. Fractions of 6 mL each are collected. Rab3 GEP appears in Fractions 52–63.
4. Hydroxyapatite-column chromatography: The active fractions of the phenyl-Sepharose-column chromatography (72 mL, 8.6 mg of protein) are collected and applied to a hydroxyapatite column (Calbiochem-Novabiochem Co., 1.0 × 30 cm) equilibrated with buffer E. Elution is performed with a 75-mL linear gradient of potassium phosphate (20–100 m*M*) in buffer E and a subsequent 75-mL linear gradient (100–300 m*M*) in buffer E, followed by a 50-mL linear gradient (300–500 m*M*) in buffer E, at a flow rate of 1 mL/min. Fractions of 2.5 mL each are collected. Rab3 GEP appears in Fractions 46–54.
5. Mono Q HR10/10-column chromatography: The active fractions of the hydroxyapatite-column chromatography (22.5 mL, 2.2 mg of protein) are collected, diluted with an equal volume of buffer F, and applied to a Mono Q HR10/10 column equilibrated with buffer F. Elution is performed with a 60-mL linear gradient of NaCl (0.2–0.5 *M*) in buffer F at a flow rate of 1 mL/min. Fractions of 1 mL each are collected. Rab3 GEP appears in Fractions 24–33.
6. HiLoad 16/60 Superdex 200-column chromatography: The active fractions of the Mono Q HR10/10-column chromatography (10 mL, 0.44 mg of protein) are collected, concentrated to approx 2 mL by a centrifugal ultrafiltration concentrator (Centriprep 30), and applied to a HiLoad 16/60 Superdex 200 column (Amersham-Pharmacia Biotech. Ltd., 1.6 × 60 cm) equilibrated with buffer G. Elution is performed with the same buffer at a flow rate of 0.25 mL/min. Fractions of 2 mL each are collected. Rab3 GEP appears in Fractions 26–30.
7. HPLC Hydroxyapatite-column chromatography: The active fractions of the Superdex 200-column chromatography (10 mL, 45 μg of protein) are collected. These collected fractions can be stored at –80°C for at least 3 mo. The other half of the synaptic soluble fraction is also subjected to the successive column chromatographies in the similar manner as described above. The active fractions of the two Superdex 200-column chromatographies are combined and applied to an HPLC hydroxyapatite column (Koken Co. Ltd. [Tokyo, Japan], 0.78 × 10 cm)

equilibrated with buffer E. Elution is performed with a 12.5-mL linear gradient of potassium phosphate (20–100 m*M*) in buffer E, followed by a 50-mL linear gradient of potassium phosphate (100–500 m*M*) in buffer B, at a flow rate of 0.5 mL/min. Fractions of 1 mL each are collected. Rab3 GEP appears in two peaks in Fractions 29–33 and 34–38. The first (5 mL, 15.5 µg of protein) and second (5 mL, 7.5 µg of protein) peaks are separately collected as Rab3 GEPI and -II, respectively. Rab3 GEPI and -II can be stored at –80°C for at least 6 mo.

3.2.2.2. Recombinant Rab3 GEP

The pCMV5 construct containing the cDNA of Rab3 GEP (pCMV5-Rab3 GEP) is transfected to COS7 cells with the DEAE-dextran method ***(20)***. Recombinant Rab3 GEP overexpressed in the cells is then purified by use of Mono Q PC 1.6/5-column chromatography. Recombinant Rab3 GEP show similar properties, including the substrate specificity, the requirement of the lipid-modifications of Rab3A, and the ineffectiveness to Rab3A complexed with Rab GDI, to those of native Rab3 GEPI and -II.

1. COS7 cells (1 × 10^6 cells) are plated on a 10-cm dish and cultured for 1 d in Dulbecco's modified Eagle's medium (DMEM) supplemented with 10% fetal calf serum (FCS). The cells are washed with PBS and incubated at 37°C for 30 min with 44 µg of pCMV5-Rab3 GEP in 2 mL of buffer H containing 0.5 mg/mL DEAE-dextran. After the solution is aspirated, the cells are incubated at 37°C for 3 h with 100 m*M* chloroquine in DMEM supplemented with 10% FCS. The cells are washed with PBS, followed by incubation at room temperature for 2 min in DMEM containing 20% glycerol. The cells are gently washed with PBS and incubated at 37°C for 48 h in DMEM supplemented with 10% FCS.
2. The COS7 cells expressing Rab3 GEP are washed with PBS and scraped. The cells are homogenized with a Teflon-glass homogenizer in 2 mL of buffer I containing 1 µ*M* APMSF and 10 µg/mL leupeptin, and centrifuged at 100,000 *g* for 60 min.
3. Mono Q PC 1.6/5-column chromatography: The supernatant (2 mL, 4.2 mg of protein) is applied to a Mono Q PC1.6/5 column equilibrated with buffer I containing 0.1 *M* NaCl. Elution is performed with a 1.5-mL linear gradient of NaCl (0.1–0.5 *M*) in buffer I at a flow rate of 50 µL/min. Fractions of 50 µL each are collected. Recombinant Rab3 GEP appears in Fractions 19–21. The active fractions (150 µL, 0.1 mg of protein) are collected. This recombinant Rab3 GEP can be stored at –80°C for at least 3 mo.

3.3. Rab3 GAP

3.3.1. Assays for Rab3 GAP

The Rab3 GAP activity to stimulate the intrinsic GTPase activity of lipid-modified Rab3A is assayed by three methods as follows:

3.3.1.1. Standard Assay (Filter Assay)

1. [γ-^{32}P]GTP-bound Rab3A is made by incubating Rab3A (3 pmol) with 1.5 μ*M* [γ-^{32}P]GTP (1 × 10^4 cpm/pmol) at 30°C for 10 min in a reaction mixture (10 μL) containing 25 m*M* Tris-HCl, pH 8.0, 10 m*M* EDTA, 5 m*M* $MgCl_2$, 0.5 m*M* DTT, and 0.3% CHAPS.
2. The reaction is stopped by adding 2.5 μL of 80 m*M* $MgCl_2$.
3. To this mixture (12.5 μL), the sample to be assayed is added in a total volume of 50 μL and further incubated at 30°C for 5 min.
4. The reaction is stopped and the radioactivity trapped on the filters is determined by Cerenkov counting as described in **Subheading 3.1.1.1., step 4**.

3.3.1.2. Overlay Assay

1. The sample to be assayed is subjected to SDS-PAGE.
2. After semi-dry Western blotting, the nitrocellulose filter-bound proteins are renatured in PBS containing 1% BSA, 0.5 m*M* $MgCl_2$, 0.1% Triton X-100, and 5 m*M* DTT.
3. The filter is incubated at 25°C for 10 min with [γ-^{32}P]GTP-bound Rab3A (3 pmol), prepared with the same method as described in **Subheading 3.3.1.1.**, in a buffer containing 25 m*M* HEPES-NaOH, pH 7.0, 1.25 m*M* $MgCl_2$, 0.05% Triton X-100, and 2.5 m*M* DTT.
4. After the filter is washed with PBS containing 25 m*M* HEPES-NaOH, pH 7.0, 5 m*M* $MgCl_2$, and 0.05% Triton X-100, the hydrolysis of [γ-^{32}P]GTP bound to Rab3A is analyzed with a Fujix BAS 2000 Imaging Analyzer.

3.3.1.3. Thin-Layer Chromatography Assay

1. [α-^{32}P]GTP-bound Rab3A (3 pmol) is made as described in **Subheading 3.1.1.1.** except that [α-^{32}P]GTP is used instead of [γ-^{32}P]GTP.
2. The sample to be assayed is mixed with [α-^{32}P]GTP-bound Rab3A in a reaction mixture (50 μL) containing 35 m*M* Tris-HCl, pH 8.0, 12 m*M* $MgCl_2$, 2 m*M* EDTA, 0.2 m*M* EGTA, 1 m*M* DTT, and 0.06% CHAPS at 30°C for 5 min.
3. The mixture is applied to a nitrocellulose filter and then rinsed 3× with an ice-cold solution containing 20 m*M* Tris-HCl, pH 7.5, 25 m*M* $MgCl_2$, and 100 m*M* NaCl.
4. Guanine nucleotides bound to Rab3A are eluted by immersing the filter in a buffer containing 20 m*M* Tris-HCl, pH 8.0, 20 m*M* EDTA, 2% SDS, 1 m*M* GDP, and 1 m*M* GTP at 65°C for 5 min.
5. The release nucleotides are separated on a poly(ethylene)imine cellulose thin-layer chromatography plate (Macherey-Nägel) with 1 *M* KH_2PO_4 at pH 3.4.
6. After developing the thin-layer chromatography plate, the plate is dried, and the GDP and GTP spots are visualized with a short-wave ultraviolet (UV) lamp and analyzed with Fujix BAS 2000 Imaging Analyzer.

3.3.2. Purification of Rab3 GAP

3.3.2.1. Native Rab3 GAP

All the purification procedures are carried out at 0-4°C. Rab3 GAP is detected by measuring the hydrolysis of [γ-^{32}P]GTP bound to lipid-modified Rab3A (*see* **Note 5**). In the last column chromatography described in **Subheading 3.3.2.1., step 5**, Rab3 GAP shows two bands with molecular masses of approx 130,000 (p130) and 150,000 (p150) on SDS-PAGE followed by protein staining with silver.

1. The synaptic soluble fraction is prepared from 200 rat brains (*see* **Subheading 2.2.**).
2. Q-Sepharose FF-column chromatography: One-fifth of the synaptic soluble fraction (450 mL, 315 mg of protein) is directly applied to a Q-Sepharose FF column (2.6 × 23 cm) equilibrated with buffer J. After the column is washed with 600 mL of buffer J, elution is performed with a 600-mL linear gradient of NaCl (0–0.5 *M*) in buffer J, followed by 120 mL of 0.5 *M* NaCl in buffer J at a flow rate of 5 mL/min. Fractions of 8 mL each are collected. Rab3 GAP appears in Fractions 62–70. These fractions (72 mL, 36 mg of protein) are collected. The rest of the synaptic soluble fraction is subjected to the same Q-Sepharose-column chromatography 4× in a similar manner.
3. Hydroxyapatite-column chromatography: The samples of the five Q-Sepharose-column chromatographies are pooled and diluted with 720 mL of buffer K. The sample is applied to a hydroxyapatite column (2.6 × 6.6 cm) equilibrated with buffer K. After the column is washed with 350 mL of the same buffer, elution is performed with a 500-mL linear gradient of potassium phosphate (20–212 m*M*) in buffer K, followed by a 150-mL linear gradient (212–500 m*M*) and 150 mL of 500 m*M* potassium phosphate in buffer K at a flow rate of 1.25 mL/min. Fractions of 10 mL each are collected. Rab3 GAP appears in Fractions 29–40. These fractions (120 mL, 18 mg of protein) are collected.
4. Heparin-Sepharose CL-6B-column chromatography: The sample is diluted with 240 mL of buffer J and applied to a heparin-Sepharose CL-6B column (Amersham-Pharmacia Biotech. Ltd., 0.5 × 5 cm) equilibrated with buffer J. After the column is washed with 20 mL of the same buffer, elution is performed with 0.5 *M* NaCl in buffer H at a flow rate of 0.5 ml/min. Fractions of 1 mL each are collected. Rab3 GAP appears in Fractions 2–6. These fractions (5 mL, 4 mg of protein) are collected.
5. Mono Q PC 1.6/5-column chromatography (*see* **Note 6**): One-fifth of the sample is diluted with 2 mL of buffer H and applied to a Mono Q PC 1.6/5 (Amersham-Pharmacia Biotech. Ltd) column equilibrated with 280 m*M* NaCl in buffer H. After the column is washed with 2 mL of the same buffer, elution is performed with a 3-mL linear gradient of NaCl (280–500 m*M*) in buffer H, followed by a 0.5-mL linear gradient of NaCl (0.5–1 *M*) and 0.5 mL of 1 *M* NaCl in buffer H

at a flow rate of 0.1 mL/min. Fractions of 0.1 mL each are collected. Rab3 GAP appears in Fractions 10 and 11. These fractions (0.2 mL, 14 μg of protein) are collected. The rest of the heparin-Sepharose sample is subjected to the same Mono Q-column chromatography 4× in a similar manner. The samples of the five Mono Q-column chromatographies are pooled and stored at –80°C until use.

3.3.2.2. Recombinant His6-tagged Proteins of Rab3 GAP p130 (His6-p130) and p150 (His6-p150)

Like native Rab3 GAP, His6-p130 shows the Rab3 GAP activity in a dose-dependent and time-dependent manner, but the specific activity of His6-p130 is weaker than that of native Rab3 GAP. His6-p150 does not show the Rab3 GAP activity under the conditions in which native Rab3 GAP and His6-p130 show the activity. In addition, His6-p150 does not affect the Rab3 GAP activity of His6-p130.

1. Recombinant His6-p130 and -p150 are separately expressed with the pRSET constructs containing the cDNAs of p130 and p150, respectively, in *E. coli.*
2. The bacteria are harvested, suspended in 20 mL of PBS, and washed with 20 mL of PBS. The pellet is frozen at –80°C.
3. The pellet is quickly thawed at 37°C and suspended in 10 mL of buffer L containing 1 mg/mL lysozyme and 400 μ*M* APMSF, and the suspension is sonicated on ice for 10 s 6× at 1-min intervals. The homogenate is centrifuged at 100,000 *g* for 1 h.
4. Forty mL of the crude supernatant is incubated with Ni^{2+}-NTA-agarose beads (Qiagen K.K., Tokyo, Japan) on a rotating wheel for 2 h. After the incubation, the beads are spun down at 800 *g* for 5 min and washed with 10 mL of buffer L and once with 10 mL of buffer L containing 40 m*M* imidazole. Then, the beads are packed onto 5 mL-disposable syringe, washed with 10 mL of buffer L containing 40 m*M* imidazole, and eluted with 5 mL of buffer L containing 500 m*M* imidazole. The eluate is dialyzed with 1 L of buffer L 3×.

4. Notes

1. When crude samples are used for detecting the Rab GDI activity, it is better to use the [^{3}H]GDP dissociation assay than the [^{35}S]GTPγS-binding assay, because the crude samples often contain other GTPases, and these GTPases bind [^{35}S]GTPγS and interfere with the assay.
2. Of all the purification steps of Rab GDIα from bovine brain cytosol, Mono Q-column chromatography is the most important step needed to obtain a homogeneous sample. When the Mono Q column is used several times, Rab GDIα is eluted in the fractions earlier than the fractions described in **Subheading 3.1.2.1., step 4**, and Rab GDIα is contaminated with a bulk of proteins.
3. We recommend purification and use of native Rab GDIα from bovine brain rather than GST-Rab GDIα. When GST-Rab GDIα is expressed in *E. coli*, most of GST-

Rab GDIα is recovered in the pellet fraction, and even the soluble form of GST-Rab GDIα often forms an aggregate during the concentration procedure. Moreover, the specific activity of GST-Rab GDIα is lower than that of the native one.

4. Western blot analysis of the subcellular fractions of rat brain indicates that Rab3 GEP and GAP are highly enriched in the synaptic soluble fraction *(16)*. Consistently, immunofluorescence microscopic analysis of primary culture hippocampal neurons from rat embryo indicates that Rab3 GEP and GAP are localized at the synaptic release sites. Therefore, the preparation of the synaptic soluble fraction is a critical step in the purification of Rab3 GEP and GAP.
5. By use of the standard assay, many samples can be assayed for a short period of time. Therefore, this assay is selected for the purification procedures of native Rab3 GAP. In contrast, the overlay assay is useful for the detection of the protein band showing the Rab3 GAP activity. However, these two assay methods are not accurate and do not distinguish the GAP activity with the nonspecific phosphatase activity or the GTP dissociation activity. Therefore, the GAP activity should be confirmed by thin-layer chromatography assay.
6. In all the purification steps of native Rab3 GAP from the rat-brain synaptic soluble fraction, the Mono Q column chromatography by use of the SMART system (Amersham-Pharmacia Biotech. Ltd.) is the most important step needed to obtain a large amount of purified Rab3 GAP. Small, total gel volume and dead volume in this system contribute to low nonspecific adsorption, resulting in the superior recovery. In addition, because there is less dilution in this system, it is possible to achieve sample concentration, which also increases the recovery.

References

1. Takai, Y., Sasaki, T., Shirataki, H., and Nakanishi, H. (1996) Rab3A small GTP-binding protein in Ca^{2+}-dependent exocytosis. *Genes Cells* **1,** 615–632.
2. Novick, P. and Zerial, M. (1997) The diversity of Rab proteins in vesicle transport. *Curr. Opin. Cell Biol.* **9,** 496–504.
3. Schimmoller, F., Simon, I., and Pfeffer, S. R. (1998) Rab GTPases, directors of vesicle docking. *J. Biol. Chem.* **273,** 22,161–22,164.
4. Martinez, O. and Goud, B. (1998) Rab proteins. *Biochim. Biophys. Acta* **1404,** 101–112.
5. Takai, Y., Sasaki, T., and Matozaki, T. (2001) Small GTP-binding proteins. *Physiol. Rev.* **81,** 153–208.
6. Soldati, T., Shapiro, A. D., Svejstrup, A. B., and Pfeffer, S. R. (1994) Membrane targeting of the small GTPase Rab9 is accompanied by nucleotide exchange. *Nature* **369,** 76–78.
7. Ullrich, O., Horiuchi, H., Bucci, C., and Zerial, M. (1994) Membrane association of Rab5 mediated by GDP-dissociation inhibitor and accompanied by GDP/GTP exchange. *Nature* **368,** 157–160.
8. Weber, E., Berta, G., Tousson, A., St. John, P., Green, M.W., Gopalokrishnan, U., et al. (1994) Expression and polarized targeting of a rab3 isoform in epithelial cells. *J. Cell Biol.* **125,** 583–594.

9. Baldini, G., Hohl, T., Lin, H. Y., and Lodish, H. L. (1992) Cloning of a Rab3 isotype predominantly expressed in adipocytes. *Proc. Natl. Acad. Sci. USA* **89,** 5049–5042.
10. Geppert, M., Bolshakov, V. Y., Siegelbaum, S. A., Takei, K., Decamill, P., Hammer, R. E., et al. (1994) The role of Rab3A in neurotransmitter release. *Nature* **369,** 494–497.
11. Geppert, M., Goda, Y., Stevens, C. F., and Südhof, T. C. (1997) The small GTP-binding protein Rab3A regulates a late step in synaptic vesicle fusion. *Nature* **387,** 810–814.
12. Castillo, P. E., Janz, R., Südohof, T. C., Tzounopoulos, T., Malenka, R. C., and Nicoll, R. A. (1997) Rab3A is essential for mossy fiber long-term potentiation in the hippocampus. *Nature* **388,** 590–593.
13. Sasaki, T., Kikuchi, A., Araki, S., Hata, Y., Isomura, M., Kuroda, S., et al. (1990) Purification and characterization from bovine brain cytosol of a protein that inhibits the dissociation of GDP from and the subsequent binding of GTP to smg p25A, a ras p21-like GTP-binding protein. *J. Biol. Chem.* **265,** 2333–2337.
14. Nishimura, N., Nakamura, H., Takai, Y., and Sano, K. (1994) Molecular cloning and characterization of two rab GDI species from rat brain: brain-specific and ubiquitous types. *J. Biol. Chem.* **269,** 14,191–14,198.
15. Wada, M., Nakanishi, H., Satoh, A., Hirano, H., Obaishi, H., Matsuura, Y., et al. (1997) Isolation and characterization of a GDP/GTP exchange protein specific for the Rab3 subfamily small G proteins. *J. Biol. Chem.* **272,** 3875–3878.
16. Oishi, H., Sasaki, T., Nagano, F., Ikeda, W., Ohya, T., Wada, M., et al. (1998) Localization of the rab3 small G protein regulators in nerve terminals and their involvement in Ca2+-dependent exocytosis. *J. Biol. Chem.* **273,** 34,580–34,585.
17. Fukui, K., Sasaki, T., Imazumi, K., Matsuura, Y., Nakanishi, H., and Takai, Y. (1997) Isolation and characterization of a GTPase activating protein specific for the Rab3 subfamily of small G proteins. *J. Biol. Chem.* **272,** 4655–4658.
18. Nagano, F., Sasaki, T., Fukui, K., Asakura, T., Imazumi, K., and Takai, Y. (1998) Molecular cloning and characterization of the noncatalytic subunit of the Rab3 subfamily-specific GTPase-activating protein. *J. Biol. Chem.* **273,** 24,781–24,785.
19. Kikuchi, A., Nakanishi, H., and Takai, Y. (1995) Purification and properties of Rab3A. *Methods Enzymol.* **257,** 57–70.
20. Hata, Y. and Südhof, T. C. (1995) A novel ubiquitous form of Munc-18 interacts with multiple syntaxins. Use of the yeast two-hybrid system to study interactions between proteins involved in membrane traffic. *J. Biol. Chem.* **270,** 13,022–13,028.

13

Analysis and Preparation of Stable Complexes between Rab GTPases, Rab Escort Protein, and Rab Geranylgeranyl Transferase

José B. Pereira-Leal, Anita Q. Gomes, and Miguel C. Seabra

1. Introduction

Rab proteins are small Ras-like GTPases that regulate vesicular trafficking events in the cell. More than 50 Rabs have been described in mammalian cells *(1)*, each with a specific subcellular localization reflecting the functional specificity of Rabs to specific trafficking steps *(2–5)*. Rabs contain two cysteine residues at or near the carboxyl terminus, arranged in a variety of motifs. Both cysteine residues are modified by the attachment of geranylgeranyl groups via thioether bonds, in a reaction catalyzed by Rab geranylgeranyl transferase (also known as GGTase type II, RGGT) *(6)*. This enzyme is a tightly bound heterodimer, composed of a 60-kDa α-subunit and a 38-kDa β-subunit, both related to the α- and β-subunits of the other known protein prenyltransferases, farnesyl transferase and caax geranylgeranyl transferase (also known as GGTase type I).

RGGT is unique among prenyl transferases. It cannot catalyze prenylation of Rabs on its own, but requires the presence of an additional component designated as Rab escort protein (REP). REP binds unprenylated Rabs, and the complex REP:Rab is the true substrate for the enzyme. Unlike other prenyl transferases, substrate recognition is not mediated by the prenylation motif of the substrate. RGGT binds the REP:Rab complex and then catalyzes the covalent addition of two geranylgeranyl groups to the carboxyl terminal cysteines of the Rab in a sequential process that is unique for RGGT. After prenylation, RGGT dissociates from diGG-Rab which remains associated with REP *(7)*, presumably because the addition of two GG groups renders the

From: *Methods in Molecular Biology, vol. 189: GTPase Protocols: The Ras Superfamily*
Edited by: E. J. Manser and T. Leung © Humana Press Inc., Totowa, NJ

protein too hydrophobic to exist in aqueous solution. DiGG-Rab can then be directly delivered to their target membranes by REP *(8)*. REP has the dual function of acting as an essential activator of the prenylation reaction and escorting newly prenylated Rabs to the membranes.

This chapter describes in vitro methods to form and purify unprenylated and prenylated complexes of Rab and REP, as well as prenylated Rab, REP, and RGGT complexes. The methods presented in this chapter were optimized for REP1 and Rab1a, but they should adapt easily to other Rabs and REPs. These complexes are used in our laboratory to study the structural and biochemical factors that determine the assembly of both REP:Rab and the catalytic complexes, and to analyze membrane delivery of Rab proteins.

2. Materials

2.1. Geranylgeranyl Pyrophosphate

1. Tritium-labeled and unlabeled geranylgeranyl pyrophosphate (GGPP) are commercially available from several sources and are always stored at –20°C. We typically use:
2. Unlabeled all-trans-geranylgeranyl pyrophosphate (Sigma, G-6025).
3. [1-^{3}H] all-*trans*-GGPP, 15-30 Ci/mmol (DuPont-New England Nuclear, NET-1052).

2.2. Analytical Scale Preparation and Purification of REP:Rab Complexes

1. Stock solutions: 1 *M* sodium HEPES, pH 7.2, 1 *M* $MgCl_2$, 1 *M* dithiothreitol (DTT), recombinant 6xHisREP1, and recombinant 6xHisRab1a.
2. SMART system gel-filtration buffer (GF buffer): 50 m*M* sodium HEPES, pH 7.2, 1 m*M* DTT.
3. SMART system chromatography system (Amersham Pharmacia) and gel-filtration column Superdex 200 3.2/30 (Amersham Pharmacia).

2.3. Analytical Scale Preparation and Purification of REP:GG-Rab Complex

1. Stock solutions: 1 *M* sodium HEPES, pH 7.2, 1 *M* $MgCl_2$, 1 *M* DTT, 175 m*M* (10%) NP-40 protein grade (Calbiochem, 492017), 1 *M* Tris-HCl, pH 8.0, 4 *M* NaCl, recombinant 6xHisREP1, recombinant 6xHisRab1a, recombinant RGGT.
2. SMART system monoQ buffer A: 50 m*M* Tris-HCl, pH 8.0, 1 m*M* DTT.
3. SMART system monoQ buffer B: 50 m*M* Tris-HCl, pH 8.0, 1 m*M* DTT, 1 *M* NaCl.
4. SMART system chromatographic system (Amersham Pharmacia) and ion-exchange column monoQ PC 1.6/5 (Amersham Pharmacia).

2.4. Analytical Scale Preparation and Purification of REP:GG-Rab:RGGT Complex

1. Stock solutions: 1 *M* sodium HEPES, pH 7.2, 1 *M* $MgCl_2$, 1 *M* DTT, 175 m*M* (10%) NP-40 protein grade (Calbiochem, 492017), 1 *M* Tris-HCl pH 8.0, 4 *M* NaCl, recombinant 6×HisREP1, recombinant 6×HisRab1aCS (mutant Rab1a with the C-terminal cysteine residue mutated to a serine), recombinant RGGT.
2. SMART system monoQ buffer A: 50 m*M* Tris-HCl, pH 8.0, 1 m*M* DTT.
3. SMART system monoQ buffer B: 50 m*M* Tris-HCl, pH 8.0, 1 m*M* DTT, 1 *M* NaCl.
4. SMART system chromatography system (Amersham Pharmacia) and ion-exchange column monoQ PC 1.6/5 (Amersham Pharmacia).

2.5. Analytical Scale Preparation and Purification of REP:[^{3}H]GG-Rab Complex

1. Stock solutions: 1 *M* sodium HEPES, pH 7.2, 2 *M* $MgCl_2$, 1 *M* DTT, 175 m*M* (10%) NP-40 protein grade (Calbiochem, 492017), recombinant 6×HisREP1, recombinant 6×HisRab1a, recombinant RGGT.
2. G75 Sephadex gel-filtration buffer (G75): 50 m*M* sodium HEPES, pH 7.2, 50 m*M* $MgCl_2$, 1 m*M* DTT, 0.05 m*M* NP40, 100 m*M* NaCl.
3. Disposable 2-mL Bio-Rad columns, 1-mL G75 Sephadex matrix (Amersham Pharmacia, 17-0051-01).
4. India His-Probe-HRP (Pierce, 15165ZZ).
5. 1.2-µm glass fiber filters (Whatman, 1822025).

2.6. Large-scale Preparation and Purification of REP:GG-Rab:RGGT Complex

1. Stock solutions: 1 *M* sodium HEPES, pH 7.2, 1 *M* $MgCl_2$, 1 *M* DTT, 175 m*M* (10%) NP-40 protein grade (Calbiochem, 492017), 1 *M* Tris-HCl, pH 8.0, 4 *M* NaCl, recombinant 6×HisREP1, recombinant 6×HisRab1aCS, recombinant RGGT.
2. Fast performance liquid chromatography (FPLC) monoQ buffer A: 50 m*M* Tris-HCl, pH 8.5, 1 m*M* DTT.
3. FPLC monoQ buffer B: 50 m*M* Tris-HCl, pH 8.5, 1 m*M* DTT, 1 *M* NaCl.
4. FPLC chromatography system (Amersham Pharmacia) and ion-exchange column monoQ HR 5/5 (Amersham Pharmacia).

3. Methods

3.1. Expression and Purification of Recombinant Proteins

We have previously described in detail the protocols used in our lab to express and purify recombinant Rabs, REPs, and RGGT ***(9,10)***. Rab1a is expressed as an N-terminal 6-histidine-tagged fusion protein in *Escherichia coli* (*E. coli*) and purified by affinity chromatography using a Ni^{2+}-Sepharose resin. RGGT is expressed in Sf9 insect cells after co-infection with baculovirus coding for both subunits, purified by cation-exchange chromatography followed by gel-filtration chromatography (*see* **Note 1**). REP is also expressed in Sf9 insect cells, produced as a C-terminal 6-histidine-tagged fusion protein and purified by Ni^{2+} affinity chromatography.

3.2. Analytical Scale Preparation and Purification of REP:Rab Complexes

REP and Rab form stable complexes, which are held together mainly by electrostatic interactions ***(11,12)***. An important consideration is therefore the concentration of salt used in the preparation of the complex and the methodology used to purify the complex away from the free recombinant proteins. We have developed a method based on gel-filtration chromatography. REP migrates upon gel filtration with an apparent mol wt of 150 kDa, both in the free and complexed forms. As such, it is advisable to always use limiting REP concentrations to avoid contamination of the complex with free REP. This is a simple way to address the binding of the two molecules in a semiquantitative manner, but it is unlikely to be useful for preparative purposes.

1. Mix, in a final volume of 50 µL, 50 m*M* sodium HEPES, pH 7.2, 5 m*M* $MgCl_2$, 1 m*M* DTT, 4 µ*M* Rab1a, 2 µ*M* REP-1.
2. Incubate for 15 min at 37°C.
3. Load the reaction mixture onto a Superdex 200 3.2/30 column (Amersham Pharmacia) pre-equilibrated with GF buffer, and run at a flow rate of 50 µL/min. Fractions of 100 µL are collected between 0.75 and 1.95 mL. Elution of proteins from the column is monitored by absorbance at 280 nm. The complex elutes at approx 1.6 mL, corresponding to a predicted Mw of 150 kDa.
4. Run an aliquot of each fraction on SDS-gel electrophoresis and visualize proteins by silver (or Coomassie) staining. In the presence of REP1, Rab1a co-elutes with REP1, indicating the formation of the complex.

3.3. Analytical Scale Preparation and Purification of REP:GG-Rab Complex

After prenylation, the resulting REP: diGG-Rab complex is more stable than the nonprenylated complex ***(11,13)***. The REP: diGG-Rab interaction involves both ionic (as in the REP: Rab complex) and hydrophobic interactions probably caused by interactions between the prenyl groups and hydrophobic amino acids in REP.

The first step in the formation of the REP:diGG-Rab complex is in vitro prenylation. Rab geranylgeranylation occurs upon incubation of REP and Rab

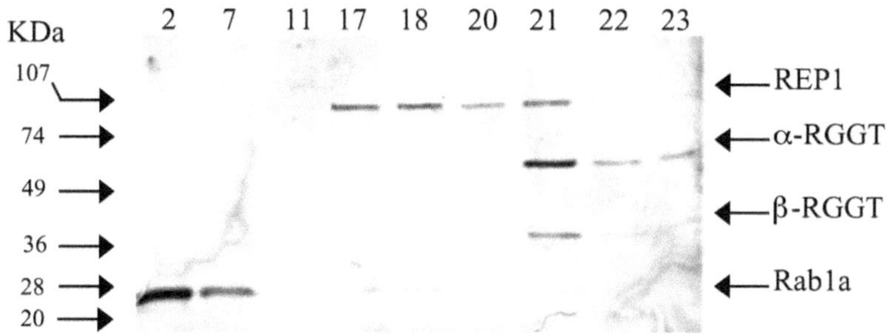

Fig. 1. Separation of the REP:GG-Rab and REP:GG-Rab:RGGT complexes by ion-exchange chromatography. See text for chromatography details. Proteins were visualized by silver-staining following SDS-gel electrophoresis on a 4–15% gradient gel according to the manufacturer's protocol (Bio-Rad).

with GGPP and RGGT. Detailed protocols for in vitro geranylgeranylation of Rabs were published in a previous book in this series *(see* **ref.** ***10)*** (*see* **Note 2**). After the reaction, the mixture is separated by ion-exchange chromatography and Rab1a co-elutes with REP1, indicating the formation of the prenylated complex. This separation procedure offers the advantage of separating the REP:diGG-Rab complex from the uncomplexed proteins, as well as separating the complex from any nonprenylated REP:Rab complex, as the presence of salt in the elution buffer promotes the dissociation of the latter complex. REP elution does not change significantly between free and complexed forms; thus, the use of limiting concentrations of REP in the reaction mixture as suggested in **Subheading 3.2.** is recommended. Also, one should use optimal conditions for the in vitro prenylation reaction to maximize the amount of diGG-Rab obtained. The majority of the reaction product is REP:diGG-Rab complex, but some ternary-complex REP:diGG-Rab:RGGT is produced. However, the ternary complex elutes differentially, and is therefore easily separated away (**Fig. 1**).

1. Mix the following components on ice, in a final volume of 50 µL: 50 m*M* sodium HEPES, pH 7.2, 5 m*M* $MgCl_2$, 1 m*M* DTT, 0.05 m*M* NP-40, 20 µ*M* Rab1a, 5 µ*M* REP1, 5 µ*M* RGGT, 50 µ*M* GGPP.
2. Incubate at 30°C for 1 h.
3. Load the reaction mixture onto a 0.1-mL monoQ column pre-equilibrated in monoQ (SMART system) buffer A at a flow rate of 0.3 mL/min, collecting 0.2-mL fractions.
4. The column is washed with 0.3 mL of monoQ (SMART system) buffer A, followed by elution with a 4-mL linear gradient from 0–35% monoQ (SMART system) buffer B and then a 0.7-mL linear gradient from 35–100%.
5. Run an aliquot of each fraction on a 12% SDS-PAGE gel and visualize proteins by silver (or Coomassie) staining.

6. Free Rab elutes in the flowthrough and/or binds weakly to the column, and is eluted at very low salt concentrations (fractions 2 and 7). The dimeric complex REP:diGG-Rab elutes in fractions 17–20 and RGGT (and the ternary complex) elutes in fractions 21–22 (**Fig 1**).

3.4. Analytical Scale Preparation and Purification of REP:GG-Rab:RGGT Complex

The ternary REP:diGG-Rab:RGGT complex is not stable, and is therefore difficult to purify. We have shown previously that a mutant Rab1a, Rab1aCS (C-terminal cysteine residue mutated to a serine residue) which can only undergo single geranylgeranylation, forms a tighter complex with REP than wild-type doubly geranylgeranylated Rab1a, Rab1aCC *(13)*. By using this mutant as the substrate of the prenylation reaction, we favor the formation of the ternary complex. Apparently, RGGT has a higher affinity for the REP:monoGG-Rab complex than for the REP:Rab complex, and the equilibrium is shifted to the formation of ternary complex *(13)*. The procedure is identical to the one described in **Subheading 3.3.** When Rab1aCS is used, we observe that most REP and Rab co-elute with RGGT in fractions 21–22, while some REP:monoGG-Rab complex elute in fractions 17–20.

3.5. Analytical Scale Preparation and Purification of REP:[^{3}H]GG-Rab Complexes

REP:[^{3}H]GG-Rab complexes can be formed by modifying in vitro Rab in the presence of tritiated GGPP. Radiolabeling allows to monitor the formation of the reaction product and to follow the prenylated Rab in subsequent studies. The prenylated REP:[^{3}H]GG-Rab is purified by manual gel filtration, because it is undesirable to contaminate the chromatography systems with radioactivity.

In order to ensure that all prenylated Rabs are bound to REP, the same principles apply. We perform in vitro reactions by incubating excess Rab with RGGT and GGPP under limiting REP conditions. The free Rab that is not prenylated can then be easily separated from the REP:Rab[^{3}H]GG complex by gel filtration. It would also be possible to use limiting amounts of Rab and excess REP to ensure that all Rabs would be prenylated and bound to REP. There are two main reasons to avoid this procedure. First, it has been shown that excess REP inhibits the geranylgeranylation reaction, probably as a result of the formation of nonproductive REP:RGGT complexes *(11)*. Second, the separation of free REP from the prenylated Rab:REP complex would be impossible with gel filtration, as both free REP and REP:diGG-Rab complex elute with the same apparent mol wt on gel-filtration chromatography.

1. Mix on ice, in a final volume of 50 µL, the following reagents: 50 m*M* sodium, HEPES, pH 7.2, 5 m*M* $MgCl_2$, 1 m*M* DTT, 0.05 m*M* NP-40, 10 µ*M* Rab,

2.5 μ*M* REP1, 0.7 μ*M* Rab RGGT, 20 μ*M* GGPP (0.2 μ*M* [^{3}H]GGPP mixed with 19.8 μ*M* unlabeled GGPP).
2. Incubate at 37°C for 45 min.
3. At 4°C, load the reaction mixture in a G75 Sephadex column (1 mL) previously equilibrated in G75 buffer (as in **ref. *10***). Run the column by adding 20 × 50 μL of G75 buffer and collecting the respective 20 × 50 μL fractions (*see* **Note 3**).
4. Quantify each fraction as described previously ***(10)***. Briefly, precipitate the proteins in a mixture of EtOH/HCl (10:1), filter the precipitable radioactivity on 1.2-μm glass fiber filters, add 5 mL of scintillation fluid, and count in the scintillation counter.
5. Run aliquots from each fraction on 12% SDS-gel electrophoresis. Visualize the proteins by silver staining on one gel. Use a replicate gel for Western blotting with the India-His probe (use as instructed by the manufacturer). This probe will allow visualization of histidine-tagged Rab and REP proteins. If necessary, use a third gel for autoradiography to visualize Rab[^{3}H]GG (*see* **Note 4**).
6. REP:[^{3}H]GG-Rab complex elutes in fractions 8–10 away from the free Rab that elutes in fractions 12–15. Note that RGGT co-elutes with the complex, but its stoichiometry is purposely reduced so that the total amount of ternary complex in the mixture is low. These fractions are pooled and used in subsequent in vitro studies (*see* **Notes 5** and **6**).

3.6. Large-scale Preparation and Purification of REP:GG-Rab:RGGT Complexes

This method is essentially a scale-up of **Subheadings 3.3.** and **3.4.** As discussed above, both ternary and dimeric complexes are produced. The choice of wild-type Rab1a (RabCC) versus Rab1aCS will determine the enrichment in either the ternary or binary complex. If RabCC is used, the formation of the dimeric complex is favored. If Rab1aCS is used, the ternary complex is favored. Here, we present conditions to preferentially form the ternary-complex REP:monoGG-Rab:RGGT.

1. Mix on ice, in a final volume of 500 μL, the following reagents: 50 m*M* sodium HEPES, pH 7.2, 5 m*M* $MgCl_2$, 1 m*M* DTT, 0.05 m*M* NP-40, 20 μ*M* Rab, 5 μ*M* REP1, 5 μ*M* RGGT, 50 μ*M* GGPP.
2. Incubate at 30°C for 1 h.
3. Load reaction mixture in a monoQ column previously equilibrated in monoQ (FPLC) buffer A at a flow rate of 1 mL/min, collecting fractions of 1.5 mL.
4. The column is washed with 3 mL of monoQ (FPLC) buffer A, then eluted with a 20-mL linear gradient from 0–30% of monoQ (FPLC) buffer B.
5. Run an aliquot of each fraction on a 12% SDS-PAGE gel, and visualize proteins by silver (or Coomassie) staining. The ternary complex typically elutes approx 24 mL under these conditions, and no further purification is usually needed. The complex is stable enough to allow a further purification/concentration step if required for specific applications (*see* **Note 7**).

4. Notes

1. Alexandrov et al. have successfully expressed geranylgeranyl transferase in bacteria (K. Alexandrov, personal communication). It is potentially a simpler way of expressing RGGT, but a direct comparison of the properties of both recombinant enzymes is not yet available.
2. It is advisable to use the utmost care when working with Rab proteins to constantly maintain the proteins on ice and avoid freeze-thawing cycles. Rab proteins will undergo partial proteolysis in the carboxyl terminus, and will generate a truncated protein without the prenylatable cysteines. These truncated proteins will effectively act as a competitor to the binding of REP in the prenylation reaction.
3. The equilibration of the column and subsequent purification steps should be performed at 4°C to avoid aggregation of the complex.
4. Blotting with India His-Probe allows a direct comparison between REP and Rab to determine whether fractions have equimolar amounts of both proteins.
5. The reaction can be scaled up to the amount of product desired, but the separation should always be performed as mentioned (50 µL of reaction mixture per 1 mL-G75 column).
6. The complexes should be stored at 4°C with 1 mg/mL of BSA as a carrier protein. They are stable for 5 d under these conditions.
7. The ternary complex is stable to freezing at –80°C, although repeated freezing and thawing is not advisable.

References

1. Pereira-Leal, J. and Seabra, M. C. (2000) The mammalian Rab family of small GTPases: definition of family and subfamily sequence motifs suggests a mechanism for functional specificity in the Ras superfamily. *J. Mol. Biol.* **301,** 1077–1087.
2. Lazar, T., Gotte, M., and Gallwitz, D. (1997) Vesicular transport: how many Ypt/Rab-GTPases make a eukaryotic cell? *Trends Biochem. Sci.* **22,** 468–472.
3. Novick, P. and Zerial, M. (1997) The diversity of Rab proteins in vesicle transport. *Curr. Opin. Cell Biol.* **9,** 496–504.
4. Schimmoller, F., Simon, I., and Pfeffer, S. R. (1998) Rab GTPases, directors of vesicle docking. *J. Biol. Chem.* **273,** 22,161–22,164.
5. Chavrier, P. and Goud, B. (1999) The role of ARF and Rab GTPases in membrane transport. *Curr. Opin. Cell Biol.* **11,** 466–475.
6. Seabra, M. C. (2000) Biochemistry of Rab geranylgeranyl transferase, in *The Enzymes* (Tamanoi, F. and Sigman, D., eds.), vol. XXI, Academic Press, New York, NY, pp. 131–154.
7. Andres, D. A., Seabra, M. C., Brown, M. S., Armstrong, S. A., Smeland, T. E., Cremers, F. P., et al. (1993) cDNA cloning of component A of Rab geranylgeranyl transferase and demonstration of its role as a Rab escort protein. *Cell* **73,** 1091–1099.

8. Alexandrov, K., Horiuchi, H., Steele-Mortimer, O., Seabra, M. C., and Zerial, M. (1994) Rab escort protein-1 is a multifunctional protein that accompanies newly prenylated Rab proteins to their target membranes. *EMBO J.* **13,** 5262–5273.
9. Armstrong, S. A., Brown, M. S., Goldstein, J. L., and Seabra, M. C. (1995) Preparation of recombinant Rab geranylgeranyltransferase and Rab escort proteins. *Methods Enzymol.* **257,** 30–41.
10. Seabra, M. C. and James, G. L. (1998) Prenylation assays for small GTPases. *Methods Mol. Biol.* **84,** 251–260.
11. Anant, J. S., Desnoyers, L., Machius, M., Demeler, B., Hansen, J. C., Westover, K. D., et al. (1998) Mechanism of Rab geranylgeranylation: formation of the catalytic ternary complex. *Biochemistry* **37,** 12,559–12,568.
12. Alexandrov, K., Simon, I., Iakovenko, A., Holz, B., Goody, R. S., and Scheidig, A. J. (1998) Moderate discrimination of REP-1 between Rab7 × GDP and Rab7 × GTP arises from a difference of an order of magnitude in dissociation rates. *FEBS Lett.* **425,** 460–464.
13. Shen, F. and Seabra, M. C. (1996) Mechanism of digeranylation of Rab proteins. *J. Biol. Chem.* **271,** 3692–3698.

II

Specific Methods

Arf Family GTPases

14

Preparation of Myristoylated Arf1 and Arf6 Proteins

Paul A. Randazzo and Henry M. Fales

1. Introduction

ADP-ribosylation factor (Arf) proteins were first identified as cofactors for cholera toxin-catalyzed ADP-ribosylation of the heterotrimeric G-protein Gs. Subsequent cloning led to the discovery that Arfs were part of a group of GTP-binding proteins that is a subfamily of the ras superfamily. The Arf family is comprised of the Arf proteins and related Arf-like (Arl) proteins. Activity as cofactors for cholera toxin distinguishes the Arf proteins from the Arls. The Arf proteins can be further subdivided into class I (comprised of Arf1 and Arf3 in humans), class II (comprised of Arf4 and Arf5), and class III (comprised of Arf6). Arfs have been found in all eukaryotes studied. Multiple Arf proteins occur within single organisms and within single cells, and Arf orthologs are highly conserved between species. The most extensively studied mammalian Arfs are Arf1 and Arf6, which have been found to function as regulators of membrane traffic and the actin cytoskeleton ***(1–9)***. In biochemical studies, Arfs have also been found to activate phospholipase D (PLD) and phosphatidylinositol 4-phosphate 5-kinase, and to bind to several putative effectors such as arfaptin, arfophilin, and the GGAs ***(10–20)***. Unlike other ras family members which are prenylated at the C-terminus, Arfs are myristoylated on an N-terminal glycine in a cotranslational event catalyzed by N-myristoyl transferase (NMT) ***(21–24)***. This lipid modification is crucial for activity.

Myristoylation of Arf has been shown to be important in at least three respects. First, myristoylation has dramatic effects on nucleotide-binding kinetics ***(25–27,42)***. Myristoylated Arf, in the presence of lipids, has a higher affinity for GTP than for GDP. The opposite is true for unmodified Arf, which has a 5-10 fold higher affinity for GDP than for GTP. Apart from its effects on nucleotide exchange, myristoylation affects physiological function and protein

From: *Methods in Molecular Biology, vol. 189: GTPase Protocols: The Ras Superfamily*
Edited by: E. J. Manser and T. Leung © Humana Press Inc., Totowa, NJ

interactions. Regulation of membrane traffic is the most extensively studied role of the Arf proteins ***(1–4)***. Membrane traffic is mediated through transport vesicles. Coat proteins, such as clathrin or coatomer, are recruited to a membrane where their assembly deforms the membrane driving the formation of a bud. A protein-coated transport vesicle is formed upon fission at the neck of the bud. The coat then dissociates, allowing the vesicle to dock and fuse with an acceptor membrane. This process is regulated by linking GTP binding and hydrolysis on Arf. Arf-GTP recruits coat proteins to membranes. The hydrolysis of GTP to form Arf-GDP is necessary for subsequent dissociation of the coat. Myristoylation is required for this function of Arf. Nonmyristoylated Arf has no effect on coat proteins in vivo, does not affect in vitro assays of membrane traffic, and does not recruit coat proteins in vitro ***(1,21–23)***. Similarly, the activation of PLD is dependent upon myristoylated Arf. Although activation is observed with nonmyristoylated protein, the affinity of myristoylated Arf-GTP for PLD is much higher than that of unmodified Arf ***(28)***. Therefore, studies of the biochemical mechanisms involved in Arf action are most likely to yield relevant results when using myristoylated Arf proteins.

Two general approaches can be used to prepare myristoylated Arf proteins. One is to purify the proteins from mammalian tissues. Purifications of Arf1 and Arf3 from the bovine brain have been described ***(29–33)***. These yield 100% myristoylated protein of high purity. The drawbacks are that the methods are fairly labor-intensive, and it is difficult to exclude the presence of other Arfs. A second strategy is to coexpress Arf with N-myristoyltransferase in bacteria and to purify the recombinant protein from the bacterial extract ***(34,35)***. This approach offers the advantage of less labor, and the preparation is known to contain a single Arf isozyme. However, the system is limited because the Arf protein is not completely modified, necessitating the separation of myristoylated and nonmyristoylated Arf. Over the last few years, several groups have developed straightforward methods to separate Arf1 from its myristoylated form ***(27,36)*** and Arf6 from myristoylated Arf6 ***(37)***. This chapter describes methods to prepare myristoylated Arf1 and Arf6. Each requires a different purification strategy. For Arf1, the protein is partially purified as a mixture of myristoylated and nonmyristoylated forms, which are subsequently separated by hydrophobic interaction chromatography. Myristoylated Arf6 is separated from the nonmyristoylated form by partitioning onto the cellular membranes. MyrArf 6 is extracted from the cell-particulate fraction with detergent. The protein is purified in a single step upon anion exchange. We and others have successfully used these preparations of Arf1 and Arf6 to study Arf GAP ***(36–38)***, Arf GEF ***(39)***, coatomer binding to Golgi-enriched membranes (V. W. Hsu, unpublished observations), AP3 binding to TGN ***(40)***, and Arf interaction with the newly identified effector GGA ***(17)***.

2. Materials

2.1. Protein Expression

Recombinant Arf is myristoylated by co-expression of Arf with NMT in *Escherichia coli* (*E. coli*). This system has been previously described (*see* **refs. *34,35***). Briefly, the Arf open reading frame is ligated into a pET3 plasmid. BL21(DE3) bacteria are cotransformed with pET3, which is under ampicillin selection, and pBB131, a plasmid containing the open reading frame for yeast NMT. Arf and NMT are simultaneously induced with IPTG. Sodium myristate is added just before induction to enrich myristoyl CoA.

2.2. GTPase Extraction

1. For Arf1, bacteria are lysed using a probe sonicator, equivalent to the "Vibracell" (Sonics and Materials, Danbury, CT) following treatment with lysozyme.
2. Arf6 is extracted by lysing cells with a French press or equivalent pressure-based lysis, such as the Piranha Press from Tesla Inc, Paxton, IL.
3. Lysis buffer: 20 m*M* Tris-HCl, pH 8.0, 100 m*M* NaCl, 1 m*M* $MgCl_2$, 1 m*M* dithiothreitol (DTT).

2.3. GTPase Purification

For both Arf1 and Arf6, a HiTrap Q (Amersham/Pharamacia) column operated with a peristaltic pump is employed. For Arf1, a HiPrep 26/60 Sephacryl S-100 (Amersham/Pharmacia) column is used. A 200–300-mL column of AcA54 is an alternative. The preparation of Arf1 also requires a 1-mL HiTrap phenylSepharose (Amersham/Pharmacia) column and a chromatography system that can generate a reliable gradient in a volume of 15–20 mL. The proteins are concentrated, and buffers are exchanged by ultrafiltration using an Amicon YM10 membrane (Danvers, MA).

1. Buffer A: 20 m*M* Tris-HCl, pH 8.0, 100 m*M* NaCl, 1 m*M* $MgCl_2$, 1 m*M* DTT, 10% v/v glycerol.
2. Buffer B: 20 m*M* Tris-HCl, pH 8.0, 25 m*M* NaCl, 1 m*M* $MgCl_2$, 1 m*M* DTT, 1% Triton X-100, 10% v/v glycerol.
3. Buffer C: 20 m*M* Tris-HCl, pH 8.0, 3 *M* NaCl, 1 m*M* $MgCl_2$, 1 m*M* DTT.
4. Buffer D: 20 m*M* Tris-HCl, pH 8.0, 100 m*M* NaCl, 1 m*M* $MgCl_2$, 1 m*M* DTT.
5. Buffer E: 20 m*M* Tris-HCl, pH 8.0, 4 *M* NaCl, 1 m*M* $MgCl_2$, 1 m*M* DTT.

3. Methods

3.1. Protein Expression

1. 100 mL of Luria broth containing 100 μg/mL ampicillin and 25 μg/mL kanamycin is inoculated with a single colony of the transformed bacteria. The culture is grown at 37°C to an $A \approx 0.6$, and is refrigerated overnight. The bacteria are

collected by centrifugation and are used to innoculate a 1–2 L culture with the same antibiotic selection. The bacteria are grown at 37°C to an $A \approx 0.6$. Sodium myristate is added to the culture to achieve a concentration of 50 µ*M*, and the incubation is continued for an additional 20–30 min. Then isopropylthiogalactoside (IPTG) is added to a final concentration of 1 m*M*. Cultures expressing Arf1 are shifted to 25°C, and the incubation is continued for an additional 12–16 h. For Arf6, the incubation is continued for 3–4 h. Cells are harvested by centrifugation and are stored at –80°C. One-mL samples of the induced and uninduced cultures are taken to confirm expression of Arf. The cells are collected by centrifugation and lysed in 20–100 µL of Laemli's sample buffer, and 10–25 µL are fractionated on a polyacrylamide gel. The gels are stained with Coomassie blue dye.

2. As seen in **Fig. 1**, Arf expression is not completely dependent on the presence of IPTG when using the pET3-based constructs. However, expression without the addition of IPTG is more variable. In the particular induction for Arf6 shown in **Fig. 1**, Arf6 expression was substantially increased by the addition of IPTG, and the bacterial lysate was enriched in the desired protein.

3.2. Cell Lysis

1. *Arf1:* The bacterial pellet is thawed at room temperature and resuspended in 10 mL of lysis buffer/2 L of culture. Lysozyme is added to a final concentration of 1 mg/mL and the suspension is incubated at room temperature for 30 min. The cells are lysed by 3–10 s bursts with a probe sonicator. The suspension is clarified by centrifugation at 100,000*g* for 60 min at 4°C. The supernatant is used for further purification.
2. *Arf6:* The bacterial pellet is thawed at room temperature and resuspended in 10 mL of lysis buffer. The suspension is lysed in a French press using 12,000 psi. The lysate is centrifuged at 100,000*g* at 4°C for 60 min. The pellet contains the myristoylated Arf6, and is saved for further purification.

3.3. Protein Purification

1. *Arf1:* The supernatant is loaded onto a 5-mL HiLoad Q column (Pharmacia) previously equilibrated in buffer A. The column eluate is collected in 2-mL steps. The column is further developed with 20 mL of buffer A. Arf adheres poorly or not at all. The fractions containing Arf can be assessed by SDS-PAGE, stained with Coomassie blue dye, and pooled. This is the major peak of protein in the isocratically developed column. The pooled fractions, which should be approx 10 mL, are loaded onto the Sephacryl column, which has been equilibrated in buffer A at room temperature. The column is developed at 1.2 mL/min, collecting 3.6 mL fractions. The fractions containing Arf are determined by SDS-PAGE, detecting proteins with Coomassie blue. The pooled fractions are concentrated by ultrafiltration to 2–5 mL, and an equal volume of buffer E is added (i.e., Add 2 mL of buffer E to 2 mL of pool). This is repeated once to bring the NaCl concentration to 3 *M*. The protein is then loaded at a rate of 1 mL/min onto the

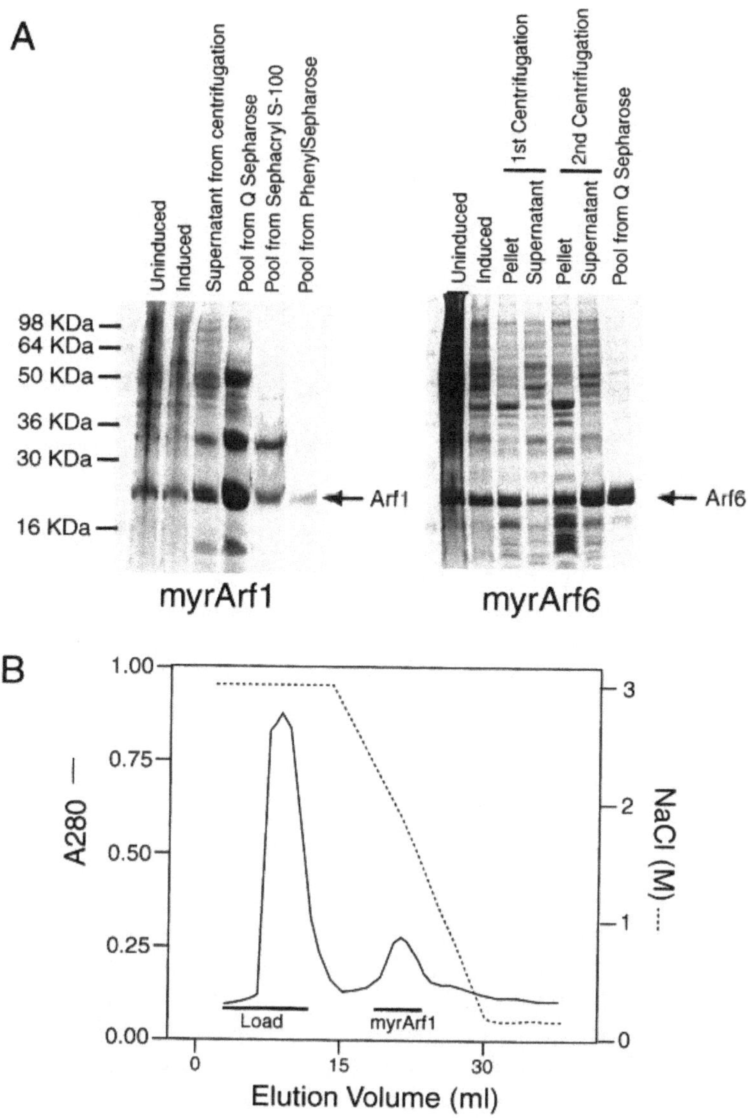

Fig. 1. Purification of myristoylated Arf1 and Arf6. (**A**) SDS-PAGE analysis of pools of protein at the indicated points in the purification. The proteins are visualized with Coomassie blue dye. (**B**) Chromatogram of final phenyl-Sepharose fractionation step in Arf1 purification. The peaks of myrArf1 and unmodified non-myrArf are indicated.

phenyl Sepharose HP column that had been equilibrated in buffer C. The column is developed in a gradient of 3 *M* to 100 m*M* NaCl (going from buffer C to buffer D) in 15 mL, collecting 1-mL fractions. Nonmyristoylated Arf does not adhere to the column, whereas myristoylated Arf1 elutes in a peak centering at 1.5 *M* NaCl (**Fig. 1B**). No nonmyristoylated Arf is detected in the second peak. Fractions containing Arf are determined by SDS-PAGE, pooled, and both concentrated and exchanged into buffer A by ultrafiltration. Aliquots are snap-frozen and stored at –80°C. **Figure 1A** shows proteins from each step of the purification fractionated on a 10–20% polyacrylamide gel, run in SDS.

2. *Arf6:* The particulate fraction is resuspended in buffer A and centrifuged. The particulate fraction is then resuspended in buffer B. This buffer contains Triton X-100 and extracts Arf6 from the particulate fraction. The suspension is passed through a French press and centrifuged at 100,000*g* × 60 min at 4°C. Guanosine diphosphate (GDP) is added to a final concentration of 10 μ*M* in the supernatant and the proteins are precipitated with 35% ammonium sulfate. The precipitated protein is collected by centrifugation and dissolved in 10 mL buffer B containing 10 μ*M* GDP. The sample is dialyzed 2× against 1 L of buffer B with GDP. The sample is then applied to a 5-mL HiTrap Q column, and the material that does not adhere to the column is collected. The protein is greater than 95% Arf6. SDS-PAGE analysis of the preparation at each step of the purification is shown in **Fig. 1A**. It is critical to snap-freeze and store the Arf6 immediately. The protein denatures, as assessed by nucleotide binding, with a $t_{1/2}$ of approximately 1 d when stored at 4°C.

3.4. Characterization of Myristoylated Arf

1. The quality of the proteins can be checked in two respects. First, the stoichiometry of nucleotide binding should be examined. More than 0.5 mol nucleotide/mol Arf1 should bind, and the binding should be highly dependent on the presence of lipids or detergents. Arf6 preparations typically bind 0.25-mol nucleotide/mol protein. For myristoylated Arf1, vesicles and mixed liposomes of cholate and phosphatidylcholine support exchange to 3–10x greater than a simple detergent such as Triton X-100. For Arf6, we have not observed any differences. Nucleotide-binding assays have been previously described *(35)*. Briefly, incubate Arf at 1 μ*M* in 20 m*M* HEPES, pH 7.4, 100 m*M* NaCl, 0.5 m*M* $MgCl_2$, 1 m*M* ethylenediamine-tetraacetic acid (EDTA) with 10 μ*M* [^{35}S]GTPγS, and one of the following lipids and/or detergents: 0.1% Triton X-100; 3 m*M* dimyristoylphosphatidylcholine (DMPC) and 0.1% sodium cholate; 700 μ*M* phosphatidylcholine and 300 μ*M* phosphatidylserine in a vesicle formed either by sonication or extrusion. Incubate at 30°C for 0–60 min, taking samples at appropriate intervals. Stop the reactions by diluting samples (10–25 μL) into 2 mL 20 m*M* Tris-HCl, pH 8.0, 100 m*M* NaCl, 10 m*M* $MgCl_2$ at 4°C. Trap the protein on nitrocellulose *(35)*, and wash 6× with 2 mL of the stop buffer by vacuum filtration. Quantify radionuclide on the filter by scintillation spectroscopy. An example binding-progress curve is shown in **Fig. 2**.

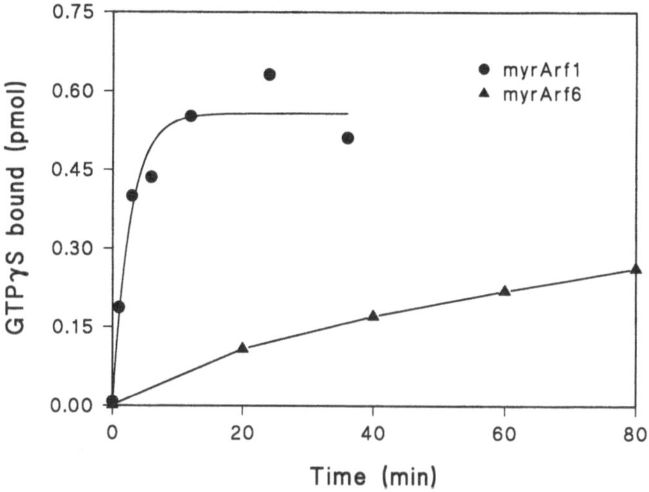

Fig. 2. Nucleotide-binding kinetics of myrArf1 and myrArf6. Binding was performed as described in the text, using mixed micelles containing 3 m*M* DMPC and 0.1% sodium cholate for myrArf1 and 0.1% Triton X-100 for myrArf6. Reactions contained 1 pM Arf and 10 pM GTPγS.

2. The extent of myristoylation can be determined in a number of ways, including HPLC chromatography using a C8 reverse-phase column developed with acetonitrile/TFA or by electrospray mass spectrometry. The former is well-described in several studies ***(33,34)***, and the latter is particularly useful if myristoylated and nonmyristoylated proteins are not available for standards. Mass spectrometric analysis is carried out as follows: Solutions of the protein (5–20 pmol/μL) are prepared in acetic acid/methanol/water (5:20:80, v/v/v). Lower concentrations can also be used, but longer signal-averaging is required. The solutions are admitted to a Finnigan TSQ 700 mass spectrometer (Sunnyside, CA) by direct infusion at 0.3 μL/min, using a modified inlet system ***(41)***. Signals are averaged in the profile mode until a steady spectrum is obtained. Deconvolution then yields the mass spectrum.
3. When the sample is partially myristoylated, an overlapping series of peaks separated by approx 27–8 amu will be obtained. These can be resolved by operating the spectrometer at an increased resolution (approx 0.5 amu half-width), at the cost of lower sensitivity. Alternatively, the lower resolution data can usually be successfully resolved by analyzing the raw data with the maximum entropy algorithm provided by MaxENT Solutions Ltd. (Cambridge, United Kingdom).
4. Matrix-assisted laser desorption/ionization (MALDI) mass spectrometry using sinapinic acid or other matrices is an alternative form of mass spectrometry that is particularly useful if the quantity of protein is more limited. Although it is quite successful with the Arf proteins, external or better internal standards

should be used to ensure accuracy. Of course, when variable myristoylation is encountered, only an average mass will be obtained with most of today's MALDI spectrometers.

4. Notes

Currently, no methods are available for the separation of modified Arf3 or Arf5 from the nonmodified proteins. Protocols have been published in which greater than 50% of the protein is modified ***(18,20,42)***. Although not ideal, these preparations of Arf3 and Arf5 can be used in many cases to compare myristoylated and nonmyristoylated proteins. A mutant of Arf, [3-7LFASK]Arf1, has been identified that is better than 90% modified by this system, but the protein interacted with an Arf GAP with very different properties than did the wild-type protein and the modification of the mutant was not restricted to myristate. Instead, acyl groups ranging from C2 to C18 were added ***(25)***. Therefore, we have chosen not to use this protein for biochemical studies.

References

1. Donaldson, J. G. and Klausner, R. D. (1994) ARF: a key regulatory switch in membrane traffic and organelle structure. *Curr. Opin. Cell Biol.* **6(4),** 527–532.
2. Donaldson, J. G., Radhakrishna, H., and Peters, P. J. (1995) The ARF GTPases: defining roles in membrane traffic and organelle structure. *Cold Spring Harb. Symp. Quant. Biol.* **60,** 229–234.
3. Moss, J. and Vaughan, M. (1998) Molecules in the Arf orbit. *J. Biol. Chem.* **273,** 21,431–21,434.
4. Moss, J. and Vaughan, M. (1995) Structure and function of ARF proteins: activators of cholera toxin and critical components of intracellular vesicular transport processes. *J. Biol. Chem.* **270(21),** 12,327–12,330.
5. Boman, A. L. and Kahn, R. A. (1995) Arf proteins: the membrane traffic police? *Trends Biochem. Sci.* **20(4),** 147–150.
6. Rothman, J. E. (1996) The protein machinery of vesicle budding and fusion. *Protein Sci.* **5(2),** 185–194.
7. Scales, S. F., Gomez, M., and Kreis, T. E. (2000) Coat proteins regulating membrane traffic. *Int. Rev. Cytol.* **195,** 67–144.
8. Wieland, F. and Hunter, C. (1999) Mechanisms of vesicle formation: insights from the COP system. *Curr. Opin. Cell Biol.* **11,** 440–446.
9. Lowe, M. and Kreis, T. E. (1998) Regulation of membrane traffic in animal cells by COPI. *Biochim. Biophys. Acta* **1404,** 53–66.
10. Godi, A., Pertile, P., Meyers, R., Marra, P., DiTullio, G., Iurisci, C., et al. (1999) ARF mediates recruitment of PtdIns-4-OH kinase-beta and stimulates synthesis of PtdIns(4,5)P2 on the golgi complex. *Nature Cell Biol.* **1,** 280–287.
11. Honda, A., Nogami, M., Yokozeki, T., Yamazaki, M., Nakamura, H., Watanabe, H., et al. (1999) Phosphatidylinositol 4-phosphate 5-kinase alpha is a downstream effector of the small G protein ARF6 in membrane ruffle formation. *Cell* **99,** 521–532.

12. Jones, D. H., Morris, J. B., Morgan, C. P., Kondo, H., Irvine, R. F., and Cockcroft, S. (2000) Type I phosphatidylinositol 4-phosphate 5-kinase directly interacts with ADP-ribosylation factor 1 and is responsible for phosphatidylinositol 4,5-bisphosphate synthesis in the Golgi compartment. *J. Biol. Chem.* **275,** 13,962–13,966.
13. Brown, H. A., Gutowski, S., Moomaw, C. R., Slaughter, C., and Sternweis, P. C. (1993) ADP-ribosylation factor, a small GTP-dependent regulatory protein, activates phospholipase C. *Cell* **75,** 1137–1144.
14. Cockcroft, S., Thomas, G. M. H., Fensome, A., Geny, B., Cunningham, E., Gout, I., et al. (1994) Phospholipase D: a downstream effector of Arf in granulocytes. *Science* **263,** 523–526.
15. Hammond, S. M., Altshuller, Y. M., Sung, T. C., Rudge, S. A., Rose, K., Engebrecht, J., et al. (1995) Human ADP-ribosylation factor-activated phosphatidylcholine-specific phospholipase D defines a new and highly conserved gene family. *J. Biol. Chem.* **270,** 29,640–29,643.
16. Hirst, J., Lui, W. W. Y., Bright, N. A., Totty, N., Seaman, M. N. J., and Robinson, M. S. (2000) A family of proteins with gamma-adaptin and VHS domains that facilitate trafficking between the trans-Golgi network and the vacuole/lysosome. *J. Cell Biol.* **149,** 67–79.
17. Dell'Angelica, E. C., Puertollano, R., Mullins, C., Agular, R. C., Vargas, J. D., Hartnell, L. M., et al. (2000) GGAs: a family of ADP ribosylation factor-binding proteins related to adaptors and associated with the Golgi complex. *J. Cell Biol.* **149,** 81–93.
18. Kanoh, H., Williger, B.-T., and Exton, J. H. (1997) Arfaptin 1, a putative cytosolic target protein of ADP-ribosylation factor, is recruited to Golgi membranes. *J. Biol. Chem.* **272,** 5421–5429.
19. Van Aelst, L., Joneson, T., and BarSagi, D. (1996) Identification of a novel Rac1-interacting protein involved in membrane ruffling. *EMBO J.* **15,** 3778–3786.
20. Shin, O. H., Ross, A. H., Mihai, I., and Exton, J. H. (1999). Identification of arfophilin, a target protein for GTP-bound class II ADP-ribosylation factors. *J. Biol. Chem.* **274,** 36,609–36,615.
21. Donaldson, J. G., Kahn, R. A., Lippincott-Schwartz, J., and Klausner, R. D. (1991) Binding of ARF and beta-COP to Golgi membranes: possible regulation by a trimeric G protein. *Science* **254(5035),** 1197–1199.
22. Donaldson, J. G., Cassel, D., Kahn, R. A., and Klausner, R. D. (1992) ADP-ribosylation factor, a small GTP-binding protein, is required for binding of the coatomer protein beta-COP to Golgi membranes. *Proc. Natl. Acad. Sci. USA* **89(14),** 6408–6412.
23. Haun, R. S., Tsai, S. C., Adamik, R., Moss, J., and Vaughan, M. (1993) Effect of myristoylation on GTP-dependent binding of ADP-ribosylation factor to Golgi. *J. Biol. Chem.* **268(10),** 7064–7068.
24. Duronio, R. J., Jackson-Machelski, E., Heuckeroth, R. O., Olins, P. O., Devine, C. S., Yonemoto, W., et al. (1990) Protein N-myristoylation in Escherichia coli: reconstitution of a eukaryotic protein modification in bacteria. *Proc. Natl. Acad. Sci. USA* **87,** 1506–1507.

25. Randazzo, P. A., Terui, T., Sturch, S., Fales, H. M., Ferrige, A. G., and Kahn, R. A. (1995) The myristoylated amino terminus of ADP-ribosylation factor 1 is a phospholipid- and GTP-sensitive switch. *J. Biol. Chem.* **270(24),** 14,809–14,815.
26. Franco, M., Chardin, P., Chabre, M., and Paris, S. (1996) Myristoylation-facilitated binding of the G protein Arf1GDP to membrane phospholipids is required for its activation by a soluble nucleotide exchange factor. *J. Biol. Chem.* **271,** 1573–1578.
27. Franco, M., Chardin, P., Chabre, M., and Paris, S. (1995) Myristoylation of ADP-ribosylation factor 1 facilitates nucleotide exchange at physiological Mg2+ levels. *J. Biol. Chem.* **270,** 1337–1341.
28. Serafini, T., Orci, L., Amherdt, M., Brunner, M., Kahn, R. A., and Rothman, J. E. (1991) ADP-ribosylation factor is a subunit of the coat of Golgi-derived COP-coated vesicles: a novel role for a GTP-binding protein. *Cell* **67(2),** 239–253.
29. Moss, J., Tsai, S. C., Price, S. R., Bobak, D. A., and Vaughan, M. (1991) Soluble guanine nucleotide-dependent ADP-ribosylation factors in activation of adenylyl cyclase by cholera toxin. *Methods Enzymol.* **195,** 243–256.
30. Tsai, S. C., Noda, M., Adamik, R., Chang, P. P., Chen, H. C., Moss, J., et al. (1988) Stimulation of choleragen enzymatic activities by GTP and two soluble proteins purified from bovine brain. *J. Biol. Chem.* **263,** 1768–1772.
31. Han, J.-S., Chung, J.-K., Kang, H.-S., Donaldson, J., Bae, Y. S., and Rhee, S. G. (1996) Multiple forms of phospholipase D inhibitor from rat brain cytosol. *J. Biol. Chem.* **271,** 11,163–11,169.
32. Kahn, R. A. and Gilman, A. G. (1986) The protein cofactor necessary for ADP-ribosylation of Gs by cholera toxin is itself a GTP binding protein. *J. Biol. Chem.* **261(17),** 7906–7911.
33. Taylor, T. C., Kahn, R. A., and Melancon, P. (1992) Two distinct members of the ADP-ribosylation factor family of GTP-binding proteins regulate cell-free intra-Golgi transport. *Cell* **70(1),** 69–79.
34. Randazzo, P. A. and Kahn, R. A. (1995) Myristoylation and ADP-ribosylation factor function. *Methods Enzymol.* **250,** 394–405.
35. Randazzo, P. A., Weiss, O., and Kahn, R. A. (1995) Preparation of recombinant ADP-ribosylation factor. *Methods Enzymol.* **257,** 128–135.
36. Randazzo, P. A. (1997) Resolution of two ADP-ribosylation factor 1 GTPase-activating proteins from rat liver. *Biochem. J.* **324(1 Jun Pt. 2),** 413–419.
37. Brown, M. T., Andrade, J., Radhakrishna, H., Donaldson, J. G., Cooper, J. A., and Randazzo, P. A. (1998) ASAP1, a phospholipid dependent Arf GTPase-activating protein that associates with and is phosphorylated by Src. *Mol. Cell. Biol.* **18,** 7038–7051.
38. Andreev, J., Simon, J. P., Sabatini, D., Kam, J., Plowman, G., Randazzo, P. A., et al. (1999) Identification of a New Pyk2 Target Protein with Arf GAP activity. *Mol. Cell. Biol.* **19,** 2338–2350.
39. Frank, S., Upender, S., Hansen, S. H., and Casanova, J. E. (1998) ARNO is a guanine nucleotide exchange factor for ADP-ribosylation factor 6. *J. Biol. Chem.* **273,** 23–27.

40. Ooi, C. E., Dell'Angelica, E. C., and Bonifacino, J. S. (1998) ADP-ribosylation factor 1 (ARF1) regulates recruitment of the AP-3 adaptor complex to membranes. *J. Cell Biol.* **142,** 391–402.
41. Konig, S., Fales, H.M., and Haegele K. D. (1998) Comment on the cylindrical capacitor interface. *Anal. Chem.* **70,** 4453–4455.
42. Claude, A., Zhao, B.-P., Kuziemsky, C. E., Dahan, S., Berger, S. J., Yan, J. P., et al. (1999) GBF1: a novel Golgi-associated BFA-resistant guanine nucleotide exchange factor that displays specificity for ADP-ribosylation factor 5. *J. Cell Biol.* **146,** 71–84.

15

ARF-Directed Guanine-Nucleotide-Exchange (GEP) Proteins

Gustavo Pacheco-Rodriguez, Joel Moss, and Martha Vaughan

1. Introduction

The ADP-ribosylation factor proteins (ARFs) are ~20-kDa guanine-nucleotide-binding proteins found in all eukaryotic cells, which are a critical part of the minimum machinery required for vesicle formation at the Golgi and other membranes *(1–3)*. The six mammalian ARFs have been grouped into three classes *(4)* based on molecular size, amino-acid sequence, and gene structure: class I (ARF1-3), class II (ARF4, 5), and class III (ARF6). Proteins closely related to ARFs include ARF-like proteins (ARLs), ARF-related protein (ARP), and ARF-domain protein 1 (ARD1) *(5)*. ARFs function as activators of cholera-toxin ADP-ribosyltransferase and phospholipase D (PLD) *(6,7)*, and have been recently implicated in the activation of phosphatidylinositol 4-phosphate 5-kinase *(8)*. ARF, like other regulatory GTPases, is active when GTP is bound. The inactivation of ARF-GTP requires its interaction with a GTPase-activating protein (GAP), which accelerates GTP hydrolysis. Activation of the inactive GDP-bound ARF requires the action of a guanine nucleotide-exchange protein or GEP, which accelerates the release of bound nucleotide via stabilization of the guanine nucleotide-free protein *(9)*.

The ARF GEPs can be grouped on the basis of sensitivity to inhibition by brefeldin A (BFA), a fungal metabolite that inhibits protein secretion and blocks guanine-nucleotide exchange on ARFs *(10)*. Initial purifications yielded only BFA-insensitive GEPs of ~50-kDa *(11,12)*. Later, a cytosolic ~670-kDa protein complex was isolated that contained a BFA-sensitive GEP of ~200-kDa. The BFA-sensitive GEPs now known include mammalian BFA-inhibited GEP 1 (BIG1 or p200) and BIG2 (or p190), and the yeast proteins Sec7, Gea1, and

From: *Methods in Molecular Biology, vol. 189: GTPase Protocols: The Ras Superfamily*
Edited by: E. J. Manser and T. Leung © Humana Press Inc., Totowa, NJ

Gea2. The BFA-insensitive GEPs are represented by cytohesin-1, cytohesin-2 (ARNO), cytohesin-3 (GRP1) and cytohesin-4, EFA6, GBF1 (for reviews *see* **refs.** ***6,9,13,14***). All of these ARF GEPs which are BFA-sensitive or -insensitive and contain a so-called Sec7 domain, which was recognized in the sequence of the *Saccharomyces cerevisiae* Sec7 gene product ***(15)*** before it was shown to possess BFA-sensitive ARF-GEP activity ***(16)***. The Sec7 domains exhibit a high degree of sequence identity ***(17)*** and structural similarity ***(18–20)***. They are composed of two subdomains, each with five α-helices. The active site is in the C-terminal portion. Its residues are critical for catalysis and are included in two sequences termed motif 1 and motif 2, which are highly conserved among Sec7 domains. Specific amino acids required for BFA inhibition are present in the Sec7 domain, but not within motifs 1 and 2 ***(9,10,21)***.

The ~50-kDa BFA-insensitive cytohesin family of GEPs contain—in addition to Sec7 domains—pleckstrin homology domains (PH) involved in phospholipid binding and/or protein-protein interactions (**Fig. 1**). Recognition of the role of phospholipid binding in the action of cytohesin family of GEPs emphasizes the importance of these kinds of interactions for integration in signal transduction and other cell functions. Identification of structural similarities among other GEPs should help to further our understanding of their multiple functions and regulatory properties.

Here we describe some procedures used to study the action of BFA-sensitive and -insensitive ARF GEPs.

2. Materials

2.1. Synthesis and Purification of Recombinant Proteins

1. The preparation of recombinant myristoylated ARFs presents many problems. For this reason, nonmyristoylated ARFs are frequently studied. N-terminal myristoylation, however, is critical for ARF action. Native ARF1 and ARF3 are relatively easily purified from tissue, whereas other ARFs are not—at least in part because they are much less abundant. Some of these procedures for ARF preparation were described and compared by Patton et al. ***(22)***. Several groups have used the method of Franco et al. ***(23)*** to prepare recombinant myristoylated ARF1, despite variability in the extent of acylation and difficulty in separating the modified and unmodified proteins. Randazzo ***(24)*** has recently optimized procedures for the synthesis in *Escherichia coli* (*E. coli*) and purification of 100% myristoylated ARF1 and ARF6, which should be widely adopted. ARFs in solution in ARF buffer are usually stored frozen at –20 or –80°C in small amounts to avoid repeated freezing and thawing. Activity of ARF preparations is routinely assayed by the ability to stimulate cholera toxin-catalyzed synthesis of ADP-ribosylagmatine ***(25)***.
2. Cytohesin-1 and cytohesin-4 proteins are prepared with N-terminal His6-tags (Qiagen) ***(17,22)***. Removal of the imidazole used for elution of His-tagged

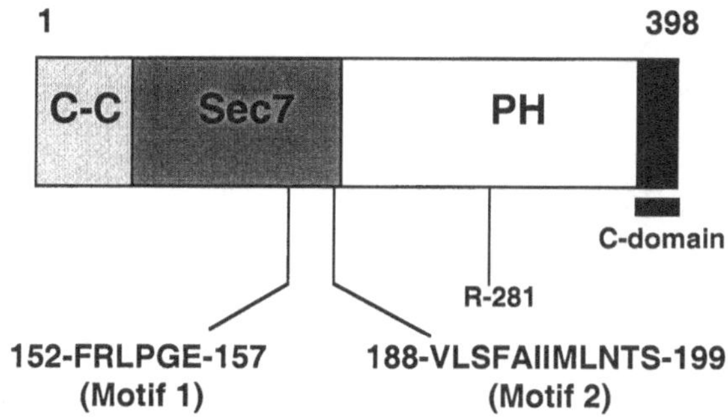

Fig. 1. Structural elements of cytohesin-1. The N-terminal region of cytohesin contains the coiled-coil (C-C) region involved in protein-protein interactions. The centrally located Sec7 domain (Sec7) contains residues critical for exchange activity in motifs 1 and 2. The Sec7 domain also contains sites responsible for BFA inhibition of BFA-sensitive GEPs. The pleckstrin homology (PH) domain in the C-terminal region can bind phospholipids, and the distal C-terminus contains a highly basic segment (C-domain).

proteins from Ni^{2+}-nitrilotriacetate resin requires dialysis before storage of the purified protein in either TENDS or phosphate buffer containing 10% glycerol at –80°C in small portions to avoid repeated freezing and thawing.

3. Recombinant BIG1 *(26)* and BIG2 *(27)* are synthesized as His6-tagged fusion proteins using baculovirus (Pharmingen) constructs in Sf9 cells, and are stored in small portions at –20°C.

2.2. Buffers and Other Reagents

1. TENDS buffer: 20 m*M* Tris-HCl, pH 8.0, containing 1 m*M* EDTA, 1 m*M* NaN_3, 1 m*M* dithiothreitol (DTT), and 0.25 *M* sucrose. Buffer with twice these concentrations can be prepared without DTT and stored at 4°C.
2. ARF buffer: TENDS buffer containing 100 m*M* NaCl and 5 m*M* $MgCl_2$.
3. Protease inhibitors: usually added for GEP and ARF preparations are leupeptin, aprotinin, soybean- and lima-bean inhibitors (1 µg/mL each), and 0.5 m*M* AEBSF (4-(2-aminoethyl)-benzene sulfonyl fluoride hydrochloride).
4. Phosphate buffer: 50 m*M* sodium phosphate, pH 7.8, containing 200 m*M* NaCl. For storage of cytohesin proteins, 10% glycerol is added. A 5x concentrated solution without glycerol can be prepared for storage.
5. Wash buffer: 25 m*M* Tris-HCl, pH 8.0, containing 5 m*M* $MgCl_2$, 100 m*M* NaCl, 1 m*M* DTT, 1 m*M* EDTA.
6. Brefeldin A (10 mg/mL) is freshly prepared in methanol or ethanol. Control incubations contain the same amount of alcohol that is added with BFA.
7. L-α-phosphatidyl-L-serine (PS) is prepared at a concentration of 4 mg/mL (~5.6 m*M*) in 10 m*M* HEPES or Tris-HCl, pH 8.0, containing 1 m*M* $MgCl_2$. PS (8 mg) in an organic solvent such as chloroform and/or methanol is dried under

a stream of nitrogen before addition of 2 mL of buffer and sonification until no particulate material is visible.

3. Methods

3.1. [^{35}S] GTPγS-Binding Assay

The effect of GEP on GTPγS binding to ARF ***(11,12,25)*** is assayed in a volume of 50 μL containing 20 m*M* Tris-HCl, pH 8.0, 40 m*M* NaCl, 250 m*M* sucrose, 2 m*M* DTT, 13.3 μ*M* bovine serum albumin (BSA) (0.8 mg/mL), and 280 μ*M* PS (0.2 mg/mL). The concentration of free magnesium has a major effect on nucleotide binding, and is, therefore adjusted according to the assay; 2 m*M* $MgCl_2$ with 1 m*M* EDTA is often used. Usual steps to evaluate the GEP activity of a protein preparation in an assay with a class I ARF as substrate are outlined here.

1. Add 10 μL of a mixture containing 40 μg of BSA protease inhibitors, 10 μg of PS, and 10 m*M* DTT to each ice-cold assay tube.
2. Addition of ~1 μg of recombinant ARF (or the indicated amount of native ARF), in 20 μL of ARF buffer is followed by addition of the GEP preparation to be tested (or vehicle).
3. To assess BFA inhibition, a concentration of 200 μ*M* is first used. Because of the mechanism of inhibition, the inhibitory concentration of BFA is influenced by the concentrations of ARF and GEP ***(10,14)***.
4. To initiate the assay, [^{35}S]GTPγS (2–4 × 10^6 cpm) is added with $MgCl_2$ and EDTA to obtain the desired concentrations (usually 4–10 μ*M* GTPγS, 2 m*M* $MgCl_2$, and 1 m*M* EDTA for class I ARFs). Nucleotide binding by various ARFs is differently affected by the concentration of free Mg^{2+}.
5. Reaction mixtures are incubated at 37°C or another indicated temperature. To assay GEP activity, it is necessary to establish a constant rate of GTPγS binding throughout the incubation. GEP proteins can differ greatly in stability, which determines incubation time and sometimes necessitates incubation at temperatures far below 37°C ***(16,21)***.
6. Incubations are terminated at the indicated time by collection of proteins on nitrocellulose filters equilibrated with wash buffer.
7. Each incubation tube is washed with 8 mL and then 4 mL of ice-cold wash buffer; washings are successively transferred to each filter.
8. Dried filters are dissolved in scintillation fluid for radioassay.
9. GEP activity is the difference between GTPγS bound to protein in incubations without and with GEP after subtraction of GTPγS bound in incubations with GEP and without ARF. BFA inhibition is the difference between binding with and without BFA after subtraction from each of [^{35}S]GTPγS bound in the absence of added GEP.

3.2. Specific Measurement of Activation of ARF by GEP

The action of ARFs as allosteric activators of cholera toxin ADP-ribosyltransferase can be used to assay ARF GEP activity in preparations that contain other GTP-binding proteins ***(11)***. Active ARF with GTPγS bound is assayed by its ability to increase the rate of cholera toxin A subunit (CTA)-catalyzed ADP-ribosylagmatine synthesis. For this purpose, GTPγS binding to ARF is carried out with or without the GEP preparation and/or BFA with all of the same controls used in the binding assay. In a second incubation, the effect of activated ARF on CTA-catalyzed ADP-ribosylagmatine synthesis is measured.

1. GTPγS binding to ARF in a 50-µL reaction mixture is carried out in a manner similar to that described for the assay of [^{35}S]GTPγS binding (*see* **Subheading 3.1.**). TENDS buffer with 30–60 µg of BSA, and 10–20 µg of L-α-phosphatidylserine (PS) is added to the GEP preparation or its control solution (vehicle), followed by the addition of 4–20 µ*M* GTPγS with or without 50 pmol (1 µg) of ARF (native or recombinant). After incubation, usually at 37°C for 40 min, tubes are placed in an ice bath and components for assay of CTA activity are added.
2. Assay of CTA-catalyzed ADP-ribosylagmatine synthesis is carried out in a total vol of 300 µL containing (final concentrations), 10 m*M* agmatine, 0.2 m*M* [^{14}C]NAD (10^5 cpm), 50 m*M* potassium phosphate buffer (pH 7.5), 20 m*M* DTT, 5 m*M* $MgCl_2$, 0.5 m*M* ATP, 20 µ*M* Cibachrome blue, 30 µg of ovalbumin, 20 µg of PS, and 2 µg of CTA, with or without the ARF, BFA, and/or GEP from the first incubation.
3. Assays are initiated with CTA addition and incubated for 1 h at 30°C before transfer of 150 µL of the reaction mixture to 1 mL of anion-exchange resin AG 1-X2 (BIO-RAD) previously equilibrated with deionized water. This is followed by five washes, each with 1 mL of water, while the effluent is collected for radioassay of [^{14}C]-labeled ADP-ribosylagmatine by liquid scintillation counting.
4. After subtraction of ADP-ribosylagmatine synthesized in the presence of ARF, but without GEP, the effect of GEP on ARF activation of CTA—and its inhibition by BFA—is calculated.

3.3. Release of Bound Guanine Nucleotides from Recombinant Proteins

Acceleration of GTPγS binding by an ARF GEP results from the acceleration of the release of bound nucleotide, which can be assessed directly by using ARF with radiolabeled guanine nucleotide bound as substrate for the GEP ***(22,28)***. In the first step, binding of radiolabeled nucleotide is carried out with a low concentration of free Mg^{2+} (usually with 1 m*M* $MgCl_2$ and 2 m*M* EDTA) and no GEP. The rate of release of bound nucleotides is then measured with and without added GEP.

1. Each assay tube contains 4 μ*M* [^{35}S]GTPγS or [^{3}H]GDP and 1 μg (50 pmol) of ARF in 50 μL of TENDS containing 10 μg of PS, 40 μg of BSA, protease inhibitors (as described in **Subheading 2.2.**), 40 m*M* NaCl, and 1 m*M* EDTA.
2. After incubation for 40 min at 37°C, tubes are placed on ice, followed by additions to bring the total vol to 100 μL with final concentrations of 1 m*M* unlabeled nucleotide, and 3 m*M* $MgCl_2$ and of other components as described in **Subheading 3.1.** for [^{35}S]GTPγS binding.
3. Release of bound [^{35}S]GTPγS or [^{3}H]GDP with or without GEP is the difference between nucleotide bound to protein at the beginning and end of the incubation period, usually measured at several times up to 20 min at 37°C.

3.4. Protein–Protein Interactions

The catalysis of guanine-nucleotide exchange on ARF appears to involve the formation of a stable complex of GEP with nucleotide-free ARF ***(29)***. These protein-protein complexes can be demonstrated by using an immobilized GST-fusion protein to bind the interacting protein ***(30)***, or by observing alteration of elution positions of interacting proteins on gel filtration ***(29)***. These complexes are more easily demonstrated between the isolated Sec7 domain of the GEP and N-terminally truncated forms of ARF than between the intact proteins. These observations are probably in accord with the findings that multiple regions of both ARF and GEP are required for functional interactions ***(31,32)***.

4. Notes

1. Although we have used recombinant nonmyristoylated ARFs in many assays, it is preferable to carry out experiments with myristoylated ARFs. Their behavior in several types of assays appears to differ qualitatively, as well as quantitatively, from that of the nonmyristoylated recombinant ARF preparations. Many published reports do not specify the extent of myristoylation of the recombinant ARF studied. There is, in any case, no information regarding the activities of mixtures containing different percentages of myristoylated and nonmyristoylated ARF.
2. To compare the activities of different classes of ARFs, we usually use ARF1 to represent Class I, ARF5 for Class II, and ARF6 for Class III. ARF1 appears to be the substrate most often used for assay of ARF GEP activity, although it is not clear that the in vitro interactions are always reflective of a biologically relevant function.
3. In guanine-nucleotide-exchange assays, the concentration of free Mg^{2+} is critical. Although no systematic comparisons have been published, it is clear that nucleotide binding by different ARFs is differently affected by the concentration of free Mg^{2+}.

4. Specific phospholipids also have major effects on guanine-nucleotide binding by ARFs and its acceleration by a GEP. The ~50 kDA cytohesins contain PH domains that specifically bind phosphatidylinositol phosphates, resulting in dramatically increased GEP activity. The structural basis of phospholipid stimulation of activity of other types of GEPs has not been established.

References

1. Rothman, J. E. and Wieland, F. T. (1996) Protein sorting by transport vesicles. *Science* **272,** 227–234.
2. Chavrier, P. and Goud, B. (1999) The role of ARF and Rab GTPases in membrane transport. *Curr. Opin. Cell Biol.* **11,** 466–475.
3. Springer, S., Spang, A., and Schekman, R. (1999) A primer on vesicle budding. *Cell* **97,** 145–149.
4. Tsuchiya, M., Price, S.R., Tsai, S.-C., Moss, J., and Vaughan, M. (1991) Molecular identification of ADP-ribosylation factor mRNAs and their expression in mammalian cells. *J. Biol. Chem.* **266,** 2772–2777.
5. Moss, J. and Vaughan M. (1995) Structure and function of ARF proteins: activators of cholera toxin and critical components of intracellular vesicular transport processes. *J. Biol. Chem.* **270,** 12,327–12,330.
6. Moss, J. and Vaughan, M. (1998) Molecules in the ARF orbit. *J. Biol. Chem.* **273,** 21,431–21,434.
7. Exton J. H. (1999) Regulation of phospholipase D. *Biochim. Biophys. Acta* **1439,** 121–133.
8. Honda, A., Nogami, M., Yokozeki, T., Yamazaki, M., Nakamura, H., Watanabe, H., et al. (1999) Phosphatidylinositol 4-phosphate 5-kinase is a downstream effector of the small G protein ARF6 in membrane ruffle formation. *Cell* **99,** 521–532.
9. Jackson, C. L. and Casanova, J. E. (2000) Turning on ARF: the Sec7 family of guanine nucleotide-exchange factors. *Trends Cell. Biol.* **10,** 60–67.
10. Peyroche, A., Antonny, B., Robineau, S., Acker, J., Cherfils, J., and Jackson, C. L. (1999) Brefeldin A acts to stabilize an abortive ARF-GDP-Sec7 domain protein complex: involvement of specific residues of the Sec7 domain. *Mol. Cell* **3,** 275–285.
11. Tsai, S.-C., Adamik, R., Moss, J., and Vaughan, M. (1994) Identification of a brefeldin A-insensitive guanine nucleotide-exchange protein for ADP-ribosylation factor in bovine brain. *Proc. Natl. Acad. Sci. USA* **91,** 3063–3066.
12. Tsai, S.-C., Adamik, R., Moss, J., and Vaughan, M. (1996) Purification and characterization of a guanine nucleotide-exchange protein for ADP-ribosylation factor from spleen cytosol. *Proc. Natl. Acad. Sci. USA* **93,** 305–309.
13. Chardin, P. and McCormick, F. (1999) Brefeldin A: the advantage of being uncompetitive. *Cell* **97,** 153–155.
14. Cherfils, J. and Chardin, P. (1999) GEFs: structural basis for their activation of small GTP-binding proteins. *Trends Biochem. Sci.* **24,** 306–311.

15. Achstetter, T., Franzusoff, A., Field, C., and Schekman, R. (1988) SEC7 encodes an unusual, high molecular weight protein required for membrane traffic from the yeast Golgi apparatus. *J. Biol. Chem.* **263,** 11,711–11,717.
16. Sata, M., Donaldson, J.G., Moss, J., and Vaughan, M. (1998) Brefeldin A-inhibited guanine nucleotide-exchange activity of Sec7 domain from yeast Sec7 with yeast and mammalian ADP ribosylation factors. *Proc. Natl. Acad. Sci. USA* **95,** 4204–4208.
17. Ogasawara, M., Kim, S.-C., Adamik, R., Togawa, A., Ferrans, V.J., Takeda, K., et al. (2000) Similarities in function and gene structure of cytohesin-4 and cytohesin-1, guanine nucleotide-exchange proteins for ADP-ribosylation factors. *J. Biol. Chem.* **275,** 3221–3230.
18. Mossessova, E., Gulbis, J. M., and Goldberg, J. (1998) Structure of the guanine nucleotide exchange factor Sec7 domain of human ARNO and analysis of the interaction with ARF GTPase. *Cell* **92,** 415–423.
19. Betz, S. F., Schnuchel, A., Wang, H., Olejniczak, E. T., Meadows, R. P., Lipsky, B. P., et al. (1998) Solution structure of the cytohesin-1 (B2-1) Sec7 domain and its interaction with the GTPase ADP-ribosylation factor 1. *Proc. Natl. Acad. Sci. USA* **95,** 7909–7914.
20. Cherfils, J., Ménétrey, J., Mathieu, M., Le Brass, G., Robineau, S., Beraud-Dufour, S., et al. (1998) Structure of the Sec7 domain of the ARF exchange factor ARNO. *Nature* **392,** 101–105.
21. Sata, M., Moss, J., and Vaughan, M. (1999) Structural basis for the inhibitory effect of brefeldin A on guanine nucleotide-exchange proteins for ADP-ribosylation factors. *Proc. Natl. Acad. Sci. USA* **96,** 2752–2757.
22. Patton, W. A., Zhang, G.-F., Moss, J., and Vaughan, M. (1997) Cholera toxin and Escherichia coli heat-labile enterotoxin: biochemical methods for assessing enzymatic activities, in *Bacterial Toxins: Tools in Cell Biology and Pharmacology* (Aktories, K., ed.), Chapman and Hall, Weinheim, Germany, pp. 15–36.
23. Franco, M., Chardin, P., Chabre, M., and Paris, S. (1995) Myristoylation of ADP-ribosylation factor 1 facilitates nucleotide exchange at physiological Mg^{2+} levels. *J. Biol. Chem.* **270,** 1337–1341.
24. Randazzo, P. A. and Fales, H. M. (2002) Preparation of myristoylated ARF1 and ARF6 proteins. *Methods Mol. Biol.: GTPase Protocols* **189,** 169–180.
25. Pacheco-Rodriguez, G., Meacci, E., Vitale, N., Moss, J., and Vaughan, M. (1998) Guanine nucleotide exchange on ADP-ribosylation factors catalyzed by cytohesin-1 and its Sec7 domain. *J. Biol. Chem.* **273,** 26,543–26,548.
26. Morinaga, N., Moss, J., and Vaughan, M. (1997) Cloning and expression of a cDNA encoding a bovine brain brefeldin A-sensitive guanine nucleotide-exchange protein for ADP-ribosylation factor. *Proc. Natl. Acad. Sci. USA* **94,** 1226–1231.
27. Togawa, A., Morinaga, N., Ogasawara, M., Moss, J., and Vaughan, M. (1999) Purification and cloning of a brefeldin A-inhibited guanine nucleotide-exchange protein for ADP-ribosylation factors. *J. Biol. Chem.* **274,** 12,308–12,315.
28. Meacci, E., Tsai, S.-C., Adamik, R., Moss, J., and Vaughan, M. (1997) Cytohesin-1, a cytosolic guanine nucleotide-exchange protein for ADP-ribosylation factor. *Proc. Natl. Acad. Sci. USA* **94,** 1745–1748.

29. Paris, S., Béraud-Dufour, S., Robineau, S., Bigay, J., Antonny, B., Chabre, M., et al. (1997) Role of protein-phospholipid interactions in the activation of ARF1 by the guanine nucleotide exchange factor ARNO. *J. Biol. Chem.* **272,** 22,221–22,226.
30. Vitale, N., Pacheco-Rodriguez, G., Ferrans, V. J., Riemenschneider, W., Moss, J., and Vaughan, M. (2000) Specific functional interaction of human cytohesin-1 and ADP-ribosylation factor domain protein (ARD1). *J. Biol. Chem.* **275,** 21,331–21,339.
31. Pacheco-Rodriguez, G,, Patton, W. A,, Adamik, R., Yoo H. S., Lee, F.-J., Zhang, G.-F., et al. (1999) Structural elements of ADP-ribosylation factor 1 required for functional interaction with cytohesin-1. *J. Biol. Chem.* **274,** 12,438–12,444.
32. Morinaga, N., Adamik, R., Moss, J., and Vaughan, M. (1999) Brefeldin A-inhibited activity of the Sec7 domain of p200, a mammalian guanine nucleotide-exchange for ADP-ribosylation factors. *J. Biol. Chem.* **274,** 17,417–17,423.

16

Arf6 and its Role in Cytoskeletal Modulation

Julie G. Donaldson

1. Introduction

The Arf family of GTP-binding proteins is highly conserved and widely expressed in eukaryotic cells. Arfs can be divided into three classes based on amino-acid sequence. Class I Arfs include mammalian Arfs 1-3, Class II includes Arfs 4, and 5, and Class III Arfs have only one member—Arf6 *(1)*. All Arf proteins have been shown to share a number of biochemical properties in vitro including: serving as cofactors in the ADP-ribosylation of Gs-alpha subunits induced by cholera toxin, activating phospholipase D (PLD), and activating phosphatidylinositol 4-phosphate 5-kinase *(2)*. However, since each Arf is broadly expressed, they must serve specific functions in cells. Our goal has been to define these functions in cells. Of these proteins, Arf1 is the most widely studied in cells and in biochemical assays. It is involved in the regulation of membrane traffic between the endoplasmic reticulum (ER) and Golgi complex and in the maintenance of the organization of the Golgi complex *(3)*. Arf1 serves this dual function through the modification of Golgi-membrane lipid composition and the assembly of cytosolic coat-protein complexes onto Golgi membranes.

After Arf1, Arf6 has been the most studied, and probably serves a related function in the periphery: regulating membrane traffic and organelle structure. Arf6 localizes to the plasma membrane (PM) and recycling endosomal membrane system, and studies from a number of groups indicate that it regulates membrane traffic between these two compartments. Arf6-GTP is predominantly associated with the PM, whereas Arf6-GDP is associated with endosomal membranes *(4,5)*. Activation of Arf6 is required for the endosomal membrane

From: *Methods in Molecular Biology, vol. 189: GTPase Protocols: The Ras Superfamily*
Edited by: E. J. Manser and T. Leung © Humana Press Inc., Totowa, NJ

to recycle its contents back to the PM, since the dominant negative mutant of Arf6, T27N, inhibits this membrane recycling *(6,7)*.

In addition to the regulation of membrane trafficking, many studies have implicated Arf6 in modulating the cortical actin cytoskeleton. This was originally shown in HeLa cells overexpressing wild-type Arf6 where addition of aluminum fluoride (AlF), a complex known to activate heterotrimeric G proteins, results in the formation of F-actin-rich protrusive structures *(8)*. Intriguingly, AlF treatment of HeLa cells expressing wild-type Rac induces formation of PM ruffles that are indistinguishable from those induced by constitutively active Rac, Q61L *(9)*. Recent studies suggest that the AlF treatment may result in the activation of Arf6 by sequestration of the endogenous Arf6 GAP in an Arf6-GAP-AlF complex, thereby allowing the free Arf6 to remain in the GTP-bound state *(10)*. Studies have shown that the apparent Rac activation by AlF treatment may also be conducted through this mechanism *(11)*. The AlF-induced Arf6-dependent protrusions and Rac-dependent ruffles can be distinguished from one another (*see* **Fig. 1**). Through the expression of dominant-negative Arf and Rho proteins, it can be shown that Arf6 protrusions occur independently of Rho proteins, whereas Rac ruffling is dependent upon Arf6 function *(9)*. Although the protrusions formed in response to AlF are an exaggerated phenotype and one that requires overexpression of wild-type Arf6, it serves as a reasonable assay for Arf6 cortical actin functions. During cell spreading, cells actually make protrusions and ruffles, and this process is dependent upon Arf6 function *(12)*. Arf6 has also been shown to be required for Fc-mediated phagocytosis *(13)*. Additional evidence for Arf6 functioning on cortical actin comes from cells expressing ARNO *(14)* or EFA6 *(15)*, guanine-nucleotide-exchange factors (GEFs) specific for Arf6, where surface actin structures have been described that are not observed if the dominant-negative mutant of Arf6 (T27N) is coexpressed.

This chapter describes an approach for investigating Arf6 effects on the actin cytoskeleton. It makes use of transient transfection of wild-type and mutants of Arf6 in cells, reversible pharmacologic reagents, and immunofluorescent localization of Arf6 and actin to assess the effects on cell shape and cytoskeleton. Assays for assessment of Arf6 roles in protrusions, rac ruffling, and cell spreading are described.

2. Materials

2.1. Expression Plasmids for Arf6 and Rac

Suitable mammalian expression plasmids encoding Arf6 and Rac proteins, both wild-type, dominant-negative (Arf6/T27N, Rac1/T17N), and constitutively active (Arf6/Q67L, Rac1/Q61L) forms are required. We have used both SV40-

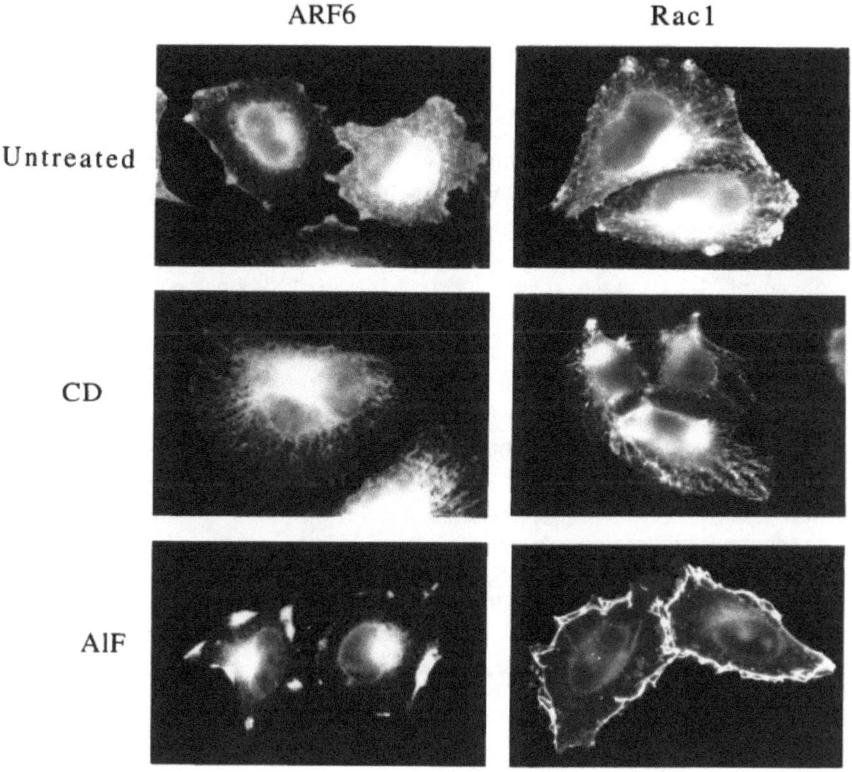

Fig. 1. Distinctive actin structures induced by Arf6 and Rac1 in HeLa cells. HeLa cells transfected with plasmids encoding either wild-type Arf6 or Rac1 were left untreated or treated with 100 nm CD or AlF (10 m*M* NaF; 50 μ*M* $AlCl_3$) for 30 min at 37°C prior to fixation and immunofluorescence localization of Arf6 or Rac. Note that in untreated cells, the localization of Arf6 and Rac are similar, and that with CD treatment both proteins redistribute on the Arf6 tubular endosomal structures. AlF treatment brings them both to the PM, but forms protrusions in cells expressing Arf6 and ruffles in cells expressing Rac. In both cases, these cortical-actin structures label with rhodamine phalloidin and many actin-associated proteins (not shown).

based *(8)* and cytomegalovirus (CMV)-promoter vectors such as pcDNA3 with success. Arf and Rac proteins can be detected by specific antibodies, if available, or an epitope tag can be appended to the sequence at the carboxy termini in Arfs and at the amino termini in Rac to avoid interference with critical post-translational lipid modifications. We have used both influenza haemogluttinin (HA) and FLAG-tagged Arf proteins with reasonable success (*see* **Note 1**).

2.2. Cells and Other Reagents

1. HeLa cells are a convenient cell type in the study of Arf6 function because they are easily transfected and the phenotypes elicited by Arf6 are well-defined. Similar activities for Arf6 have been observed in other cell types although they are not always exactly the same (*see* **Note 2**). HeLa cells are grown in Dulbecco's modified Eagle's medium (DMEM) with 10% fetal calf serum (FCS), and 1% penicillin-streptomycin. They are plated onto glass cover slips (12-mm diameter) 24 h before transfection in order to cover approx 20% of the area.
2. Cytochalasin D and latrunculin are stored at –20°C as stock solutions in dimethyl sulfoxide (DMSO). AlF is obtained by addition of 10 m*M* NaF and 50 μ*M* $AlCl_3$ into culture media. For immunofluorescence, antibodies to detect epitope tags or antibodies to Arf6 directed to the carboxyl-terminal 12 amino acids (EGLTWLTSNYKS) ***(12)*** and Rac1 (Transduction Labs, Lexington, KY) are used as primary antibodies. For secondary antibodies, we use new Alexa-coupled secondary antibodies available from Molecular Probes (Eugene, OR). Alexa 488-goat-anti-mouse or 488-goat-anti-rabbit work very well in combination with rhodamine-phalloidin to detect F-actin.
3. HEPES-buffered saline: HBS, 280 m*M* NaCl, 50 m*M* HEPES acid, 1.5 m*M* NA_2HPO_4. Adjust pH with NaOH to 7.05; this adjustment is critical.
4. AlF: mixture of 10 m*M* NaF and 50 μ*M* $AlCl_3$.

3. Methods

Two approaches for studying the Arf6 role in cortical actin are described here. The first examines effects of pharmacologic treatment on cells overexpressing either Arf6 or Rac1, particularly the ability of AlF treatment to "activate" the GTPases. The second assesses the requirement of Arf6 function in a cell-spreading assay through expression of dominant-negative Arf6 (T27N). Both assays rely on transient transfection and immunofluorescence techniques to assess these functions.

3.1. Transient Transfection and Immunofluorescence

1. Cells that had been plated on cover slips 24 h previously are placed in 6-well culture dishes, 4 cover slips/well, with 2 mL culture media. We have noted that for calcium phosphate-mediated transfections, it is very important that the culture media used contains some calcium. RPMI-1640, for example, does not contain calcium and should not be used.
2. Transient transfection is carried out using the calcium phosphate precipitation method ***(16)***. We use 2–5 μg plasmid DNA per well. In sterile tubes, 2–5 μg DNA, 16 μL of 2 *M* $CaCl_2$ solution is added and the volume is adjusted with water to 125 μL. Into this tube, 125 μL of a 2X concentration of HBS is added dropwise while gently vortexing. The final volume of this transfection solution is 250 μL with 128 m*M* $CaCl_2$. The tubes are allowed to sit at room temperature

for 10 min, and then the contents are added dropwise to each of the wells while gently swirling the dish. Alternatives to calcium phosphate, such as various lipid-based reagents, can be used. However, the relative ease and economy of the calcium-phosphate precipitation method makes it the most versatile.

3. After 12–16 h, the media and calcium-phosphate precipitate are removed, the culture wells are rinsed with sterile phosphate-buffered saline (PBS), and the media is added back. The cells can be used either 24 h or 30 h after addition of DNA.
4. Immunofluorescence staining: Following various treatment of these cells on cover slips, the cells are fixed in 2% formaldehyde solution in PBS for 10 min at room temperature. The cover slips are rinsed once with PBS and then placed in 10% FCS in PBS (FCS/PBS) for 10 min. Cover slips are inverted onto a 25-μL drop of primary antibody diluted into FCS/PBS containing 0.2% saponin on a strip of parfilm. After 1 h, the cover slips are washed with three changes over 15 min of 1 mL FCS/PBS. Cover slips are then incubated with fluorescent secondary antibodies diluted into FCS/PBS with 0.2% saponin for 1 h. The cover slips are then washed with three changes over 15 min of 1 mL FCS/PBS as before, and are finally rinsed in PBS prior to mounting on slides with Fluormount G (Southern Biotechnology, Birmingham, AL) or other mounting fluid, blotted on paper towels, and sealed with a rim of colorless nail polish.

3.2. Relationship between Arf6 and Rac Proteins

1. Cells are transfected with various Arf6 and Rac constructs alone and in combination, and cellular morphology is examined after various treatments. To test for interdependence, each wild-type and active mutant are coexpressed with the dominant-negative construct of the other.
2. Cells on cover slips are left untreated or treated with 100 nm cytochalasin D or AlF for 30 min at 37°C. They are then fixed and processed for immunofluorescence staining.

In HeLa cells, wild-type Arf6 and Rac both localize to the PM and associate with internal, juxtanuclear compartments. Upon treatment with CD, this distribution shifts from the PM to internal tubular membranes that emanate from the perinuclear region. AlF treatment results in the formation of protrusions in cells expressing Arf6 and peripheral ruffles in cells expressing Rac1 (**Fig. 1**). These structures are distinctively different from each other, and thus allow for an examination of the interdependence of the two GTPases. Indeed, it can be shown that coexpression of Arf6/T27N with wild-type Rac inhibits Rac ruffling, demonstrating that Arf6 function is required for Rac ruffling. By contrast, coexpression of Rac/T19N with wild-type Arf6 does not block Arf6 protrusions *(9)* indicating that Arf6 protrusions do not depend upon Rac activity. This assay has more recently been used to assess the ability of putative effector-domain mutants of Arf6 to form protrusions *(17)*.

3.3. Arf6 and Cell Spreading

1. For cell-spreading assays, cells that are plated in 10-cm culture dishes containing one cover slip are transfected with solutions scaled up fourfold so that 8–20 μg DNA are used in a total volume of 1 mL.
2. Prior to the replating assay, the one cover slip is removed and fixed for later staining to assess transfection efficiency. The media on the remaining cells is removed and replaced with trypsin/ethylenediamine-tetraacetic acid (EDTA) solution and cells are allowed to be released into suspension. The trypsinized is neutralized by addition of culture media, and the cells are pelleted at 300*g* for 5 min. The cells are washed in culture media and pelleted again as above and then resuspended in 10 mL culture media. The suspended cells are added to 10-cm culture dishes containing cover slips and placed back in the incubator.
3. After various periods of time, cover slips are removed, fixed, and prepared for immunofluorescence staining. To facilitate scoring of cell spreading, the cell surface can be labeled with a fluorescent lectin after fixation in the absence of saponin. After removing unbound lectin with PBS washes, the cells can be lightly fixed again, and then immunostaining for Arf6 or Rac in the presence of saponin can be done. Cells can be scored both for attachment, i.e., adhering to the cover slip, and for spreading, defined as covering a surface area equal to twice the diameter of the nucleus of the cell. The proportion of cells initially transfected as determined by examination of the initial cover slip can be determined and compared at each time-point to the proportion of cells attached and spread that are transfected.
4. Although this assessment is performed using complete media containing serum, replating assays can also be carried out in the absence of serum on cover slips coated with extracellular matrix components such as collagen, fibronectin, and laminin. For this, cover slips are coated overnight in 25 mg/mL of matrix components.

We find that expression of Arf6/T27N inhibits attachment and spreading of cells both in complete media and on defined matrix components ***(12)***. The block in cell spreading by expression of Arf6/T27N is also observed, even when spreading is stimulated by treatment of cells with phorbol esters ***(12)***.

4. Notes

1. The wide availability of good commercial antibodies to epitope tags has facilitated studies on the functions of different Arfs and mutants in cells. These antibodies are typically used for immunofluorescence, immunoblotting, and immunoprecipitation of expressed, epitope-tagged proteins. For Arfs, these tags must be appended to the carboxyl terminus because amino-terminal myristoylation is required for biological activity. The presence of a short peptide tag, however, is not entirely benign. In particular, we have noticed that the dominant-negative mutant of Arf6, T27N, is much more potent without its epitope tag. Furthermore, the ease and

popularity of tagging GTPases with green fluorescent protein (GFP) should be approached with caution. Although Arf1 and 6 tagged with GFP show remarkable localization in cells ***(18)***, it is not clear whether the activities of GFP-tagged Arfs will always be the same as untagged.

2. The AlF-induced protrusions are not observed in all cells. Notably, we do not see such structures in fibroblasts. Furthermore, in Chinese hamster ovary (CHO) cells expressing Arf6, AlF treatment or activated Arf6 results in PM ruffling that is nearly indistinguishable from that induced in CHO cells expressing Rac ***(19)***. Indeed, we have noticed that in CHO cells, although Arf6 and Rac both make ruffles in response to AlF treatment, we could still demonstrate that whereas Rac-mediated ruffling was dependent upon Arf6 activity and could be inhibited by Arf6/T27N, Arf6-mediated ruffling was not dependent upon Rac1 activity and was not inhibited by Rac/T17N. Indeed, the finding by D'Souza-Schorey et al. ***(19)*** that Rac and Arf6 share a common effector, POR1 (partner of Rac), and our observations that the trafficking of Rac and membranes is regulated by Arf6 ***(9)***, demonstrate a complexity in Arf6 and Rac functions that requires further study. The use of HeLa cells that offer the advantage of distinctive Arf6 and Rac PM structures should facilitate these studies.

References

1. Moss, J. and Vaughan, M. (1998) Molecules in the ARF orbit. *J. Biol. Chem.* **273,** 21,431–21,434.
2. Donaldson, J. G. and Jackson, C. L. (2000) Regulators and effectors of the ARF GTPases. *Curr. Opin. Cell Biol.* **12,** 475–482.
3. Lippincott-Schwartz, J., Cole, N. B., and Donaldson, J. G. (1998) Building a secretory apparatus: role of ARF1/COPI in Golgi biogenesis and maintenance. *Histochem. Cell Biol.* **109,** 449–462.
4. Peters, P. J., Hsu, V. W., Ooi, C. E., Finazzi, D., Teal, S. B., Oorschot, V., et al. (1995) Overexpression of wild-type and mutant ARF1 and ARF6: distinct perturbations of nonoverlapping membrane compartments. *J. Cell Biol.* **128,** 1003–1017.
5. van Donselaar, E., Hsu, V. W., Yang, C., Stahl, P. D., and Peters, P. J. (1998) ARF6 targets recycling vesicles to the plasma membrane: insights from an ultrstructural investigation. *J. Cell Biol.* **140,** 603–616.
6. D'Souza-Schorey, C., Li, G., Colombo, M. I., and Stahl, P. D. (1995) A regulatory role for ARF6 in receptor-mediated endocytosis. *Science* **267,** 1175–1178.
7. Radhakrishna, H. and Donaldson, J. G. (1997) ADP-ribosylation factor 6 regulates a novel plasma membrane recycling pathway. *J. Cell Biol.* **139,** 49–61.
8. Radhakrishna, H., Klausner, R. D., and Donaldson, J. G. (1996) Aluminum fluoride stimulates surface protrusions in cells overexpressing the ARF6 GTPase. *J. Cell Biol.* **134,** 935–947.
9. Radhakrishna, H., Al-Awar, O., Khachikian, Z., and Donaldson, J.G. (1999) ARF6 requirement for Rac ruffling suggests a role for membrane trafficking in cortical actin rearrangements. *J. Cell Sci.* **112,** 855–866.

10. Jackson, T. R., Brown, F. D., Nie, Z., Miura, K., Foroni, L., Sun, J., et al. (2000) ACAPs are Arf6 GTPase-activating proteins that function in the cell periphery. *J. Cell Biol.* **151,** 627–638.
11. Wolf, J., Dagher, M.C., Fuschs, A., Geiszt, M., and Ligeti, E. (1996) In vitro activation of the NADPH oxidase by fluoride—possible involvement of a factor activating GTP hydrolysis on Rac. *Eur. J. Biochem.* **239,** 369–375.
12. Song, J., Khachikian, Z., Radhakrishna, H., and Donaldson, J. G. (1998) Localization of endogenous ARF6 to sites of cortical actin rearrangement and involvement of ARF6 in cell spreading. *J. Cell Sci.* **111,** 2257–2267.
13. Zhang, Q., Cox, D., Tseng, C., Donaldson, J. G., and Greenberg, S. (1998) A requirement for ARF6 in Fcg receptor-mediated phagocytosis in macrophages. *J. Biol. Chem.* **273,** 19,977–19,981.
14. Frank, S. R., Hatfield, J. C., and Casanova, J. E. (1998) Remodeling of the actin cytoskeleton is coordinately regulated by protein kinase C and the ADP-ribosylation factor nucleotide exchange factor ARNO. *Mol. Biol. Cell* **9,** 3133–3146.
15. Franco, M., Peters, P. J., Boretto, J., van Donselaar, E., Neri, A., D'Souza-Schorey, C., et al. (1999) EFA6, a sec7 domain-containing exchange factor for ARF6, coordinates membrane recycling and actin cytoskeleton organization. *EMBO J.* **18,** 1480–1491.
16. Bonifacino, J. S., Suzuki, C. K., Lippincott-Schwartz, J., Weissman, A. M., and Klausner, R. D. (1989) Pre-Golgi degradation of newly synthesized T cell antigen receptor chains: intrinsic sensitivity and the role of subunit assembly. *J. Cell Biol.* **10,** 73–83.
17. Al-Awar, O., Radhakrishna, H., Powell, N. N., and Donaldson, J. G. (2000) Separation of membrane trafficking and actin remodeling functions of ARF6 with an effector domain mutant. *Mol. Cell. Biol.* **20,** 5998–6007.
18. Vasudevan, C., Han, W., Tan, Y., Nie, Y. Li, D., Shome, K., et al. (1998) The distribution and translocation of the G protein ADP-ribosylation factor 1 in live cells is determined by its GTPase activity. *J. Cell Sci.* **111,** 1277–1285.
19. D'Souza-Schorey, C., Boshans, R. L., McDonough, M., Stahl, P. D., and Van Aelst, L. (1997) A role for POR1, a Rac1-interacting protein, in ARF6-mediated cytoskeletal rearrangements. *EMBO J.* **16,** 5445–5454.

17

ARF GTPase-Activating Protein 1

Irit Huber, Edna Cukierman, Miriam Rotman, and Dan Cassel

1. Introduction

ARF-family GTPases function as regulators of multiple membrane trafficking processes in all eukaryotic cells. Six ARF proteins have been identified in humans, all of which contain a myristoyl residue, attached to their amino-terminal glycine by an amide bond. In their active, guanosine triphosphate-bound (GTP-bound) state ARFs associate with membranes and trigger the recruitment of cytosolic proteins to the membrane. The switch to the GTP-bound state is under the control of a family of ARF-directed guanine-nucleotide-exchange proteins containing a Sec7-homology domain (for review *see* **ref.** ***1***). The subsequent deactivation of ARFs and their dissociation from membranes depends on the hydrolysis of bound GTP. However, ARF proteins are devoid of intrinsic GTPase activity, and GTP hydrolysis depends on the action of GTPase-activating proteins (GAPs). ARF GAPs are a family of proteins which contain a conserved GAP domain of approx 130 amino acids with a unique Cys_4 zinc-finger motif near their amino-terminal part ***(2–9)***. Different ARF GAPs display a limited specificity to various ARF proteins in vitro, whereas additional specificity may be provided by targeting to distinct cellular compartments through variable domains that are present on various GAP molecules ***(10–12)***.

This chapter examines the first ARF GAP to be discovered, now known as GAP1 ***(2,13)***. The GAP1 protein has a molecular mass of 45.4 K_d, is distributed between cytosol and the Golgi, and regulates membrane traffic through this organelle. GAP1 can be purified from tissues, but this procedure is laborious and results in a very low yield ***(13)***. Thus, for most purposes the protein must be obtained from recombinant sources. Although full-length GAP1 cannot be expressed in *Escherichia coli* (*E. coli*) ***(2)***, amino-terminal fragments with full

From: *Methods in Molecular Biology, vol. 189: GTPase Protocols: The Ras Superfamily*
Edited by: E. J. Manser and T. Leung © Humana Press Inc., Totowa, NJ

GAP catalytic activity can be prepared in high yield. Most experience in our laboratory was gained with a construct encoding the first 257 amino acids, which also served as the basis for mutagenesis studies. The preparation of this protein ($GAP_{1\text{-}257}$) is described in this chapter. Full-length GAP1 can be expressed in insect cells using the baculovirus system, as described elsewhere ***(14)***.

A unique property of GAP1 and its truncated forms is the ability to refold at high efficiency following treatment with denaturing agents such as 8 *M* urea or 6 *M* guanidinium hydrochloride. Refolding at good yield is achieved by simple dilution, even after standing in the guanidine solution for several days at room temperature. This property allows the purification of GAP1 from *E. coli* under denaturing conditions, which is advantageous in two ways. First, extraction of bacteria with guanidinium hydrochloride results in the recovery of a large pool of insoluble material, resulting in much higher yield than that obtained by purification from a soluble pool. Second, under the denaturing conditions the risk of proteolysis is eliminated.

The principles of the purification of GAP1 are rather straightforward. The protein, when expressed, is fused to a hexahistidine tag, usually at the amino terminus. Expressing bacteria are extracted with guanidinium hydrochloride, and the protein is purified by Ni-NTA chromatography in the presence of the guanidinium solution. The purified material is dialyzed to remove the guanidine, and precipitated material that apparently contains denatured forms of GAP1 is removed. Finally, protein in the supernatant is purified by high-performance anion-exchange chromatography. Again, this step results not only in further purification, but also in a resolution between native and denatured forms of GAP1.

Assays of GAP activity are generally based on measurement of a single round of GTP binding and hydrolysis. This eliminates any indirect influence of the rate of GTP binding on the final readout. Thus, the G protein is loaded with radiolabeled GTP in the presence of a nanomolar concentration of free magnesium, generated by a "magnesium buffer" containing an excess of EDTA over magnesium. This low magnesium concentration facilitates the exchange of bound guanosine diphosphate (GDP) for GTP. The exchange reaction is stopped by the addition of excess magnesium ions, and the hydrolysis of bound GTP following the addition of GAP is monitored. In the case of ARFs, loading with GTP cannot be carried out in solution, because phospholipids and/or detergents are required. These agents act by stabilizing the amphipatic amino-terminal part of ARF that becomes exposed in the GTP state ***(15–17)***. Loading with GTP can be obtained in the presence of detergent alone (e.g., 0.1% Triton-X100), although at relatively low efficiency. High loading efficiency is obtained in the presence of phospholipids of various compositions or certain mixtures of phospholipid and detergent such as dimyristoylphosphatidylcholine (DMPC) and cholate, and only assays based on such loading conditions are described here. In the final stage of the GAP assay, GTP hydrolysis products are resolved and determined. When the GTP substrate is labeled in the γ phosphate, addition of a charcoal suspension results in the absorption of all nucleotides and of ARF, and radioactivity in the charcoal supernatant indicates the amount of inorganic ^{32}P released. Alternatively, [$\alpha^{32}P$]-GTP can be used as substrate, and the amount of radiolabeled GDP formed is determined after separation by anion-exchange thin-layer chromatography. This approach when combined with an energy-regenerating system can eliminate background in assays of crude preparations containing GAP activity ***(13)***, but the procedure employing [$\gamma^{32}P$]-GTP is faster and more convenient, and is the preferable one when using purified proteins that are devoid of nucleotidase activity.

2. Materials

2.1. Expression Plasmids

1. $GAP_{1\text{-}257}$ can be expressed from a number of expression plasmids. We have routinely used for expression vectors from the pET series. In these vectors the cDNA is placed under the control of the T7 promoter, and the resulting expression vector is introduced into *Escherichia coli* (*E. coli*) strain BL21 (DE3) which contains the T7 RNA polymerase under the control of the lacZ promoter. Induction of the T7 polymerase with isopropyl-β-D-thiogalactopyranozide (IPTG) triggers the expression of the cDNA. By choosing the appropriate vector, GAP1-257 can be expressed with a hexahistidine tag at either the amino or carboxy terminus, with or without a protease cleavage site.
2. Our standard $GAP_{1\text{-}257}$ preparation is produced in a pKM260 vector (**Fig. 1**). A fragment of GAP cDNA cloned in the Bluescript SK^+ vector (clone Z6, **ref.** ***2***) is amplified by PCR using the T3 primer and a sequence-specific antisense primer containing an extension creating a *Bam*H I site. The product is cloned between the *Nco* I and *Bam*H I sites of pKM260, using the endogenous *Nco* I site that includes the initiating methionine in GAP1. The product contains an amino-terminal extension of hexahistidine followed by a TEV protease cleavage site. Alternatively, GAP cDNA is amplified with a sense primer that starts at codons for Ser-3 and contains an Nhe I site (GCATATGGCTAGCCCAAGAACCAGAA) together with the antisense described here. Cloning of the product into the corresponding sites in pKM260 results in a construct encoding $GAP_{1\text{-}257}$ with hexahistidine extension but without the TEV recognition peptide.

2.2. Reagents for GAP1-257 Purification

1. Lysis buffer: 6 *M* guanidine hydrochloride, 0.1 *M* sodium phosphate, 10 m*M* Tris-HCl, pH 8.0, 5 m*M* β-mercaptoethanol (*see* **Note 1**).

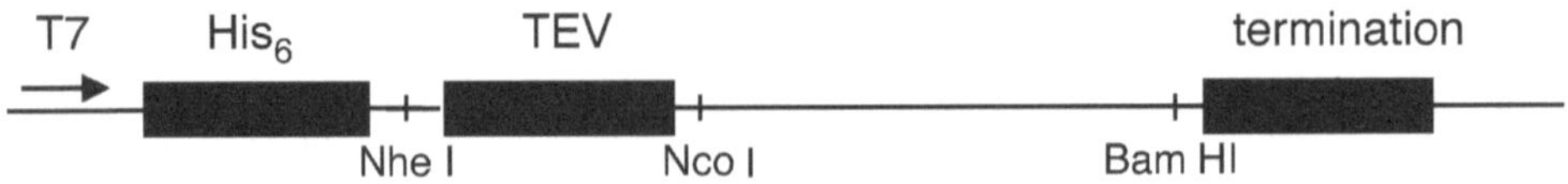

Fig. 1. Schematic presentation of cloning sites in the expression plasmid pKM260. The plasmid contains an Amp resistance marker. GAP_{257} is cloned between either *Nco* I and *Bam*H I or *Nhe* I and *Bam*H I.

2. Elution buffer: Lysis buffer containing 250 m*M* imidazole-HCl, pH 8.0.
3. Dialysis buffer: 50 m*M* NaCl, 25 m*M* Tris-HCl, pH 7.5, 1 m*M* dithiothreitol (DTT).
4. Resource-Q buffers: 25 m*M* Tris-HCl, pH 7.5, 1 m*M* DTT, and either 50 or 500 m*M* NaCl.

2.3. GAP Assay Reagents

1. ARF-loading buffer: 2 m*M* EDTA, 1 m*M* $MgCl_2$, 50 m*M* NaCl, 25 m*M* MOPS pH 7.4.
2. DMPC/cholate mixture: 6–8 mg of dimyristoylphosphatidylcholine (DMPC, from Sigma) are weighed in a 1-cm diameter thin glass test-tube and suspended in 25 m*M* MOPS pH 7.4 to give a DMPC concentration of 22 mg/mL. The mixture is sonicated in water bath sonicator for a few minutes, and sodium cholate is added from a 10% stock to yield a 1% final concentration. At this point, the suspension should become completely clear.
3. Liposomes: Unilamellar lipsomes of uniform size are prepared by filter extrusion ***(18)***. Stocks of pure phospholipids in chloroform are mixed at the desired ratio to yield 2 mg in total, and chloroform is added to a final vol of 0.3 mL. The mixture is placed in a 50-mL round or conical flask, and the solvent is removed under reduced pressure in a rotary evaporator. Traces of solvent are removed under a stream of nitrogen, and the phospholipids are resuspended by rotation in 0.5 mL of 100 m*M* KCl/50 m*M* HEPES pH 7.5. The mixture is transferred to a 2-mL microfuge tube and the tube is vortexed for 20 min, then subjected to 5 cycles of freezing with liquid nitrogen and thawing at 37°C (*see* **Note 2**). Following the final thaw, the liposomes may not be kept below room temperature. The liposomes are forced 19 times through a 0.4-μ polycarbonate filter using either a homemade 2-syringe device or a commercial device (available from Lipex Biomembranes, Vancouver, B.C.). The liposome suspension becomes less turbid at this point. It is important to ensure that the filter has remained intact at the end of this procedure, as excessive application of force may result in its disruption.
4. GAP assay buffer: 5 m*M* $MgCl_2$, 1 m*M* DTT, 20–50 m*M* NaCl, 25 m*M* MOPS pH 7.4.
5. Charcoal suspension: Fine charcoal is weighed and suspended in water in a narrow vessel, and is allowed to settle. Floating material is removed by decanta-

tion, and the procedure is repeated once more. Finally the charcoal is suspended in 50 m*M* NaH_2PO_4 (unbuffered) to yield a 5% suspension.

6. Filter-wash buffer: 10 m*M* $MgCl_2$, 50 m*M* KCl, 20 m*M* Tris-HCl, pH 7.5.

3. Methods

3.1. Expression and Purification of $GAP_{1\text{-}257}$

1. BL21/DE3 *E. coli* strain harboring the $GAP_{1\text{-}257}$ expression plasmid (*see* **Subheading 1.1.**) are grown overnight at 37°C in Luria broth (LB) medium supplemented with 50 mg/L ampicillin. Then, 4 mL of this culture are diluted into 400 mL of this medium and growth is continued with shaking at 220–250 rpm. When OD_{600} of the culture reaches 0.6–0.8 add IPTG to a final concentration of 0.4 m*M*, and continue growth for an additional 2.5 h.
2. All procedures are carried out at room temperature. Centrifuge the bacteria for 10 min at 4,000*g*. Resuspend the pellet in 16 mL lysis buffer (*see* **Subheading 2.2.**) by repeated pipetting until homogenous lysate is obtained, then mix by inversion for 20 min. Centrifuge for 15 min at 26,800*g*.
3. Add to the cleared lysate 2.5 mL (packed volume) of Ni-NTA beads (from Qiagen) prewashed with the lysis buffer, and mix by inversion for 45 min.
4. Wash the Ni-NTA beads twice by centrifugation for 1 min at 500*g* and resuspension in 20 mL of lysis buffer.
5. Transfer the Ni-NTA beads to a disposable 0.8 × 4-cm plastic column (such columns can be obtained from Bio-Rad). Wash the column with 10 mL of lysis buffer.
6. Add elution buffer, and collect 0.5-mL fractions. Test the protein concentration of each fraction by using the Bio-Rad protein assay reagent prediluted 5-fold with water, and pool the fractions containing the protein peak (usually fractions 3–5).
7. Dialyze the pooled fractions against dialysis buffer (*see* **Subheading 2.2.**) overnight with one buffer replacement. A considerable amount of denatured protein may precipitate at this stage.
8. Centrifuge the dialysate for 10 min at 25,000*g*, and load the supernatant on a 1-mL Resource Q column (a Mono-Q column may be used as well; both columns are obtained from Pharmacia-LKB) (*see* **Note 3**).
9. Develop the column with NaCl gradient (50–500 m*M*) using the Resource Q buffers (*see* **Subheading 2.2.**) over 30 min at 1 mL/min, recording OD_{280}. Collect the symmetrical peak that elutes at approx 200 m*M* salt. Check purity by SDS-PAGE (11% acrylamide). $GAP_{1\text{–}25}$ appears as a major 31 K_d band, although some low mol wt degradation products may also be present.
10. Divide the preparation into small aliquots, freeze with liquid nitrogen, and store at –80°C. Alternatively, add to the preparation an equal vol of 80% (w/v) glycerol, and store at –15°C (*see* **Note 4**).

3.2. Assay of GAP Activity

1. To preload ARF1 with radiolabeled GTP, prepare a mixture containing ARF-loading buffer (*see* **Subheading 2.3.**), 4 μ*M* myristoylated ARF1 (*see* **Note 5**)

and 1 μ*M* [γ^{32}P]-GTP (a mixture of labeled and unlabeled GTP containing 10,000–20,000 cpm/pmol; if carrier-free [γ^{32}P]-GTP is employed, its contribution to the final GTP concentration may be ignored). Bring the reaction mixture to room temperature, and add lipids (1 : 10 dilution of DMPC/cholate mixture or 1 : 5–1 : 10 dilution of liposomes, *see* **Subheading 2.3.**). Incubate for 15 min at 30°C, and stop the reaction by the addition of $MgCl_2$ from a 40-m*M* stock to give 2 m*M* (*see* **Note 6**).

2. To determine loading efficiency, remove a 2-μL aliquot into 3 mL of cold filter-wash buffer (*see* **Subheading 2.3.**) and filter through a 0.45-μ nitrocellulose membrane (such as NC45 from Schleicher and Schuell). Wash the filter with 5 mL wash buffer, place the filter (unfolded) in a scintillation vial, and add 2 μL of the loading reaction into a separate vial. Determine the percentage of ARF-bound GTP (i.e., filter-bound counts/total counts); values above 40% are acceptable, and typical values are 50–75%.
3. To assay for GAP activity, prepare reaction mixtures in a final vol of 10–20 μL containing GAP assay buffer (*see* **Subheading 2.3.**) and test samples. Bring to 30°C and add preloaded ARF (1 : 5–1 : 10 of the reaction volume). Incubate for the desired period (typically 10 min), then terminate the reaction by the addition of 0.5 mL of charcoal suspension (*see* **Subheading 2.3.**). Spin the samples in the microfuge for 1 min, and count 0.4 mL of the supernatant by using the Cherenkov radiation.
4. Determine the percentage of ARF-bound GTP that was hydrolyzed by comparing the results of the GAP assay with the total amount of radioactivity in each assay corrected for the percentage of loading as determined by the filter assay described here (*see* **Note 7**).

4. Notes

1. Guanidine solutions are highly irritating and corrosive. Use protective gloves and avoid spillage in centrifuges containing metal parts.
2. Liposomes may be stored at –80°C under nitrogen at this stage.
3. The preparation that is obtained after dialysis shows high GAP activity; however, at this stage the preparation contains denatured forms that tend to stick to hydrophobic surfaces, and is thus unsuitable for binding experiments ***(18)***.
4. The purified GAP_{1-25} is stable at –80°C for several months, and the glycerol stock is stable at –15°C for several weeks. Avoid temperatures lower than –15°C because of possible freezing of the preparation. Storage of purified GAP1 at 4°C results in a partial loss of activity within a few days.
5. Myristoylated ARF1 is prepared as described in Chapter 14.
6. The use of highly myristoylated ARF1 is recommended to avoid compromising loading efficiency. If low loading efficiency is obtained, an increase in the ARF1 concentration in the loading reaction may be attempted.
7. ARF preloaded with GTP in the presence of a DMPC/cholate mixture may be frozen with liquid nitrogen and stored in aliquots at –80°C. ARF preloaded in the

presence of liposomes should be used fresh, and therefore only limited amounts should be prepared. Typical loading is conducted in a final vol of 30–100 µL.

8. GAP_{257} concentrations that cause 50% hydrolysis of ARF1-bound GTP over a 10-min assay range from 50–500 ng/mL (16–160 nm). In addition to differences in activity reflecting the lipid composition in the assay, there is some variability that probably reflects the quality of the GAP preparation.

Acknowledgments

We thank the Israel Science Foundation and the Fund for the Promotion of Research at the Technion for support.

References

1. Jackson, C. L. and Casanova, J. E. (2000) Turning on ARF: the Sec7 family of guanine-nucleotide-exchange factors. *Trends Cell Biol.* **10,** 60–67.
2. Cukierman, E., Huber, I., Rotman, M., and Cassel, D. (1995) The ARF1 GTPase-activating protein: zinc finger motif and Golgi complex localization. *Science* **270,** 1999–2002.
3. Poon, P. P., Wang, X., Rotman, M., Huber, I., Cukierman, E., Cassel, D., et al. (1996) Saccharomyces cerevisiae Gcs1 is an ADP-ribosylation factor GTPase-activating protein. *Proc. Natl. Acad. Sci. USA* **93,** 10,074–10,077.
4. Poon, P. P., Cassel, D., Spang, A., Rotman, M., Pick, E., Singer, R. A., et al. (1999) Retrograde transport from the yeast Golgi is mediated by two ARF GAP proteins with overlapping function. *EMBO J.* **18,** 555–564.
5. Premont, R. T., Claing, A., Vitale, N., Freeman, J. L., Pitcher, J. A., Patton, W. A., et al. (1998) beta2-adrenergic receptor regulation by GIT1, a G protein-coupled receptor kinase-associated ADP ribosylation factor GTPase-activating protein. *Proc. Natl. Acad. Sci. USA* **95,** 14,082–14,087.
6. Bagrodia, S., Bailey, D., Lenard, Z., Hart, M., Guan, J. L., Premont, R. T., et al. (1999) A tyrosine-phosphorylated protein that binds to an important regulatory region on the cool family of p21-activated kinase-binding proteins. *J. Biol. Chem.* **274,** 22,393–22,400.
7. Brown, M. T., Andrade, J., Radhakrishna, H., Donaldson, J. G., Cooper, J. A., and Randazzo, P. A. (1998) ASAP1, a phospholipid-dependent arf GTPase-activating protein that associates with and is phosphorylated by Src. *Mol. Cell. Biol.* **18,** 7038–7051.
8. Andreev, J., Simon, J. P., Sabatini, D. D., Kam, J., Plowman, G., Randazzo, P. A., et al. (1999) Identification of a new Pyk2 target protein with Arf-GAP activity. *Mol. Cell. Biol.* **19,** 2338–2350.
9. Kondo, A., Hashimoto, S., Yano, H., Nagayama, K., Mazaki, Y., and Sabe, H. (2000) A new raxillin-binding protein, PAG3/papalpha/KIAA0400, bearing an ADP-ribosylation factor GTPase-activating protein activity, is involved in paxillin recruitment to focal adhesions and cell migration. *Mol. Biol. Cell* **11,** 1315–1327.

10. Huber, I., Cukierman, E., Rotman, M., Aoe, T., Hsu, V. W., and Cassel, D. (1998) Requirement for both the amino-terminal catalytic domain and a noncatalytic domain for in vivo activity of ADP-ribosylation factor GTPase-activating protein. *J. Biol. Chem.* **273,** 24,786–24,791.
11. Aoe, T., Huber, I., Vasudevan, C., Watkins, S. C., Romero, G., Cassel, D., et al. (1999) The KDEL receptor regulates a GTPase-activating protein for ADP-ribosylation factor 1 by interacting with its non-catalytic domain. *J. Biol. Chem.* **274,** 20,545–20,549.
12. Randazzo, P. A., Andrade, J., Miura, K., Brown, M. T., Long, Y. Q., Stauffer, S., et al. (2000) The Arf GTPase-activating protein ASAP1 regulates the actin cytoskeleton. *Proc. Natl. Acad. Sci. USA* **11,** 4011–4016.
13. Makler, V., Cukierman, E., Rotman, M., Admon, A., and Cassel, D. (1995) ADP-ribosylation factor-directed GTPase-activating protein. Purification and partial characterization. *J. Biol. Chem.* **270,** 5232–5237.
14. Huber, I., Rotman, M., Pick, E., Makler, V., Rothem, L., Cukierman, E., et al. (2001) *Methods Enzymol.* **329,** 307–316.
15. Antonny, B., Beraud-Dufour, S., Chardin, P., and Chabre, M. (1997) N-terminal hydrophobic residues of the G-protein ADP-ribosylation factor-1 insert into membrane phospholipids upon GDP to GTP exchange. *Biochemistry* **36,** 4675–4684.
16. Goldberg, J. (1998) Structural basis for activation of ARF GTPase: mechanisms of guanine nucleotide exchange and GTP-myristoyl switching. *Cell* **95,** 237–248.
17. Beraud-Dufour, S., Paris, S., Chabre, M., and Antonny, B. (1999) Dual interaction of ADP ribosylation factor 1 with Sec7 domain and with lipid membranes during catalysis of guanine nucleotide exchange. *J. Biol. Chem.* **274,** 37,629–37,636.
18. Antonny, B., Huber, I., Paris, S., Chabre, M., and Cassel, D. (1997) Activation of ADP-ribosylation factor 1 GTPase-activating protein by phosphatidylcholine-derived diacylglycerols. *J. Biol. Chem.* **272,** 30,848–30,851.

II

Specific Methods

Ran GTPases

18

The Use of Permeabilized Cell Systems to Study Nuclear Transport

Amy M. Brownawell, James M. Holaska, Ian G. Macara, and Bryce M. Paschal

1. Introduction

1.1. Nuclear Protein Import Assays

Nucleocytoplasmic transport occurs through nuclear pore complexes (NPC), macromolecular structures which span the nuclear envelope *(1)*. NPCs contain aqueous channels with a diameter of ~9 nm, allowing ions, metabolites, and small proteins to passively diffuse between the nucleus and the cytoplasm *(2)*. Most proteins and RNAs, however, exceed the ~60-kDa diffusion limit and are transported through the NPC by mechanisms that are saturable *(3)*, energy-dependent *(4)*, and signal-mediated *(6)*.

"Classical" nuclear localization sequences (NLS) are characterized by short amino-acid stretches enriched in basic amino acids. The NLS of the large T antigen of simian virus 40 (SV40) was identified by deletion analysis that resulted in mislocalization of the protein to the cytoplasm *(7)*. It was later defined as a seven amino-acid sequence (PKKKRKV) sufficient to confer nuclear localization when conjugated to a carrier protein *(8)*. Microinjection of excess NLS-peptide conjugates led to saturation of the protein import pathway, providing strong evidence for the existence of an NLS-receptor *(3)*.

The development of in vitro permeabilized-cell assays led to the discovery of the NLS receptor, as well as the identification of key regulatory proteins. Some assays utilized mammalian cells in which the plasma membrane was selectively permeabilized with digitonin, a detergent which binds cholesterol and perforates the lipid bilayer *(5)*. The nuclear membrane in these cells remained intact because of its low cholesterol content. A fluorescent import

From: *Methods in Molecular Biology, vol. 189: GTPase Protocols: The Ras Superfamily*
Edited by: E. J. Manser and T. Leung © Humana Press Inc., Totowa, NJ

substrate was introduced into these cells through the permeabilized plasma membrane, and substrate uptake into nuclei was monitored by fluorescence microscopy. Importantly, it was observed that nuclear import in digitonin-permeabilized cells was dependent upon the re-addition of cytosol ***(5,9)***. Subsequent fractionation of the cytosolic extract led to the characterization of soluble factors required for the import of substrates that possess a classical NLS: the small GTPase Ran ***(10,11)***, nuclear transport factor 2 (NTF2)/p10 ***(12,13)***, importin α/karyopherin α ***(14,15)***, and importin β/karyopherin β/p97 ***(14,16–18)***. It is important to note that essential factors can be detected in this assay only when they are sufficiently depleted by digitonin extraction, and that insoluble factors (such as components of the NPC) are probably involved in the translocation process ***(19)***. Since these initial observations, 6 importin α and more than 21 importin β family members have been identified in humans, reflecting the complexity of the nuclear transport system ***(2)***.

In the permeabilized-cell assay, the importin α/β heterodimer is both essential and sufficient for docking of classical NLS-containing substrates to the pore ***(9)***. Importin α binds directly to the NLS, and importin β interacts with the NPC ***(25,26)***. However, there are many exceptions to this type of docking. For example, ribosomal proteins and histones have been shown to bind directly to many different importin β family members and dock at the NPC independently of importin α ***(27,28)***.

Subsequent to formation of a stable import complex, the docked proteins translocate through the pore into the nucleus, where the complex encounters a high concentration of RanGTP. RanGTP binds directly to importin β family members and promotes the release of import substrates into the nucleoplasm ***(29–31)***. Therefore, a critical function of Ran is to regulate the loading and unloading of cargo in a compartment-specific manner, and the constituents of the Ran system are distributed in order to provide this directionality for transport.

Ran is an abundant protein that is predominantly nuclear at steady state. Like other GTPases, Ran requires cofactors to efficiently hydrolyze guanosine triphosphate (GTP) and exchange guanosine diphosphate (GDP) for GTP. The only known guanine-nucleotide-exchange factor (GEF) for Ran is the nuclear protein RCC1 ***(20)***, and the only known GTPase-activating protein (RanGAP in humans; Fug1 in mouse) is cytoplasmic ***(21,22)***. NTF2 binds RanGDP in the cytoplasm and mediates Ran import into the nucleus in an energy-dependent manner ***(23,24)***. The disparate subcellular localization of RCC1 and RanGAP, in conjunction with NTF2-dependent translocation of Ran, maintains a high nuclear—and a low cytoplasmic—RanGTP concentration.

In vitro permeabilized mammalian-cell import assays can be used to determine the import requirements of a specific substrate or to assess the role that a novel transport receptor may play in the process. Assays can be performed using rabbit reticulocyte lysate *(32)* or HeLa cell lysate *(33)* as a crude source of soluble transport factors. Alternatively, all required transport factors can be expressed as recombinant proteins and added back to the permeabilized-cell system to support import *(31,34)*.

Inhibitors of the classical import pathway have been identified. These include preventing the formation of a Ran-gradient by: the addition of nonhydrolyzable GTP analogs *(10)*, energy depletion *(5,15)*, or adding Ran mutants locked in the GTP-bound conformation *(31)* to the import assay reaction. Both wheat-germ agglutinin *(5)* and a dominant-negative importin β fragment (residues 45–462) *(35)* block import by irreversibly binding to the NPC. The importin β-binding domain (IBB), the residues of importin α that are necessary and sufficient to bind importin β, block the classical import pathway by competing for binding to importin β *(34)*. Finally, by performing the assays on ice, necessary protein-protein interactions are inhibited and import is prevented *(5)*. These inhibitors can be useful in discerning whether import proceeds via a classical import pathway. It is important to note that many nonclassical import pathways have been described. One example is the importin α-independent import of ribosomal proteins and histones *(27,28)*. Other proteins are imported by Ran-independent mechanisms. RCC1 can be imported into the nucleus via both an N-terminal classical NLS that binds importin α3 and a nonclassical pathway that is Ran- and energy-independent *(36)*. hnRNPK and β-catenin can translocate through the pore in the absence of any soluble factors *(37,38)*. Therefore, careful evaluation of in vitro import assays using available inhibitors is required to elucidate the import mechanism of a protein.

1.2. Nuclear Protein Export Assays

Permeabilized cell assays have also been devised to study nuclear-protein export. These "two-step" assays require the initial import of a fluorescent reporter protein into the nucleus, and subsequent export of the same fluorescent reporter out of the nucleus. To date, four laboratories have described permeabilized-cell assays for analyzing nuclear export *(39–42)*. The assays differ with regard to the choice of fluorescent reporter and its fusion partner, and whether the import step is performed in the intact cell or the permeabilized cell. In all four assays, the goal is to reconstitute nuclear protein export through the NPC in a signal- and soluble-factor-dependent manner. As detailed here, these assays have facilitated the characterization of both wild-type and mutant nuclear-export factors, and in one case, the discovery of a new export factor.

Our laboratory developed a nuclear-export assay based on protein kinase inhibitor (PKI). In its physiological context, PKI binds tightly to the catalytic subunit of protein kinase A (PKA) and mediates its transport from the nucleus to the cytoplasm. Export of the PKI-PKA complex is mediated by the leucine-rich nuclear-export signal (NES; **L**A**L**K**L**AG**L**D**I**N) in PKI that is in turn recognized by the export receptor Crm1 in the nucleoplasm. The interaction between PKI and Crm1 is stabilized by the presence of Ran in its GTP-bound form. Thus, a complex of PKA, PKI, Crm1, and RanGTP is believed to form in the nucleoplasm and translocate through the NPC to the cytoplasm.

In the PKI-export assay, we have substituted PKA with a fluorescent reporter protein that can be monitored in the microscope ***(39)***. The reporter protein is fluorescein isothiocyanate-labeled recombinant streptavidin that has been engineered with an NLS (denoted FITC-STV-NLS). PKI binding to FITC-STV-NLS is mediated by a biotin moiety which has been added to recombinant PKI (denoted bPKI). Nuclear loading of the FITC-STV-NLS (~80-kDa tetramer) in digitonin-permeabilized cells is followed by the introduction of bPKI, whose small size (75 amino acids) permits rapid nuclear entry by diffusion. The biotin-streptavidin interaction in the nucleus results in the formation of a complex of bPKI/FITC-STV-NLS, which is a substrate for export mediated by soluble export factors such as Crm1 and RanGTP. We have used this assay to characterize the RanGTP-binding protein NXT1 and its role in nuclear export mediated by the receptor Crm1 ***(49)***. We have also used the PKI assay to purify a novel export receptor that operates independently of the Crm1 pathway (Holaska and Paschal, unpublished observations).

2. Materials

2.1. Materials for Import Assays

2.1.1. Fluorescent NLS-Conjugated Import Substrate

The use of fluorescent import substrates allows direct visualization of nuclear import by confocal or conventional microscopy. Several classical import substrates have been used successfully. These include: GST-GFP-NLS (~60 kDa) ***(43)***, tetramethylrhodamine isothiocyanate (TRITC)-BSA-NLS (~67 kDa) ***(33)***, FITC-nucleoplasmin (~155 kDa, homopentamer) ***(15)***, and allophycocyanin-NLS (104 kDa) ***(33)***. In addition, other proteins that translocate into the nucleus in intact cell systems can be expressed in *Escherichia coli* (*E. coli*). The specific requirements for import can be determined with permeabilized-cell assays if the proteins are labeled with a fluorophore or are expressed as green fluorescent protein (GFP) fusions. This technique has been successfully utilized in our laboratory for the study of RCC1, as well as in other laboratories ***(27,36)***. An important consideration to take into account when deciding on an import substrate is the diffusion limit of the nuclear pore. Import substrates less than ~60 kDa may diffuse passively through the pore, thereby complicating differentiation between active transport and passive diffusion followed by nuclear retention.

Our laboratory has successfully used GST-GFP-NLS (SV40) expressed in and purified from E. coli for both import assays and cellular microinjections using GST-GFP as negative control. However, TRITC-BSA-NLS (SV40) can also be prepared easily, and can be used for both permeabilized-cell assays and microinjections. TRITC-maleimide is conjugated to BSA via sulfhydryl groups on the protein. The fluorescent protein is then activated with a crosslinker consisting of a NHS-ester and a maleimide group connected to a spacer arm. The NHS-ester reacts with primary amines on the protein, leaving the maleimide available to react with the N-terminal cysteine on a synthetic peptide containing the SV40 NLS (CYTPPKKKRKV). Whenever a carrier protein is used as an import substrate, the protein must be labeled with the fluorophore prior to conjugation with peptide to avoid inactivation of the NLS by modification of essential amino-acid residues. In general, TRITC or Oregon Green™ conjugates of proteins are preferable to FITC conjugates for transport studies, because of their greater resistance to photobleaching. Xenopus nucleoplasmin is a nuclear protein with a naturally occurring NLS that has been used for analysis of signal-mediated nuclear import ***(15)***. Although the recombinant protein can be purified easily, FITC-labeled nucleoplasmin prepared in our laboratory resulted in high nonspecific binding to adherent permeabilized cells. Allophycocyanin, a fluorescent protein, conjugated to SV40 NLS peptide, is also a good import substrate ***(33)***. However, the cost of allophycocyanin (Molecular Probes) is prohibitive if a large quantity of import substrate is required. The first half of the following protocol can be used for the conjugation of TRITC-maleimide to a protein that already carries a NLS.

2.1.1.1. Solutions

1. Sodium phosphate buffer: 20 m*M* NaH_2PO_4, pH 7.2.
2. Crosslinking buffer: 50 m*M* HEPES, pH 7.2.
3. Import assay buffer: 20 m*M* HEPES-KOH, pH 7.4, 110 m*M* potassium acetate, 4 m*M* magnesium acetate, 0.5 m*M* EGTA.
4. Synthetic peptide: CYTPPKKKRKV.

2.1.1.2. Preparation of a Fluorescent Import Substrate

1. Dissolve 5–20 mg of BSA or HSA (crystallized and lyophilized, Sigma) in 0.5 mL sodium phosphate buffer. Calculate the molar concentration of the protein.
2. Dissolve 1 mg per mL (2.1 m*M*) TRITC-maleimide (Molecular Probes) in dimethylformamide.
3. Mix equimolar amounts of protein and TRITC-maleimide and incubate with gentle agitation at 4°C for 1 h.

4. Separate unconjugated TRITC-maleimide from the protein using a gel-filtration column, such as a Sephadex G-25 column (NAP-5, Amersham-Pharmacia, sample volume 0.5 mL). Exchange the protein into crosslinking buffer.
5. Dissolve the crosslinker, sulfo-SMCC (Pierce Chemical) in dimethyl sulfoxide (DMSO). Activate the protein with a 20-fold molar excess of the crosslinker for 1 h at room temperature.
6. Remove unreacted crosslinker using a gel-filtration column (PD-10, Amersham-Pharmacia, sample volume 2.5 mL). Exchange the protein into crosslinking buffer.
7. Calculate a 25-fold molar excess of the peptide. Resuspend in crosslinking buffer.
8. Add the peptide to the activated protein and readjust the pH to 7.2.
9. Allow the reaction to proceed 2–3 h at 4°C with gentle agitation. Remove any unconjugated peptide using a gel-filtration column, and exchange into import assay buffer. Concentrate using a Centricon-30 (Millipore). Determine the final protein concentration using a protein assay reagent (Bio-Rad protein assay). Aliquot, snap-freeze with liquid nitrogen, and store at –80°C. Freezing and thawing of substrate is not recommended.

The number of peptides conjugated per bovine serum albumin/human serum albumin (BSA/HSA) molecule can be estimated by comparison of the mobilities of sulfo-SMCC-activated protein with peptide-conjugated protein on a SDS-polyacrylamide gel. The conditions described here generally yield 4–8 peptides conjugated per BSA molecule.

2.1.2. Cell Extracts

Import assays have been performed utilizing cell lysates as the source of soluble import factors. Assays utilizing HeLa cell lysate ***(33,44)***, *Xenopus* egg ***(45)***, and oocyte lysate ***(9)***, and rabbit reticulocyte lysate ***(32)*** have been described. Assays utilizing mammalian-cell lysates are excellent starting points for analyzing the import requirements of a particular protein. In an effort to focus on mammalian systems, this section explores the preparation of lysate from HeLa cells and discusses the use of commercially available rabbit reticulocyte lysate. In our laboratory, HeLa cell lysate has greater import capacity and exhibits less variability between preparations compared to different lots of commercial rabbit reticulocyte lysate.

2.1.2.1. Solutions

1. 10X transport buffer: 200 m*M* HEPES-KOH, pH 7.4, 1.1 *M* potassium acetate, 20 m*M* magnesium acetate, and 5 m*M* EGTA. Sterile-filter and store at 4°C.
2. 1X transport buffer: 20 m*M* HEPES-KOH, pH 7.4, 110 m*M* potassium acetate, 2 m*M* magnesium acetate, and 0.5 m*M* EGTA.

3. Cell-lysis buffer: 5 m*M* HEPES-KOH, pH 7.4, 10 m*M* potassium acetate, 2 m*M* magnesium acetate, and 1 m*M* EGTA. Store at 4°C.
4. 1X phosphate-buffered saline (PBS): For 1 L, dissolve 8 g sodium chloride, 0.2 g potassium chloride, 1.44 g sodium phosphate (dibasic), and 0.24 g potassium phosphate (monobasic) in 900 mL deionized water. Adjust the pH to 7.4, and bring the final vol to 1 L. Store at 4°C.

2.1.2.2. Preparation of HeLa Cell Lysate

1. Grow HeLa cells at a density of 2–7 × 10^5 cells per mL in a spinner flask in a 37°C incubator (CO_2 is not required). Use Joklik's modified SMEM containing 2.0 g sodium bicarbonate and 2.38 g HEPES per L of media. Adjust the pH to 7.3, sterile-filter, and store at 4°C. Supplement the medium with 10% newborn calf serum, 1% penicillin/streptomycin, and 2 m*M* glutamine before use. The cells have an estimated doubling time of 18 h, making it necessary to dilute the culture with new medium every 1–2 d.
2. HeLa cells from a 250-mL culture provide the basis for scaling up the cells to 15 L. This is carried out by sequential dilution of the culture into larger spinner flasks. The culture should not be diluted to a density below 2 × 10^5 cells/mL. The spinner flasks used for scaling up the preparation are 250 mL (1 each), 1 L (1 each), 3 L (2 each), and 7 L (2 each). This process takes approx 5 d.
3. Harvest the cells by centrifugation at 400*g* for 15 min in 500-mL bottles. This and all subsequent steps should be performed at 4°C.
4. Wash cells 2× with 1 L ice-cold PBS. Wash once with 1X transport buffer containing 2 m*M* dithiothreitol (DTT).
5. Resuspend the cells in 100 mL 1X transport buffer containing 2 m*M* DTT, transfer to 50-mL polypropylene centrifuge tubes, and collect by centrifugation. The yield from a 15-L culture should be approx 40 mL.
6. Resuspend the cell pellet in 1.5 vol lysis buffer, supplemented with 3 µg/mL each aprotinin, leupeptin, pepstatin, 0.5 m*M* PMSF, and 5 m*M* DTT. Allow the cells to swell on ice for 10 min.
7. Disrupt the cells by two to three passes in a stainless-steel homogenizer. Monitor the progress of homogenization by Trypan blue staining and phase–contrast microscopy. The goal is to obtain ~95% cell disruption. Excessive homogenization causes nuclear fragmentation and release of nuclear contents into the soluble fraction of the preparation. This should be avoided.
8. Dilute the homogenate with 0.1 vol of 10 × transport buffer and centrifuge at 40,000*g* for 30 min.
9. Filter the resulting low-speed supernatant fraction through four layers of gauze. Spin this supernatant at 150,000*g* for 60 min.
10. The resulting supernatant fraction (50 mL, protein concentration approx. 10 mg/mL) should be aliquoted into 1-mL fractions, snap-frozen in liquid nitrogen, and stored at –80°C.

2.1.2.3. Preparation of Dialyzed Rabbit Reticulocyte Lysate

Rabbit reticulocyte lysate can be prepared directly from whole blood ***(33)***, or can be purchased as an untreated lysate (Promega, L4151). Our laboratory and others have successfully used commercial lysate in in vitro import assays ***(27,36)***. It provides a readily available source of transport factors when beginning to use this technique. However, we have experienced variability in the ability of different lots of commercially available lysate to support the import reaction. The basis for this variability is not understood, but dialysis against transport buffer and clarification by centrifugation can improve transport activity in these lysates ***(46)***. Dialyze batches of reticulocyte lysate against several changes of transport buffer (20 m*M* HEPES-KOH, pH 7.4, 110 m*M* potassium acetate, 4 m*M* magnesium acetate, 0.5 m*M* EGTA, 1 m*M* DTT, 0.1 m*M* phenylmethyl sulfonyl fluoride (PMSF), 10 µg/mL aprotinin). Centrifuge the dialyzed extract at 100,000*g* for 1 h, aliquot, snap-freeze in liquid nitrogen, and store at –80°C.

2.1.3. Recombinant Proteins

wt Ran, G19V Ran, Ran BP1, importin β (residues 45–462), and RanGAP (murine Fug1) have been subcloned into pGEX-2T (Amersham-Pharmacia) and expressed as GST-fusion proteins in E. coli strain XL-1 Blue ***(47,48)***. Purification of the fusion proteins and subsequent thrombin cleavage is carried out according to the published protocol (Amersham-Pharmacia). Importin α and β lose activity when expressed as GST-fusion proteins. Instead, they are expressed with C-terminal 6× histidine tags from the vectors pQE70 ***(15)*** and pQE60 ***(35)*** (Qiagen), respectively. The nonspecific-binding blocking protein, nucleoplasmin core (lacking its NLS), is also expressed from pQE70 ***(15)***. Q69L Ran is expressed with an N-terminal 6× histidine tag from pQE32 (Qiagen) ***(35)***. The IBB (residues 1–55 of importin α) has been subcloned into pzz70 and expressed with an N-terminal zz tag and a C-terminal 6× histidine tag ***(34)***. Purification of histidine-tagged proteins is achieved using protocols published previously ***(15,35)***. All imidazole used to remove the histidine-tagged proteins from their Ni^{+2} or Co^{+2} affinity matrix must be removed by desalting and concentration of the proteins, since imidazole is a potent inhibitor of nuclear transport. NTF2 is nonfunctional if expressed with any tag tested to date. Therefore, we express untagged NTF2 from the vector pET23b in *E. coli* strain BL21(DE3) according to published protocols ***(13)***. All proteins should be exchanged into assay buffer, concentrated, aliquoted, snap-frozen in liquid nitrogen, and stored at –80°C. Proteins should not be refrozen after use and used again in the import assay system; loss of import activity results from repeated freeze/thaw of the component proteins.

2.2. Materials for Export Assays

2.2.1. Buffers and Additives

1. TNE buffer: 50 m*M* Tris-HCl, pH 8.0, 50 m*M* NaCl, and 1 m*M* EDTA.
2. Transport buffer: 20 m*M* HEPES, pH 7.4, 110 m*M* potassium acetate, 2 m*M* magnesium acetate, and 0.5 m*M* EGTA.
3. Bicarbonate coupling buffer: 0.2 *M* sodium bicarbonate, pH 8.8, and 0.5 *M* NaCl.
4. Protease inhibitors (all stored as 100-µL aliquots at –20°C): PMSF (1 *M* in DMSO), leupeptin (1 mg/mL in water), pepstatin (1 mg/mL in water), aprotinin (1 mg/mL in DMSO).
5. DTT (1 *M* in water).
6. IPTG (100 m*M* in water, sterile-filtered).

2.2.2. Biotinylated PKI (bPKI)

The α-isoform of human PKI is expressed from a T7 promoter-based plasmid (pRSET; Invitrogen) in the bacterial strain BL21(DE3)pLysS (Novagen).

1. A single colony from a fresh transformation plate is used to inoculate a 20-mL Luria broth (LB) culture containing 100 µg/mL ampicillin. The 20-mL culture is grown overnight in a shaking incubator at 37°C and used to inoculate a 1-L culture (prewarmed to 37°C).
2. The culture is grown to an optical density (OD) of 0.6–0.8, induced with 0.5 m*M* IPTG, and grown for an additional 3 h.
3. The culture is harvested by centrifugation (4200*g* × 10 min at 4°C). The bacterial pellet (~10 mL) is frozen in liquid nitrogen and stored at –80°C until use.
4. The pellet is rapidly thawed by resuspension in TNE buffer (20 mL) containing a protease inhibitor cocktail (2 µg/mL each of aprotinin, leupeptin, pepstatin, 0.5 m*M* PMSF), and 2 m*M* DTT. Although the lysozyme encoded by the pLysS plasmid is generally sufficient for cell lysis, we often supplement the sample with lysozyme (added as powder to 0.5 mg/mL), and allow the sample to mix end-over-end for 30 min at 4°C.
5. The viscous sample is homogenized using 4 strokes in a tight-fitting stainless-steel homogenizer (7-mL capacity; Wheaton #357572).
6. The homogenate is clarified by centrifugation (40,000*g* × 30 min at 4°C). The supernatant (~20 mL) is further clarified by ultracentrifugation (150,000*g* × 30 min at 4°C).
7. The high-speed supernatant is subjected to a heat treatment, which denatures bacterial proteins while leaving the heat-stable PKI in its native form. For the heat treatment, the high-speed supernatant is transferred to a polypropylene centrifuge tube, which is placed in a boiling water bath for 5 min. The tube is then plunged into an ice-water slurry and chilled for 5 min.
8. The sample is clarified by centrifugation (40,000*g* × 30 min at 4°C). The supernatant is highly enriched in PKI, which is visible after electrophoresis on a 15% gel and Coomassie blue staining.

9. The final step of PKI purification involves chromatography of the heat-stable supernatant on an S100 gel-filtration column (120-mL bed vol; Amersham-Pharmacia) in PBS operated at 15 mL/h. PKI should elute in a monodisperse peak with an apparent mol-wt of 22 kDa.
10. The purified PKI (~20 mg in 3 mL) is biotinylated with a 20-fold molar excess of the amine-specific NHS-LC-biotin (Pierce Chemical). We typically combine 22.2 mg of the amine-specific biotinylation reagent with 20 mg of PKI in a total vol of 3.5 mL for 20 min at room temperature.
11. Unincorporated biotinylation reagent is removed by desalting on a PD-10 column (Amersham-Pharmacia) equilibrated in transport buffer, dispensed into 100-µL aliquots, quick-frozen in liquid nitrogen, and stored at –80°C.
12. Three different methods can be used to confirm the biotin incorporation into PKI. The bPKI can be resolved on a 15% SDS gel, transferred to nitrocellulose, and detected with commercial streptavidin conjugated with peroxidase. Methods that can reveal the molar ratio of biotin: PKI are mass spectrometry, and biotin detection using a commercial kit (Pierce Chemical).

2.2.3. Fluorescently-Labeled Streptavidin-NLS (FITC-STC-NLS)

Streptavidin-NLS is expressed from a T7-promoter-based plasmid (pET23b; Novagen) in the bacterial strain BL21(DE3)pLysS plasmid. STV-NLS is a 189 amino-acid fusion protein containing residues 1–138 from the streptavidin ORF, 19 residues of linker sequence from the polylinker of the pET plasmid, and residues 83–136 from SV40 large T antigen that include the NLS. Because the protein appears to be toxic to bacteria after prolonged growth in culture, we have adopted a protocol that ensures that the culture does not reach the stationary phase.

1. Approximately 10 colonies from a fresh transformation plate are vortexed in 1 mL LB, which is then added to a 500-mL LB culture containing 100 µg/mL ampicillin. The culture is allowed to sit on the benchtop overnight at room temperature.
2. The flask is then transferred to a shaking incubator (37°C) and grown until the culture reaches an OD of 0.6. IPTG (0.5 m*M*) is added to induce STV-NLS expression for 3 h.
3. The culture is harvested and the pellet is stored as described above for PKI.
4. The frozen pellet containing STV-NLS is treated as described above for PKI. The pellet is resuspended in TNE buffer (20 mL) containing a protease-inhibitor cocktail and DTT, supplemented with lysozyme, and mixed end-over-end for 30 min at 4°C.
5. The sample is homogenized using the stainless-steel homogenizer, and clarified by centrifugation (40,000*g* × 30 min at 4°C).
6. Because STV-NLS partitions into the inclusion body fraction, the supernatant is discarded and the pellet is washed 2× to release bacterial contaminants. This involves resuspending the pellet in 100 mL PBS using the stainless-steel

homogenizer (4 passes), and collection of the inclusion body fraction by centrifugation (40,000*g* × 30 min at 4°C).
7. The washed inclusion body fraction is solubilized in TNE buffer containing 8 *M* urea and 2 m*M* DTT for 60 min at 4°C.
8. The sample is clarified by sequential centrifugation at 40,000*g* for 30 min and 150,000*g* for 60 min.
9. The highly clarified supernatant is applied to a Sephacryl S300 gel filtration column (Amersham-Pharmacia) equilibrated in TNE buffer containing 8 *M* urea, and chromatographed at a flow rate of 15 mL/h at room temperature. The STV-NLS protein, which is detected by Coomassie blue staining as a 22 kDa (15% SDS PAGE), elutes with an apparent mol wt of ~80 kDa. This reflects the fact that the STV-NLS fusion protein, like native streptavidin, assembles into a tetramer which is resistant to disassembly by 8 *M* urea. The STV-NLS protein yield from the S300 column is ~4 mg in a total volume of ~12 mL.
10. Our procedure for urea removal involves several steps. The first step is a rapid dilution into 7 vol of ice-cold distilled water, which reduces the urea concentration to 1 *M*.
11. The rapid dilution step should not result in visible protein precipitation. Nonetheless, as a precaution, the sample is clarified at 40,000*g* × 30 min at 4°C.
12. The relatively dilute sample (~100 mL) is concentrated 10-fold in a nitrogen pressure, stirred-cell apparatus (Amicon #8050) operated at 28 PSI at 4°C. Note that no urea is removed during this step. The time required for this concentration step is ~2 h.
13. The sample is further concentrated, and the urea is removed in a vacuum dialysis apparatus and a 25-kDa membrane collodion membrane (both from Pierce Chemical) under house vacuum against a bicarbonate coupling buffer (4 changes of 250 mL each).
14. The STV-NLS protein (~4 mg in 2 mL buffer) is labeled with fluorescein-S-isothiocyanate (FITC) (Molecular Probes). The labeling reaction is performed by end-over-end mixing (or by stirring using a micro-stir bar) at room temperature for 30 min, using a fourfold molar excess of FITC to STV-NLS. The tube is wrapped in foil to minimize exposure to light during the labeling reaction.
15. Unincorporated FITC is removed by desalting on a PD-10 column (Amersham-Pharmacia) equilibrated in transport buffer. The FITC-STV-NLS is dispensed into 100-µL aliquots, quick-frozen in liquid nitrogen, and stored at –80°C. The concentration of FITC-STV-NLS is ~0.5 mg/mL.

2.2.4. Suspension Culture HeLa Cells

The cells used for the PKI export assays are suspension culture HeLa cells, which are grown exactly as described in **Subheading 2.1.2.2.** for preparation of HeLa cytosol. In our experience, adherent cells yield unsatisfactory results with the PKI export assay because of the nonspecific interactions of FITC-STV-NLS with the cells.

3. Methods

3.1. In Vitro Nuclear Import Assays

Protocols utilizing either suspension or adherent mammalian cells for in vitro import assays have been described ***(15,33,46)***. The following protocol focuses on the use of HeLa and BHK21 cells growing on cover slips for these permeabilized-cell assays.

3.1.1. Solutions

1. Digitonin stock solution: Digitonin can be prepared in either water or DMSO. Both yield similar results. We use high-purity digitonin from Calbiochem.
   a. Dissolve 5% digitonin in boiling water. Place solution on ice for 60 min and remove any precipitate by centrifugation in a microfuge at 40,000*g* for 30 min. This supernatant is defined as the 5% stock solution. Store refrigerated for approx 1–2 wk.
   b. Dissolve 10% digitonin in DMSO. Store as aliquots at –20°C.
2. Assay buffer: 20 m*M* HEPES-KOH, pH 7.4, 110 m*M* potassium acetate, 4 m*M* magnesium acetate, 0.5 m*M* EGTA, 1 m*M* DTT (added fresh the day of the experiment). Store at 4°C.
3. Import assay soluble factors: Either cell lysate or recombinant proteins can be used as a source of required import pathway proteins.
   a. Cell lysate mix: Thaw 50 µL dialyzed rabbit reticulocyte lysate per assay to be performed (i.e., per cover slip). Alternatively, dilute HeLa cell lysate with assay buffer to a final concentration of 1 mg/mL to yield a 50 µL reaction volume.
   b. Recombinant protein mix: Prepare in assay buffer (50 µL per cover slip) the following protein solution: 1.5 µ*M* Ran, 150 nm RanBP1, 150 nm RanGAP (Fug1), 150 nm NTF2, 1 µ*M* importin α, 3 µ*M* importin β, 1 to 3 µ*M* substrate, 1 mg/mL nucleoplasmin core, and 10 mg/mL BSA (the final two components are used to block nonspecific binding).
4. Energy-regenerating system: Stocks of ATP and GTP (Sigma) can be aliquoted and frozen at –20°C. Creatine phosphokinase (Sigma) should be resuspended in 20 m*M* HEPES-KOH, pH 7.5 containing 50% glycerol and stored at –20°C. Phosphocreatine (Sigma) should be made fresh the day of the experiment in assay buffer. Combine these components and add to the cell lysate or recombinant protein mix to these final concentrations: 20 m*M* phosphocreatine, 1 m*M* GTP, 1 m*M* ATP, and 50 µg/mL creatine phosphokinase. Repeated freeze/thaw of energy components results in loss of energy-regenerating system activity and is not recommended.
5. Inhibitors: These components are used to inhibit many nuclear import pathways.
   a. Energy depletion: Do not add the energy-regenerating system. Instead, add either 100 U/mL apyrase (Sigma) stored in 20 m*M* HEPES, pH 7.5 containing 50% glycerol at –20°C, or 50 U/mL hexokinase plus 12.5 m*M* glucose (Sigma) stored in 20 m*M* HEPES, pH 7.5 containing 50% glycerol at –20°C.

b. NPC-binding proteins: Add 200 µg/mL of the lectin, wheat-germ agglutinin (Sigma), dissolved in assay buffer to a standard import mix. Alternatively, 10 µ*M* importin β (45–462) can be added.
c. Inhibition of import complex formation: Add 100 µg/mL G19V Ran or Q69L Ran. Both proteins are Ran mutants locked in the GTP-bound state. They bind directly to importin β and inhibit import-complex formation. Add 100 µg/mL of IBB, the residues of importin α that are necessary and sufficient for binding to importin β.
d. Temperature-dependent inhibition: Perform the import reaction on ice.

3.1.2. Nuclear Import of Fluorescent Import Substrate

1. 18-mm diameter cover slips (Fisher) are flame-sterilized, placed in a 12-well tissue-culture plate, and coated with poly-L-lysine. HeLa or BHK21 cells should be plated on the cover slips at 50–60% confluence approx 20–24 h before intended usage. Cells are grown in Dulbecco's Modified Eagle's medium (DMEM) containing 10% (v/v) fetal calf serum (FCS) and 1% penicillin/streptomycin and maintained in a humidified incubator at 37°C and 5% CO_2.
2. Wash cells 2× with ice-cold assay buffer.
3. Permeabilize cells on ice for 5 min in ice-cold 0.005–0.008% digitonin diluted in assay buffer with gentle rocking to improve the homogeneity of permeabilization (*see* **Note 5**). Wash 2× with assay buffer.
4. Incubate for 20 min at room temperature with 1% BSA diluted in assay buffer. BSA stops the permeabilization of the plasma membrane by scavenging residual digitonin.
5. Move the cover slips cell-side up to dental wax (Dentsply International) or Parafilm (American National Can) using jeweler's forceps.
6. Begin the import assay by adding 50 µL of import assay mixture prepared as described in **Subheading 3.1.1., step 3.** Incubate 30 min (or desired time-point) at room temperature.
7. Return cover slips to the 12-well dish and wash 3× with assay buffer.
8. Fix for 15–20 min in 4% paraformaldehyde prepared in assay buffer. Wash once with assay buffer.
9. Permeabilize with –20°C methanol for 2 min. Wash once with assay buffer. Stain nuclei with 10 ng/mL 4′,6-diamidino-2-phenylindole dilactate (DAPI; Molecular Probes; D-3571) prepared in assay buffer for 10 min at room temperature. DAPI associates with the minor groove of dsDNA and allows for easy identification of cell nuclei.
10. Wash 2× with assay buffer. Mount the cover slips onto slides using an aqueous mounting medium with anti-fading agents, such as GelMount (Biomeda). Visualize the accumulation of import substrate with a fluorescence microscope.

3.2. In Vitro Nuclear Export Assays

3.2.1. Digitonin Permeabilization of HeLa Cells

The PKI export assay involves an import phase and an export phase, as illustrated in **Fig. 1A**. The import phase is required to load the fluorescent reporter FITC-STV-NLS into the nucleus.

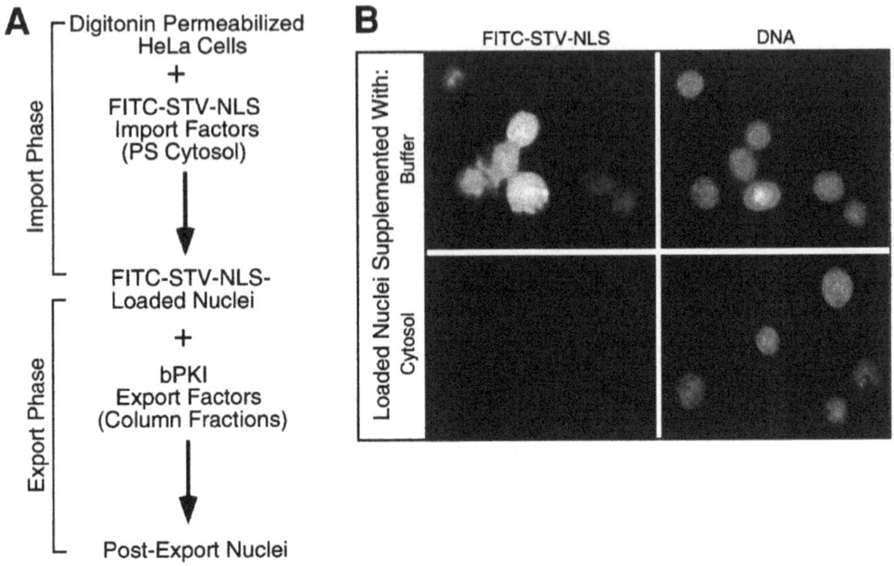

Fig. 1. Assay for nuclear export of PKI using digitonin-permeabilized cells. **(A)** Summary of the major steps of the assay. In the import phase, fluorescently labeled streptavidin engineered with a nuclear localization sequence (FITC-STV-NLS) is imported into the nuclei of permeabilized HeLa cells. After washing, biotinylated PKI (bPKI) is added, which diffuses into the nucleus and binds FITC-STV-NLS. PKI contains the nuclear export signal (LALKLAGLDIN; Wen et al., 1995). The bPKI/FITC-STV-NLS complex is exported from the nucleus upon addition of soluble factors. **(B)** Fluorescence microscopy of cells, containing the bPKI/FITC-STV-NLS complex, that were incubated with buffer (upper panels) or HeLa cell cytosol (lower panels) as a source of export factors. The fluorescent complex undergoes export in a cytosol-dependent reaction (Holaska and Paschal, 1998; Black et al., 1999).

1. To prepare HeLa cells for the import phase, harvest 50 mL of cells (density ~5×10^5/mL) by centrifugation at 400*g* × 5 min using a swinging bucket rotor.
2. The cell pellet is resuspended in transport buffer containing 2 m*M* DTT, and collected by centrifugation.
3. The washed cell pellet (~0.25 mL) is resuspended in 1 mL of the same buffer, and the cell density is measured using a hemacytometer. The cells are then diluted to exactly 5×10^6/mL in the same buffer and placed on ice.
4. Digitonin is then added to a final concentration of 50 µg/mL (from a 100 mg/mL stock in DMSO; *see* **Subheading 3.1.1.**), and the cells are gently pipetted twice and placed on ice for exactly 5 min.
5. To terminate the permeabilization reaction, dilute the sample 10-fold with transport buffer containing protease inhibitors (no PMSF) and DTT, and collect the cells by centrifugation.

6. The cell pellet (~100 μL) is resuspended gently in an equal volume of buffer and placed on ice.
7. Verify that the permeabilization was successful (~90% of the cells, permeabilized) by staining ~5 μL cells with Trypan blue and examining the preparation with a tissue-culture microscope (*see* **Note 5**).

3.2.2. Nuclear Import of FITC-STV-NLS

Because digitonin permeabilization releases soluble proteins from cells, reconstitution of nuclear import and export requires the addition of soluble transport factors to the assay. The factors can be added as recombinant proteins, or as HeLa-cell cytosol. The aim of the PKI export assay is to reconstitute nuclear export in a reaction that requires the addition of cytosolic factors. If, however, export factors are inadvertently added to the system during the import phase, then the permeabilized cells may display a reduced dependence on cytosol addition during the export phase. This is apparently because the export factors accumulate in the nuclei of permeabilized cells. We avoid introducing the major export factor Crm1 during the import phase by using cytosol that has been depleted of Crm1. This involves a simple preincubation of cytosol with Phenyl Sepharose ***(39)***. The alternative is to reconstitute import of FITC-STV-NLS using recombinant import factors, as described in **Subheading 3.1.1., step 3.**

1. Incubate HeLa cell cytosol (1.5 mL at 5 mg/mL) with Phenyl Sepharose beads (0.5 mL) by end-over-end mixing for 20 min at 4°C. The beads are sedimented by brief centrifugation at 200*g*.
2. The unbound fraction (supernatant) is collected and dispensed into aliquots (~100 μL), which are quick-frozen and stored at –80°C. Hereafter, the Phenyl-Sepharose pretreated cytosol is referred to as PS cytosol.
3. Nuclear import of FITC-STV-NLS is performed in suspension-culture HeLa cells. Each import reaction should yield enough FITC-STV-NLS-loaded nuclei for 6 export reactions. The import reaction (total vol = 50 μL) is assembled in 4 mL polystyrene, snap-top tube (Falcon #352054) on ice. Add 20 μL PS cytosol and 5 μL FITC-STV-NLS to the tube and incubate 10 min.
4. Add 5 μL energy-regenerating system (described in **Subheading 3.1.1.**) and 20 μL digitonin-permeabilized cells (~1.5×10^6), and mix by gently tapping the tube.
5. Transfer the tubes to a 30°C water bath and incubate for 15 min.
6. Add GTP to a final concentration of 5 m*M* (from a 50 m*M* stock of Mg-GTP) and continue the 30°C incubation for an additional 5 min.
7. The import reaction is terminated by transferring the tubes to an ice bucket, and by adding 4 mL ice-cold transport buffer to the cells.
8. Collect the cells by centrifugation (400*g* × 5 min) at 4°C, and carefully remove the supernatant by aspiration using a Pasteur pipet.

9. The cell pellet is resuspended in a total volume of 100 µL, and placed on ice. This completes the import phase, and its results can be visualized in the fluorescence microscope.

3.2.3. Nuclear Export of bPKI/FITC-STV-NLS

The export reaction (total volume = 60 µL) includes cells preloaded with FITC-STV-NLS, bPKI, energy-regenerating system, and the experimental sample (such as a column fraction) for measurement of export activity.

1. We assemble the export reactions by adding 37 µL of the experimental sample (or buffer) to a 4-mL polystyrene tube on ice.
2. Add 5 µL energy-regenerating system and 15 µL cell loaded with FITC-STV-NLS.
3. Add 3 µL of bPKI, and allow the reaction to incubate on ice for 4 min. During this incubation, bPKI diffuses into the nucleus and binds to FITC-STV-NLS.
4. The samples are transferred to a 30°C water bath and incubated for 8 min.
5. The export reaction is terminated by transferring the tubes to an ice bucket, and by adding 4 mL ice-cold transport buffer. The transport buffer is supplemented with DAPI (0.25 µg/mL) for DNA staining.
6. The cells are collected by centrifugation (400*g* × 5 min) at 4°C, and the supernatant is removed by aspiration.
7. The cells are chemically fixed by gentle resuspension in 2 mL 3.7% formaldehyde (made in PBS) for at least 30 min at room temperature.
8. The formaldehyde-cell suspension is diluted with 2 mL transport buffer and collected by centrifugation as in **Subheading 3.2.3.**
9. After removing the supernatant by aspiration, the cell pellet (barely visible) is resuspended in a small volume (~20 µL) of transport buffer.
10. Pipet 6 µL of cell suspension and 3 µL of the mounting media Vectashield (Molecular Probes) onto a glass microscope slide, invert a #1 cover slip (18 mm^2) on the sample, and examine in the fluorescence microscope.

3.2.4. Quantitation of Nuclear Export

We have developed a method for quantitating export of the bPKI/FITC-STV-NLS complex. In brief, the method involves using digital fluorescence microscopy to measure the mean nuclear fluorescence of 25–50 cells from each transport reaction. By comparing the nuclear fluorescence before and after the export reaction, it is possible to quantify the export activity of a protein or biochemical fraction of interest.

1. Capture the images using a CCD camera (Hamamatsu ORCA) mounted on a Nikon Microphot-SA microscope, using OpenLab software (2.0.6) running on a Power MacIntosh computer. Using a 40× dry objective (N.A. = 0.95), cells are randomly selected using phase contrast or DAPI staining, and recorded as

8-bit images using ~1-s exposure. Multiple fields from a single cover slip are sampled in this manner.
2. The mean fluorescence from a sample is determined by measuring the average pixel intensity/mm^2 in each nuclei of 25–50 cells. These values are then averaged to yield a mean and standard deviation (SD) for the sample.
3. Scale the data from 1 to 0, with 1 corresponding to the mean pixel intensity of a sample supplemented with buffer alone (no transport factors) during the export phase, and 0 corresponding to the mean pixel intensity of a region of the cover slip that is free of cells.

The results of a typical experiment, as shown in **Fig. 1B**, illustrate the level of nuclear export promoted by the addition of HeLa cell cytosol. Addition of buffer alone reduces the nuclear fluorescence by ~15%, and addition of cytosol, which contains export factors, reduces nuclear fluorescence by ~90%. We have used this assay to characterize the export factor NXT1, a RanGTP-binding protein that regulates the Crm1-dependent export pathway *(49)*. The assay has also been used to identify a novel export activity present in HeLa cytosol *(39)*.

4. Notes

1. When considering various import substrates, keep in mind that the diffusion limit of the nuclear pore is ~60 kDa for a globular protein.
2. GST-GFP-NLS and TRITC-BSA(or HSA)-NLS are good import substrates. Using fluorescently labeled nucleoplasmin results in high nonspecific background in in vitro import assays performed with adherent cells plated on cover slips.
3. HeLa cell lysate provides the most consistent source of cellular extract for these import assays. Commercial rabbit reticulocyte lysate can be used, but different lots vary in their import activity. Extensive dialysis followed by clarification by centrifugation is recommended to improve the activity of these commercial lysates.
4. Do not freeze/thaw proteins used in the recombinant protein mix. This leads to loss of import assay activity. Also, remove all traces of imidazole used in the purification of 6x His-tagged proteins. Imidazole strongly inhibits import.
5. Titrate the digitonin concentration for your particular cell system. Monitor the degree of permeabilization with Trypan-blue staining and phase-contrast microscopy. Underpermeabilization as well as overpermeabilization can lead to the failure of these assays.
6. Do not freeze/thaw the components of the energy-regenerating system. Storing the creatine phosphokinase in a 50% glycerol solution at –20°C (as described in **Subheading 3.1.1.**) prolongs the life of this enzyme, and is recommended.
7. Both 1 mg/mL recombinant nucleoplasmin core domain and 10 mg/mL BSA have been recommended to block nonspecific binding of the import substrate to cells. However, the nucleoplasmin core blocks nonspecific binding more

efficiently than BSA. Therefore, problems with nonspecific binding may be resolved by increasing the concentration of nucleoplasmin core.

References

1. Davis, L. I. (1995) The nuclear pore complex. *Annu. Rev. Biochem.* **64,** 865–896.
2. Gorlich, D. and Kutay, U. (1999) Transport between the cell nucleus and the cytoplasm. *Annu. Rev. Cell Dev. Biol.* **15,** 607–660.
3. Goldfarb, D. S., Gariepy, J., Schoolnik, G., and Kornberg, R. D. (1986) Synthetic peptides as nuclear localization signals. *Nature* **322,** 641–644.
4. Newmeyer, D. D., Finlay, D. R, and Forbes, D. J. (1986) In vitro transport of a fluorescent nuclear protein and exclusion of non-nuclear proteins. *J. Cell Biol.* **103,** 2091–2102.
5. Adam, S. A., Marr, R. S., and Gerace, L. (1990) Nuclear protein import in permeabilized mammalian cells requires soluble cytoplasmic factors. *J. Cell Biol.* **111,** 807–816.
6. Dingwall, C., Sharnick, S. V., and Laskey, R. A. (1982) A polypeptide domain that specifies migration of nucleoplasmin into the nucleus. *Cell* **30,** 449–458.
7. Lanford, R. E. and Butel, J. S. (1984) Construction and characterization of an SV40 mutant defective in nuclear transport of T antigen. *Cell* **37,** 801–813.
8. Lanford, R. E., Kanda, P., and Kennedy, R. C. (1986) Induction of nuclear transport with a synthetic peptide homologous to the SV40 T antigen transport signal. *Cell* **46,** 575–582.
9. Moore, M. S. and Blobel, G. (1992) The two steps of nuclear import, targeting to the nuclear envelope and translocation through the nuclear pore, require different cytosolic factors. *Cell* **69,** 939–950.
10. Melchior, F., Paschal, B., Evans, J., and Gerace, L. (1993) Inhibition of nuclear protein import by nonhydrolyzable analogues of GTP and identification of the small GTPase Ran/TC4 as an essential transport factor. *J. Cell Biol.* **123,** 1649–1659.
11. Moore, M. S. and Blobel, G. (1993) The GTP-binding protein Ran/TC4 is required for protein import into the nucleus. *Nature* **365,** 661–663.
12. Moore, M. S. and Blobel, G. (1994) Purification of a Ran-interacting protein that is required for protein import into the nucleus. *Proc. Natl. Acad. Sci. USA* **91,** 10,212–10,216.
13. Paschal, B. M. and Gerace, L. (1995) Identification of NTF2, a cytosolic factor for nuclear import that interacts with nuclear pore complex protein p62. *J. Cell Biol.* **129,** 925–937.
14. Adam, E. J. and Adam, S. A. (1994) Identification of cytosolic factors required for nuclear location sequence-mediated binding to the nuclear envelope. *J. Cell Biol.* **125,** 547–555.
15. Gorlich, D., Prehn, S., Laskey, R. A., and Hartmann, E. (1994) Isolation of a protein that is essential for the first step of nuclear protein import. *Cell* **79,** 767–778.

16. Chi, N. C., Adam, E. J., and Adam, S. A. (1995) Sequence and characterization of cytoplasmic nuclear protein import factor p97. *J. Cell Biol.* **130,** 265–274.
17. Gorlich, D., Kostka, S., Kraft, R., Dingwall, C., Laskey, R. A., Hartmann, E., et al. (1995) Two different subunits of importin cooperate to recognize nuclear localization signals and bind them to the nuclear envelope. *Curr. Biol.* **5,** 383–392.
18. Radu, A., Blobel, G., and Moore, M. S. (1995) Identification of a protein complex that is required for nuclear protein import and mediates docking of import substrate to distinct nucleoporins. *Proc. Natl. Acad. Sci. USA* **92,** 1769–1773.
19. Wente, S. R. (2000) Gatekeepers of the nucleus. *Science* **288,** 1374–1377.
20. Bischoff, F. R. and Ponstingl, H. (1991) Catalysis of guanine nucleotide exchange on Ran by the mitotic regulator RCC1. *Nature* **354,** 80–82.
21. Bischoff, F. R., Krebber, H., Kempf, T., Hermes, I., and Ponstingl, H. (1995) Human RanGTPase-activating protein RanGAP1 is a homologue of yeast Rna1p involved in mRNA processing and transport. *Proc. Natl. Acad. Sci. USA* **92,** 1749–1753.
22. Hopper, A. K., Traglia, H. M., and Dunst, R. W. (1990) The yeast RNA1 gene product necessary for RNA processing is located in the cytosol and apparently excluded from the nucleus. *J. Cell Biol.* **111,** 309–321.
23. Ribbeck, K., Lipowsky, G., Kent, H. M., Stewart, M., and Gorlich, D. (1998) NTF2 mediates nuclear import of Ran. *EMBO J.* **17,** 6587–6598.
24. Smith, A., Brownawell, A., and Macara, I. G. (1998) Nuclear import of RanGDP is mediated by NTF2. *Curr. Biol.* **8,** 1403–1406.
25. Gorlich, D., Vogel, F., Mills, A. D., Hartmann, E., and Laskey, R. A. (1995) Distinct functions for the two importin subunits in nuclear protein import. *Nature* **377,** 246–248.
26. Moroianu, J., Hijikata, M., Blobel, G., and Radu, A. (1995) Mammalian karyopherin alpha 1 beta and alpha 2 beta heterodimers: alpha 1 or alpha 2 subunit binds nuclear localization signal and beta subunit interacts with peptide repeat-containing nucleoporins. *Proc. Natl. Acad. Sci. USA* **92,** 6532–6536.
27. Jakel, S. and Gorlich, D. (1998) Importin β, transportin, RanBP5 and RanBP7 mediate nuclear import of ribosomal proteins in mammalian cells. *EMBO J.* **17,** 4491–4502.
28. Jakel, S., Albig, W., Kutay, U., Bischoff, F. R., Schwamborn, K., Doenecke, D., et al. (1999) The importin beta/importin 7 heterodimer is a functional nuclear import receptor for histone H1. *EMBO J.* **18,** 2411–2423.
29. Izaurralde, E., Kutay, U., Vonkobbe, C., Mattaj, I. W., and Gorlich, D. (1997) The asymmetric distribution of the constituents of the Ran system is essential for transport into and out of the nucleus. *EMBO J.* **16,** 6535–6547.
30. Rexach, M. and Blobel, G. (1995) Protein import into nuclei: association and dissociation reactions involving transport substrate, transport factors, and nucleoporins. *Cell* **83,** 683–692.

31. Gorlich, D., Pante, N., Kutay, U., Aebi, U., and Bischoff, F. R. (1996) Identification of different roles for RanGDP and RanGTP in nuclear protein import. *EMBO J.* **15,** 5584–5594.
32. Adam, S. A. and Gerace, L. (1991) Cytosolic proteins that specifically bind nuclear location signals are receptors for nuclear import. *Cell* **66,** 837–847.
33. Adam, S. A., Sterne-Marr, R., and Gerace, L. (1992) Nuclear protein import using digitonin-permeabilized cells. *Methods Enzymol.* **219,** 97–110.
34. Gorlich, D., Henklein, P., Laskey, R. A., and Hartmann, E. (1996) A 41 amino acid motif in importin-alpha confers binding to importin-beta and hence transit into the nucleus. *EMBO J.* **15,** 1810–1817.
35. Kutay, U., Izaurralde, E., Bischoff, F. R., Mattaj, I. W., and Gorlich, D. (1997) Dominant-negative mutants of importin-beta block multiple pathways of import and export through the nuclear pore complex. *EMBO J.* **16,** 1153–1163.
36. Nemergut, M. E. and Macara, I. G. (2000) Nuclear import of the Ran exchange factor, RCC1, is mediated by at least two distinct mechanisms. *J. Cell Biol.* **149,** 835–850.
37. Michael, W. M., Eder, P. S., and Dreyfuss, G. (1997) The K nuclear shuttling domain: a novel signal for nuclear import and nuclear export In the hnRNP K protein. *EMBO J.* **16,** 3587–3598.
38. Yokoya, F., Imamoto, N., Tachibana, T., and Yoneda, Y. (1999) Beta-catenin can be transported into the nucleus in a Ran-unassisted manner. *Mol. Biol. Cell* **10,** 1119–1131.
39. Holaska, J. M. and Paschal, B. M. (1998) A cytosolic activity distinct from Crm1 mediates nuclear export of protein kinase inhibitor in permeabilized cells. *Proc. Natl. Acad. Sci. USA* **95,** 14,739–14,744.
40. Kehlenbach, R. H., Dickmanns, A., and Gerace, L. (1998) Nucleocytoplasmic shuttling factors including Ran and Crm1 mediate nuclear export of NFAT in vitro. *J. Cell Biol.* **141,** 863–874.
41. Love, D. C., Sweitzer, T. D., and Hanover, J. A. (1998) Reconstitution of HIV-1 rev nuclear export: independent requirements for nuclear import and export. *Proc. Natl. Acad. Sci. USA* **95,** 10,608–10,613.
42. Englmeier, L., Olivo, J. C., and Mattaj, I. W. (1999) Receptor-mediated substrate translocation through the nuclear pore complex without nucleotide triphosphate hydrolysis. *Curr. Biol.* **9,** 30–41.
43. Plafker, K. and Macara, I. G. (2000) Facilitated nucleocytoplasmic shuttling of the Ran binding protein RanBP1. *Mol. Cell. Biol.* **20,** 3510–3521.
44. Paschal, B. M. (1998) Assay of nuclear protein import in permeabilized cells using flow cytometry, in *Cell Biology: A Laboratory Handbook* (Academic Press, San Diego, CA), pp. 305–313.
45. Newmeyer, D. D. and Wilson, K. L. (1991) Egg extracts for nuclear import and nuclear assembly reactions. *Methods Cell Biol.* **36,** 607–634.
46. Dingwall, C. and Palacios, I. (1998) In vitro systems for the reconstitution of snRNP and protein nuclear import. *Methods Cell Biol.* **53,** 517–543.

47. Lounsbury, K. M., Richards, S. A., Carey, K. L., and Macara, I. G. (1996) Mutations within the Ran/TC4 gtpase—effects on regulatory factor interactions and subcellular localization. *J. Biol. Chem.* **271,** 32,834–32,841.
48. Welch, K., Franke, J., Kohler, M., and Macara, I. G. (1999) RanBP3 contains an unusual nuclear localization signal that is imported preferentially by importin-alpha3. *Mol. Cell. Biol.* **19,** 8400–8411.
49. Black, B. E., Levesque, L., Holaska, J. M., Wood, T. C., and Paschal, B. M. (1999) Identification of an NTF2-related factor that binds Ran-GTP and regulates nuclear protein export. *Mol. Cell. Biol.* **19,** 8616–8624.

19

Analysis of Nuclear Protein Import and Export In Vitro Using Fluorescent Cargoes

Ralph H. Kehlenbach and Larry Gerace

1. Introduction

Nucleocytoplasmic transport of protein is central to the regulation of many cellular functions. Nuclear import and export takes place through nuclear pore complexes (NPCs; reviewed in **refs.** ***1*** and ***2***), large supramolecular structures that span the nuclear envelope. Cargo proteins contain specific signals for nuclear transport, which are often short stretches of amino acids. Nuclear localization signals (NLSs) direct the nuclear import of proteins, whereas nuclear export sequences (NESs) specify protein export from the nucleus. Both types of signals are recognized by nucleocytoplasmic shuttling receptor proteins belonging to the importin/karyopherin β family (reviewed in **refs.** ***3*** and ***4***). After binding to cargo, these receptors interact with proteins of the NPC to mediate cargo translocation into or out of the nucleus. The best-characterized nuclear-import pathway utilizes the major import-receptor importin β, which mediates the transport of proteins with basic amino-acid-rich NLSs. Importin β interacts with cargo molecules either directly ***(5)*** or via an adapter protein, importin α ***(3)***. Export of many proteins from the nucleus is mediated by the export receptor chromosome region maintenance 1 (CRM1), which recognizes leucine-rich NESs ***(6–10)***.

The small GTPase Ran, which is a member of the Ras superfamily, plays a fundamental role in regulating nuclear transport. One of the major functions of Ran is to control the binding of cargo molecules to transport receptors. Ran is believed to be concentrated in the nucleus as RanGTP under steady state conditions (*see* **ref.** ***3*** for review). In the case of nuclear export, RanGTP cooperatively binds to export receptors together with NES-containing cargo,

From: *Methods in Molecular Biology, vol. 189: GTPase Protocols: The Ras Superfamily*
Edited by: E. J. Manser and T. Leung © Humana Press Inc., Totowa, NJ

and thus promotes the formation of the export complex that is translocated out of the nucleus *(6)*. RanGTP also promotes the targeting of the export complex to the cytoplasmic side of the NPC *(11)*. In nuclear import, however, the binding of RanGTP to the import complex in the nucleus releases the cargo molecule from the receptor.

The activity of both nuclear import and export pathways can be regulated either positively or negatively by phosphorylation of cargo molecules (*see* **ref. *12*** for review). One example of such regulation is seen with the transcription factor NFAT (nuclear factor of activated T cells). NFAT is imported into the nucleus upon dephosphorylation by the calcium-activated phosphatase calcineurin. Rephosphorylation of NFAT by certain nuclear kinases then initiates its CRM1-dependent nuclear export *(12)*.

In vitro approaches have provided a basis for detailed biochemical analysis of nuclear transport. The first in vitro system for studying nuclear import involved nuclei that were assembled or resealed using Xenopus egg extracts *(13,14)*. Another in vitro system was subsequently developed, which involved cultured cells treated with digitonin to permeabilize the plasma membrane *(15)*. Digitonin-permeabilized cells are now widely used to study nuclear transport because this system is simple to use, and allows the analysis of nuclear transport in a variety of cell types under conditions that closely mimic various in vivo situations. Furthermore, under appropriate conditions, transport is dependent on exogenously added cytosol or shuttling nuclear transport factors, including Ran.

Nuclear import in permeabilized cells is often analyzed by fluorescence microscopy using adherent cells (or cells attached to a slide after the transport reaction). Permeabilized cells are incubated at 20–30°C in the presence of an ATP-regenerating system, a fluorescently labeled import substrate containing an NLS, and exogenous cytosol or recombinant import factors. After the transport reaction, cells are imaged by fluorescence microscopy to visualize the nuclear accumulation of substrate. For quantification, a digital fluorescence image can be obtained and the average nuclear fluorescence of 50–100 individual cells can be determined (*see* **ref. *16***). A much more rapid method for the quantification of nuclear import in permeabilized cells uses flow cytometry *(17)*, allowing the analysis of a large number of cells (~10,000) in a short period of time (<1 min). It is therefore particularly useful for detailed biochemical analysis of transport (*see* **refs. *17*** and ***18***).

Here, we describe an efficient in vitro system for the analysis of CRM1-dependent export of green-fluorescent-protein-labeled NFAT (GFP-NFAT) from the nuclei of permeabilized cells *(10)*. The molecular characteristics of NFAT allow tight control of its import and export signals in vivo and in vitro, which is important for activating its export and preventing its re-import in vitro. The nuclear import of fluorescently labeled protein substrates can be analyzed concurrently in the same cells, allowing a comprehensive analysis of nuclear import/export in a single reaction. This assay has been used to identify Ran, CRM1, and RanBP1 as major soluble factors in nuclear export in permeabilized cells ***(10,11)***. We have also used it for the characterization of novel Ran mutants ***(19)*** and for the analysis of the effects of protein phosphorylation on the nuclear import and export machinery ***(20)***.

2. Materials

2.1. HeLa Cells Stably Transfected with GFP-NFAT

The coding sequence of human NFAT was PCR-amplified from pSH107c (kindly provided by G.R. Crabtree, Stanford, CA), and cloned via *Hind*III and *Bam*HI into a pS65T-C1 (Clontech)-based eukaryotic expression vector, behind the coding sequence of a modified version of GFP with enhanced fluorescence characteristics (S65T). HeLa cells were grown in Dulbecco's Modified Eagle's medium (DMEM), containing 10% fetal bovine serum (FBS), penicillin (100 U/mL) and streptomycin (100 µg/mL) on plastic dishes. Transfections were performed with Lipofectamine (Gibco-BRL), using 25 µg AseI-linearized plasmid per ~2×10^6 adherent cells. Approximately 30 G418 (Gibco-BRL) resistant clones were pooled and induced overnight with 250 nm trichostatin A (TSA; Wako BioProducts) to express GFP-NFAT (*see* **Note 1**). Cells expressing high levels of GFP-NFAT were enriched by fluorescence-activated cell sorting (FACS; *see* **Note 2**). Positive cells were expanded and frozen in aliquots.

2.2. Preparation of Cytosol

a. Lysis buffer: 5 mM Hepes-KOH, pH 7.3, 10 m*M* KOAc, 2 m*M* $Mg(OAc)_2$, 2 m*M* dithiothreitol (DTT), 1 m*M* PMSF, 1 µg/mL of each leupeptin, pepstatin, and aprotinin.
b. Transport buffer: 20 m*M* HEPES-KOH, pH 7.3, 110 m*M* KOAc, 2 m*M* $Mg(OAc)_2$, 1 m*M* EGTA, DTT and protease inhibitors as in **item a** above (*see* **Note 3**).

1. Grow HeLa cells to mid-log phase in suspension culture in Joklik's modified S-MEM (Gibco-BRL), containing 10% fetal bovine serum (FBS), 100 U/mL penicillin and 100 µg/mL streptomycin, and collect them by centrifugation at 300*g* for 15 min (3–10 L can easily be handled). All of the following steps are performed on ice or at 4°C.
2. Wash 2× with phosphate-buffered saline (PBS) and once with transport buffer.
3. Resuspend in 1 vol of lysis buffer and swell for 10 min on ice.
4. Lyse the cells in a dounce homogenizer (*see* **Note 4**). After centrifugation at 1,500*g* for 15 min, the supernatant is cleared by centrifugation at 120,000*g* for 1 h.
5. Dialyze overnight against transport buffer (10,000 K_d cutoff), freeze the resulting cytosol (~10 mg/mL) in aliquots in liquid nitrogen and store at –80°C.

2.3. Preparation of Recombinant Transport Factors

2.3.1. Ran

a. Ran buffer: 50 m*M* Tris-HCl, pH 8.0, 75 m*M* NaCl, 1 m*M* MgC12, 0.1 m*M* phenylmethyl sulfonyl fluoride (PMSF), 1 µg/mL of each leupeptin, pepstatin, and aprotinin.

Various methods for the preparation of Ran have been described in the literature (*see* **refs. *21*** and ***22***). We typically use a modified version of the one described by Melchior et al. ***(23)***. Wild-type Ran is expressed from pET11d in BL21-(DE3) cells.

1. Grow 2 L of a bacterial culture to an OD_{600} of 0.6 and induce expression with 0.5 m*M* isopropylthio-β-Δ-galactside (IPTG) for 3 h at 37°C. All subsequent steps are performed on ice or at 4°C.
2. Harvest the cells and resuspend them in 100 mL of Ran-buffer containing 1 mg/mL Lysozyme (Sigma) and 1 µg/mL DNase I (Sigma). After one freeze-thaw cycle, leave the cells on ice for 20–30 min (*see* **Note 5**) and then centrifuge the lysates at 100,000*g* for 30 min.
3. Incubate the supernatant in batch with 70 mL DEAE Sepharose (Amersham Pharmacia Biotech), equilibrated in Ran buffer, for 1 h with gentle agitation. Pour the slurry into a column and collect the flowthrough and approx 100 mL of the subsequent wash (Ran buffer; wash until no protein can be detected).
4. Precipitate Ran from the DEAE flowthrough with ammonium sulfate: add salt gradually to 55% saturation, stir for 2 h, and centrifuge 20 min at 100,000*g*. The precipitate is then resuspended in 6–8 mL transport buffer containing 250 µ*M* GDP, and clarified by centrifugation at 14,000*g* for 15 min.
5. Load onto a preparative S200 column (Amersham Pharmacia Biotech), equilibrated in transport buffer. Collect 5-mL fractions and analyze them by SDS-PAGE.
6. Pool Ran-containing fractions, freeze in aliquots in liquid nitrogen, and store at –80°C.

The final concentration of Ran (ca. 95% pure) should be 1–2 mg/mL. Some Ran mutants (e.g., RanQ69L, *see* **ref. *11***) can be expressed and purified using the same method.

2.3.2. CRM1

a. CRM1 buffer: 50 m*M* HEPES, pH 8, 500 m*M* NaCl, 2 mM MgC12, 1 m*M* PMSF, 5 µg/mL each of aprotinin, leupeptin, and pepstatin.

The preparation of His-tagged CRM1 from pQE60-CRM1 (kindly provided by I. Mattaj, EMBL, Heidelberg, Germany) has been described in detail (*see* **ref. *24***).

1. Express the protein at 37 °C in TG1 cells without induction by IPTG.
2. Harvest bacteria, resuspend them in cold CRM1 buffer, and lyse them by sonification. All following steps are performed on ice or at 4°C.
3. Clear the lysate by centrifugation at 100,000*g* for 45 min.
4. Precipitate CRM1: add saturated ammonium sulfate (ca. 4.1 *M*) to the supernatant to a final concentration of 1.4 *M*. Stir for 2 h. Collect the precipitate by centrifugation (10,000*g* for 20 min) and resuspend pellet in CRM1 buffer containing 1 m*M* imidazole and 20 µg/mL DNase I (Sigma).
5. Incubate with Talon beads (Clontech) for 1 h at 4°C and wash the beads with lysis buffer containing 10 m*M* imidazole.
6. Elute CRM1 with CRM1 buffer containing 50–100 m*M* imidazole, dialyze against transport buffer, and freeze in aliquots in liquid nitrogen. Store at –80°C.

2.4. Cy5-BSA-NLS

1. Couple BSA (fatty-acid free; 2.5 mg in 1 mL 0.1 *M* Na_2CO_3) to activated Cy5 (monoreactive CyDye FluoroLink, 1 vial, as provided by Amersham Pharmacia Biotech; **Note 6**) for 40 min at room temperature.
2. Separate the Cy5-BSA conjugate from the free dye by chromatography on a PD-10 column (Amersham Pharmacia Biotech), equilibrated with PBS.
3. Activate Cy5-BSA by incubating with 2 m*M* sulfo-SMCC (Pierce Chemical Company; prepare 10 m*M* stock in DMSO) for 30 min at room temperature. Remove free crosslinker using a PD-10 column, as in **item 2** above.
4. Reduce the NLS-peptide derived from the sequence of the SV40 large T antigen (CGGGPKKKRKVED; ***25,26***): add 10 mg of DTT to 1 mL of peptide (10 mg/mL in 50 m*M* HEPES-KOH, pH 7.0). Incubate for 1 h at room temperature and separate the peptide from DTT, using a G10 column (10 mL), equilibrated with 1% acetic acid. Fractions containing reduced peptide (as determined by a sensitive peptide/protein detection system) are lyophilized in 10 aliquots, assuming 100% recovery, and stored at –20°C (*see* **Note 7**).
5. Redissolve 1 mg of peptide with activated Cy5-BSA, incubate the solution overnight at 4°C, and remove free peptide from the Cy5-BSA-NLS conjugate by chromatography on a PD-10 column, equilibrated with transport buffer.
6. Freeze the conjugate in liquid nitrogen and store at –80°C. After thawing, the import substrate can be kept at 4°C in the dark for a few weeks. Other fluorescent import ligands (such as FITC-BSA-NLS and Cy5-GST-NLS) may be prepared similarly ***(10,17)***.

2.5. Additional Reagents for Transport Assays

1. ATP-regenerating system: 100 m*M* ATP (Sigma) in 20 m*M* HEPES, 100 m*M* $Mg(OAc)_2$. Adjust pH to 7.4 with NaOH (*see* **Note 8**). 80 mg/mL creatine phosphate (Calbiochem) in H_2O. 2000 U/mL creatine phosphokinase (Calbio-

chem) in 50% glycerol, 20 m*M* HEPES, pH 7.4. Store at –20°C and mix 2:2:1 (ATP:creatine phosphate:creatine phosphokinase) before use.

2. Trichostatin A (TSA, Wako BioProducts; *see* **Note 1**): dissolve at 1 m*M* in DMSO. Dilute to 100 μ*M* with sterile PBS and freeze in aliquots. Store at –80°C.
3. Ionomycin (Calbiochem): prepare stock (1 m*M* in DMSO) and store at –80°C.
4. Digitonin (Calbiochem): prepare stock (1% in DMSO), store at –20°C.
5. Trypan blue (Sigma). Use 1:1 with cell suspension.
6. Oligonucleotide (*see* **Note 9**): dissolve desalted oligonucleotides (5′AGAG GAAAATTTGTTTCATA and 5′ TATGAAACAAATTTTCCTCT), each at 200 μ*M*, in 40 m*M* Tris-HCl, pH 7.4, 20 m*M* $MgCl_2$, 50 m*M* NaCl, and anneal by heating to 65°C for 5 min and slow cooling to room temperature. Freeze in aliquots and store at –20°C.
7. Wheat-germ agglutinin (WGA, Sigma): WGA, which inhibits most nucleocytoplasmic transport pathways, is dissolved in transport buffer at 1 mg/mL and stored at –80°C.
8. NES-peptide: The NES-peptide of the minute virus of mice (CVDEMTKKFGTLTIHDTEK *(27)* is dissolved in transport buffer at 1 mg/mL and stored at –20°C.
9. Leptomycin B (LMB). (LMB; obtained from B. Wolff, Novartis, Vienna, Austria) is dissolved in ethanol (1 m*M*) and stored at –20°C. It is now commercially available from Sigma (L2913).
10. If cells are to be preincubated in order to release nuclear transport factors, normal transport buffer (*see* **Subheading 2.2.**) is substituted by transport buffer containing 30 m*M* LiOAc and 80 m*M* KOAc instead of 110 m*M* KOAc (*see* **Note 10**).

3. Methods

The principles of the nuclear export/import assay with digitonin-permeabilized cells are outlined schematically in **Fig. 1**. HeLa cells, stably transfected with GFP-NFAT, are grown as in **Subheading 2.1.** in plastic dishes (containing cover slips or multitest slides (ICN), if analysis is to be done by fluorescence microscopy), and are stimulated to express GFP-NFAT by the addition of 100–200 nm trichostatin A (TSA). After incubation overnight, nuclear import of the reporter protein is induced with 1 μ*M* ionomycin and 30 m*M* LiOAc for 30 min (*see* **Note 10**). Subsequently, the cells are permeabilized with digitonin and nuclear export of GFP-NFAT, as well as nuclear import of a fluorescently labeled NLS-containing cargo, are achieved with an appropriate in vitro incubation.

3.1. Analysis of Transport by Fluorescence Microscopy

See also **refs. *10*** and ***15***.

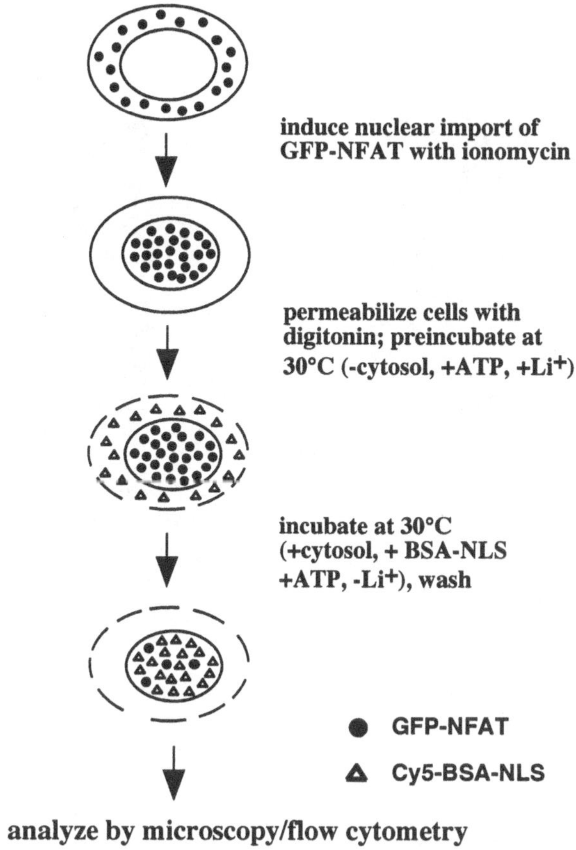

Fig. 1. Flow diagram depicting steps in the analysis of nuclear import and export in vitro. Transport of GFP-NFAT *(filled circles)* and the synthetic substrate Cy5-BSA-NLS *(empty triangles)* can be analyzed concurrently in the same cells using either fluorescence microscopy or flow cytometry, since these two reporters differ in their fluorescence emission.

1. Wash cells with cold transport buffer and permeabilize with digitonin (final concentration of approx 30 μg/mL in transport buffer) on ice. The required amount should be tested on a separate slide for each cell type, using Trypan blue; *see* **Note 4**).
2. Subject cells to preincubation and transport reaction, using reagents as outlined in **Subheading 3.2., steps 3** and **4**. Use ca. 12 μL solution per well of a 10x multitest slide and incubate at 30°C in a moisturized chamber.
3. Wash with cold transport buffer. Fix cells with 3.7% formaldehyde in PBS. You may also do an indirect immunofluorescence labeling (for example, for CRM1 or Ran) at this step.
4. Analyze cells by fluorescence microscopy, using appropriate filters.

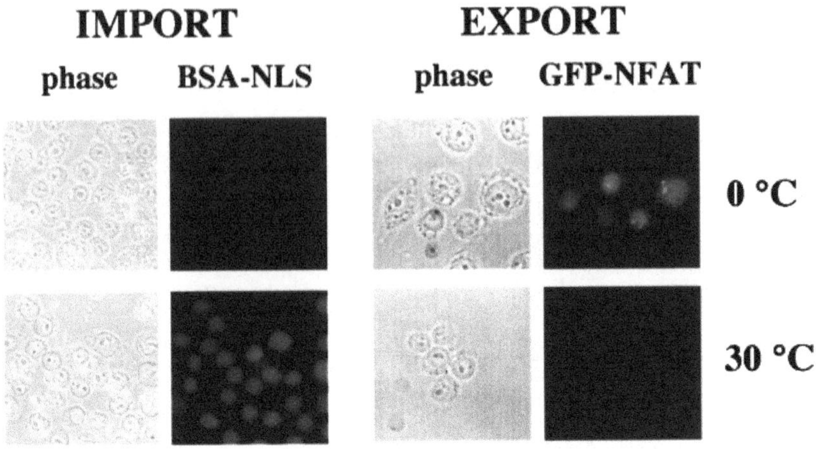

Fig. 2. Analysis of nuclear import and nuclear export by fluorescence microscopy. Left panels: import of a synthetic substrate, FITC-BSA-NLS. Right panels: export of GFP-NFAT. Reactions at 0°C *(top)*, where most transport is blocked, and at 30°C *(bottom)*, which supports efficient transport, are shown for comparison.

An example of nuclear import and export reactions analyzed by fluorescence microscopy on permeabilized adherent HeLa cells is shown in **Fig. 2**.

3.2. Analysis of Transport by Flow Cytometry

See also **refs. *10*** and ***17***.

1. Trypsinize cells: wash with PBS, add trypsin containing 1 µ*M* ionomycin and 30 m*M* LiOAc (2 mL for a 15-cm dish). Remove most of the trypsin solution immediately to minimize carryover of the protease. Incubate at 37°C until cells detach. Collect cells in cold transport buffer containing 5% fetal bovine serum (FBS). Centrifuge for 5 min at 300*g* at 4°C and wash once in transport buffer.
2. Resuspend cells in transport buffer or transport buffer containing LiOAc (*see* **Note 10**) at 10^7/mL. Add digitonin to 100 µg/mL (1 µL of a 1% stock per 10^6 cells). Leave on ice for 3 min and check permeabilization with Trypan blue, using 5 µL of cells. Add more digitonin if less than 90–95% of the nuclei appear blue. Wash with transport buffer (+/– LiOAc) and centrifuge as in **step 1**.
3. Preincubation (optional): Resuspend cells in transport buffer (+ LiOAc) at 10^7/mL and add 25 µL ATP-regenerating system per mL of cell suspension. Incubate 15 min in a 30°C water bath. Wash cells with transport buffer (–LiOAc). The preincubation results in depletion of transport factors (CRM1) that initially remain associated with the permeabilized cells. Without the preincubation, Ran will be the only rate-limiting factor for CRM1-dependent export ***(10)***.
4. Resuspend cells in transport buffer (–LiOAc) at 3×10^7/mL. The transport reaction (40 µL in a 5-mL fluorescence-activated cell sorting (FACS) tube)

should contain: 10 µL of cells (300,000), 1 µL ATP-regenerating system, 0.2 µL annealed oligonucleotide (final concentration: 1 µ*M*), ca. 1 µL Cy5-BSA-NLS (*see* **Note 6**). Add cytosol (2.5 mg/mL usually yields optimal transport) and/or recombinant transport factors. For nuclear import, the amount of cytosol or transport factors affects the background signal obtained in the 0°C control, as the fluorescent import cargo may stick to cytoplasmic structures (*see* **Note 11**).

5. Incubate at 0°C or in a 30°C water bath. Transport should be linear for approx 30 min.
6. Add 4.5 mL cold transport buffer. Centrifuge for 5 min at 400*g* and 4°C. Remove most of supernatant by aspiration, leaving ca. 200 µL behind.
7. Analyze transport by flow cytometry. We use the FACS Calibur configuration (Becton Dickinson), detecting GFP-NFAT in FL1 and Cy5-BSA-NLS in FL4 (*see* http://facs.scripps.edu/facslab.html; **Note 12**) and typically count 10,000 cells. In most cases, the median fluorescence is best suited for statistical analysis of transport. Export reactions may be standardized by assigning a GFP-NFAT fluorescence value of 100 to a 0°C control. For nuclear import, the 30°C reaction with the highest Cy5-BSA-NLS signal is usually assigned a fluorescence value of 100.

The set up and analysis of up to 100 reactions should take about 3–4 h. Typical results for nuclear transport reactions as analyzed by flow cytometry are shown in **Figs. 3** and **4**. The physiological significance of the results obtained for nuclear export can be validated by using specific inhibitors of nuclear transport. Wheat-germ agglutinin (WGA) is a lectin that binds to various O-glycosylated nucleoporins *(28)* and is known to inhibit the majority of nuclear transport pathways *(29,30)*. WGA inhibits both nuclear import and export at similar concentrations (**Fig. 3A**). Leptomycin B (LMB) is a fungal metabolite that covalently binds to the export receptor CRM1, inhibiting its interaction with NES-cargo molecules *(31)*. It therefore inhibits CRM1-dependent export *(32)*, but not importin β-dependent import (**Fig. 3B**; *see* **Note 13**). CRM1 has different affinities for different leucine-rich NESs *(27)*. The NES of the NS2-peptide of minute virus of mice (which in fact contains only a single leucine) has a relatively high affinity for CRM1 *(27)*. It is therefore a potent inhibitor of CRM1-dependent export, but not of nuclear import (**Fig. 3C**; *see* **Note 13**). Taken together, these controls validate our in vitro assay as a transport system that faithfully reconstitutes nuclear transport in permeabilized cells.

Digitonin-permeabilized cell assays have proven to be very useful for identifying and analyzing shuttling nuclear transport factors by biochemical reconstitution. Ran and CRM1 are required for nuclear export in vivo *(6,7,33)*. Recombinant Ran and highly purified CRM1 obtained from HeLa cytosol promote nuclear export in vitro in permeabilized cells that are preincubated to deplete cytosolic factors *(10)*. Recombinant CRM1 also stimulates nuclear

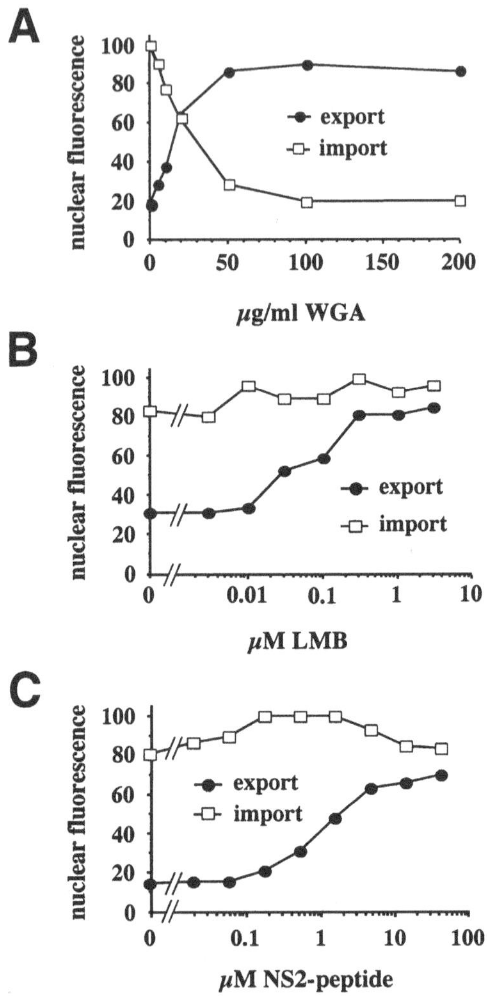

Fig. 3. Effects of various inhibitors on concurrent nuclear import and export in permeabilized cells, as monitored by flow cytometry. The import substrate is Cy5-BSA-NLS, and the export substrate is GFP-NFAT. Note that nuclear import is represented by an increase in nuclear fluorescence, whereas nuclear export is denoted as a decrease in nuclear fluorescence. Wheat-germ agglutinin (WGA; **A**), which binds to proteins of the NPC, strongly inhibits both nuclear import and export. By contrast, leptomycin B (LMB; **B**) and a peptide derived from the NES of NS2 protein from minute virus of mice **(C)** selectively inhibit nuclear export and not import.

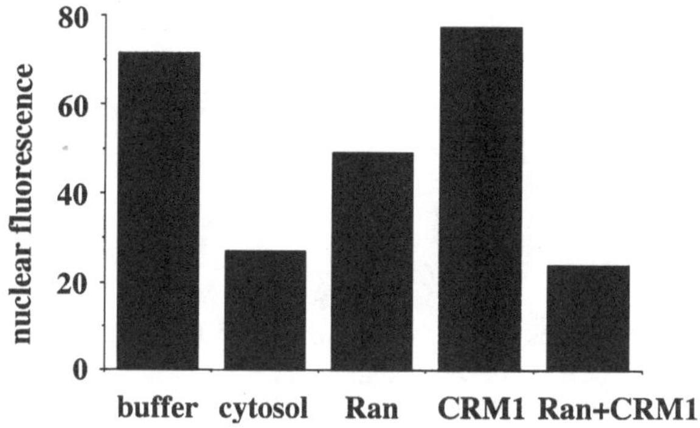

Fig. 4. Reconstitution of nuclear export with recombinant transport factors. Permeabilized cells were preincubated at 30°C in the presence of an ATP-regenerating system to deplete shuttling nuclear transport factors. Nuclear export in a subsequent transport reaction is strongly stimulated by cytosol, which contains both Ran and CRM1, or by a combination of Ran (25 μg/mL) and CRM1 (35 μg/mL), but not by CRM1 alone. Ran alone stimulates export to an intermediate level.

export of GFP-NFAT, together with Ran (**Fig. 4**). Ran without CRM1 stimulates export to an intermediate level, probably because some CRM1 remains associated with the nuclear fraction, even after preincubation of the permeabilized cells. Similar reconstitution approaches can be used with mutant versions of nuclear transport factors to analyze the specific steps and requirements for individual factors ***(11)***.

4. Notes

1. TSA is a histone deacetylase inhibitor that stimulates the expression of certain transfected genes ***(34)***. In HeLa cells, it induces the expression of GFP-NFAT from the CMV-promotor half-maximally at ~100 nm. Sodium butyrate at much higher concentrations (~10 m*M*) is much cheaper and has similar effects, although it has a noxious odor.
2. We used a FACStar Plus sorter (Becton Dickinson; *see* http://facs.scripps.edu/facslab.html; here you may also find useful links for cell sorting and flow cytometry). Typically, we sort the 10% brightest cells, expand and resort them, and freeze them in aliquots. These cells retain a high level of fluorescence (after induction with TSA) for ~3 wk in culture. Sorting of transfected cells results in a mixed population, with cells originating from various clones. This avoids possible problems with clonal variations.
3. EGTA inhibits the cellular phosphatase calcineurin, which dephosphorylates NFAT, resulting in nuclear import of the protein ***(35)***. Thus, re-import of the reporter protein, which would complicate the analysis, is prevented under our conditions.

4. Check permeabilization with Trypan blue, which stains the nuclei. Alternatively, the plasma membrane can be permeabilized with digitonin. The nuclear membrane remains intact, because of the lower level of cholesterol, as compared to the plasma membrane. Add 0.5–1 μL of a 1% solution of digitonin in DMSO for 10^6 cells.
5. Leave on ice and/or add more DNaseI until the DNA is sheared. Sonication may lead to aggregation of Ran.
6. Cy5 has maxima of absorption and fluorescence emission at 649 and 670 nm, respectively. Other dyes such as Cy2 (489 and 506 nm) or Cy3 (550 and 570 nm) may be useful, depending on the available cytometer configuration.
7. The reduction step may be optional, depending on the preparation of the peptide. In any case, the efficiency of coupling should be checked by SDS-PAGE, comparing coupled and uncoupled Cy5-BSA. Coupling results in a shift in molecular mobility, allowing the estimation of the average number of peptides per BSA-molecule. This number should not be higher than 10, as too many fluorophore-containing peptides increases the stickiness of the transport substrate, resulting in high background values.
8. Use sodium-ATP. Lithium-ATP may interfere with export of GFP-NFAT (*see* **Note 10**).
9. This double-stranded oligonucleotide mimics the DNA-binding site of NFAT. At 1 μ*M*, it stimulates export of GFP-NFAT about twofold, probably by promoting the release of GFP-NFAT from chromatin.
10. Li^+ inhibits one of the kinases, GSK-3 ***(36)***, which is involved in the rephosphorylation of nuclear NFAT. Rephosphorylation is required for the efficient export of NFAT. LiOAc may therefore be included under certain conditions (for example, during preincubation of permeabilized cells to deplete transport factors from the nucleus), in order to prevent concomitant export of the reporter protein GFP-NFAT.
11. The optimal amount of Cy5-BSA-NLS should be determined as follows: prepare two transport reactions containing 2.5 mg/mL of cytosol with 0.2, 0.4, 0.7, 1, 2 and 3 μL of import substrate each. Incubate at 0°C or 30°C for 30 min. Determine the amount of import substrate that yields the maximal fold stimulation of nuclear import (30°C vs. 0°C), as determined by flow cytometry. In the absence of cytosol, the background signal tends to be higher, because of unspecific cytoplasmic labeling.
12. Other combinations are possible. For example, one may use propidium iodide (PI; detected in FL3) to determine the DNA content of the cells. This can be done for nuclear import or export, allowing analysis of nuclear transport with respect to the cell cycle ***(20)***.
13. Inhibitors of nuclear export such as LMB or the NS2-peptide may have a small effect on nuclear import (in our case a stimulatory), as the different transport pathways may converge at some point at the NPC. Here, various transport receptors may compete for binding sites at certain nucleoporins ***(37)***.

Acknowledgments

We are grateful to Dr. Gerald Crabtree and Dr. Iain Mattaj for providing the cDNAs for NFAT and CRM1, respectively. We also thank Dr. Barbara Wolff for the gift of Leptomycin-B.

References

1. Stoffler, D., Fahrenkrog, B., and Aebi, U. (1999) The nuclear pore complex: from molecular architecture to functional dynamics. *Curr. Opin. Cell. Biol.* **11,** 391–401.
2. Ryan, K. J. and Wente, S. R. (2000) The nuclear pore complex: a protein machine bridging the nucleus and cytoplasm. *Curr. Opin. Cell. Biol.* **12,** 361–371.
3. Görlich, D. and Kutay, U. (1999) Transport between the cell nucleus and the cytoplasm. *Annu. Rev. Cell. Dev. Biol.* **15,** 607–660.
4. Mattaj, I. W. and Englmeier, L. (1998) Nucleocytoplasmic transport: the soluble phase. *Annu. Rev. Biochem.* **67,** 265–306.
5. Henderson, B. R. and Percipalle, P. (1997) Interactions between HIV Rev and nuclear import and export factors: the Rev nuclear localisation signal mediates specific binding to human importin-beta. *J. Mol. Biol.* **274,** 693–707.
6. Fornerod, M., Ohno, M., Yoshida, M., and Mattaj, I. W. (1997) CRM1 is an export receptor for leucine-rich nuclear export signals. *Cell* **90,** 1051–1060.
7. Fukuda, M., Asano, S., Nakamura, T., Adachi, M., Yoshida, M., Yanagida, M., et al. (1997) CRM1 is responsible for intracellular transport mediated by the nuclear export signal. *Nature* **390,** 308–311.
8. Ossareh-Nazari, B., Bachelerie, F., and Dargemont, C. (1997) Evidence for a role of CRM1 in signal-mediated nuclear protein export. *Science* **278,** 141–144.
9. Stade, K., Ford, C. S., Guthrie, C., and Weis, K. (1997) Exportin 1 (Crm1p) is an essential nuclear export factor. *Cell* **90,** 1041–1050.
10. Kehlenbach, R. H., Dickmanns, A., and Gerace, L. (1998) Nucleocytoplasmic shuttling factors including Ran and CRM1 mediate nuclear export of NFAT In vitro. *J. Cell Biol.* **141,** 863–874.
11. Kehlenbach, R. H., Dickmanns, A., Kehlenbach, A., Guan, T., and Gerace, L. (1999) A role for RanBP1 in the release of CRM1 from the nuclear pore complex in a terminal step of nuclear export. *J. Cell Biol.* **145,** 645–657.
12. Kaffman, A. and O'Shea, E. K. (1999) Regulation of nuclear localization: a key to a door. *Annu. Rev. Cell. Dev. Biol.* **15,** 291–339.
13. Newmeyer, D. D., Finlay, D. R., and Forbes, D. J. (1986) In vitro transport of a fluorescent nuclear protein and exclusion of non-nuclear proteins. *J. Cell Biol.* **103,** 2091–2102.
14. Newmeyer, D. D., Lucocq, J. M., Burglin, T. R., and De Robertis, E. M. (1986) Assembly in vitro of nuclei active in nuclear protein transport: ATP is required for nucleoplasmin accumulation. *EMBO J.* **5,** 501–510.
15. Adam, S. A., Marr, R. S., and Gerace, L. (1990) Nuclear protein import in permeabilized mammalian cells requires soluble cytoplasmic factors. *J. Cell Biol.* **111,** 807–816.

16. Chi, N. C., Adam, E. J., Visser, G. D., and Adam, S. A. (1996) RanBP1 stabilizes the interaction of Ran with p97 nuclear protein import. *J. Cell Biol.* **135,** 559–569.
17. Paschal, B. M. and Gerace, L. (1995) Identification of NTF2, a cytosolic factor for nuclear import that interacts with nuclear pore complex protein p62. *J. Cell Biol.* **129,** 925–937.
18. Mahajan, R., Delphin, C., Guan, T., Gerace, L., and Melchior, F. (1997) A small ubiquitin-related polypeptide involved in targeting RanGAP1 to nuclear pore complex protein RanBP2. *Cell* **88,** 97–107.
19. Kehlenbach, R.H., Assheuer, R., Kehlenbach, A., Becker, J., and Gerace, L. Stimulation of nuclear export and inhibition of nuclear import by a Ran mutant deficient in binding to Ran-binding Protein 1. *J. Biol. Chem.* **276(17),** 14,524–14,531.
20. Kehlenbach, R. H. and Gerace, L. (2000) Phosphorylation of the nuclear transport machinery down-regulates nuclear protein import in vitro. *J. Biol. Chem.* **275,** 17,848–17,856.
21. Klebe, C., Nishimoto, T., and Wittinghofer, F. (1993) Functional expression in Escherichia coli of the mitotic regulator protein $p24^{ran}$ and $p45^{rcc1}$ and fluorescence measurements of their interaction. *Biochemistry* **32,** 11,923–11,928.
22. Bischoff, F. R. and Ponstingl, H. (1995) Catalysis of guanine nucleotide exchange of Ran by RCC1 and stimulation of hydrolysis of Ran-bound GTP by Ran-GAP1. *Methods Enzymol.* **257,** 135–144.
23. Melchior, F., Sweet, D. J., and Gerace, L. (1995) Analysis of Ran/TC4 function in nuclear protein import. *Methods Enzymol.* **257,** 279–291.
24. Guan, T., Kehlenbach, R. H., Schirmer, E. C., Kehlenbach, A., Fan, F., Clurman, B. E., et al. (2000) Nup50, a nucleoplasmically oriented nucleoporin with a role in nuclear protein export. *Mol. Cell. Biol.* **20,** 5619–5630.
25. Kalderon, D., Roberts, B. L., Richardson, W. D., and Smith, A. E. (1984) A short amino acid sequence able to specify nuclear location. *Cell* **39,** 499–509.
26. Dingwall, C. and Laskey, R. A. (1991) Nuclear targeting sequences—a consensus? *Trends Biochem. Sci.* **16,** 478–481.
27. Askjaer, P., Bachi, A., Wilm, M., Bischoff, F. R., Weeks, D. L., Ogniewski, V., et al. (1999) RanGTP-regulated interactions of CRM1 with nucleoporins and a shuttling DEAD-box helicase. *Mol. Cell. Biol.* **19,** 6276–6285.
28. Hanover, J. A., Cohen, C. K., Willingham, M. C., and Park, M. K. (1987) O-linked N-acetylglucosamine is attached to proteins of the nuclear pore. Evidence for cytoplasmic and nucleoplasmic glycoproteins. *J. Biol. Chem.* **262,** 9887–9894.
29. Yoneda, Y., Imamoto-Sonobe, N., Yamaizumi, M., and Uchida, T. (1987) Reversible inhibition of protein import into the nucleus by wheat germ agglutinin injected into cultured cells. *Exp. Cell Res.* **173,** 586–595.
30. Dargemont, C. and Kuhn, L. C. (1992) Export of mRNA from microinjected nuclei of Xenopus laevis oocytes. *J. Cell Biol.* **118,** 1–9.
31. Kudo, N., Matsumori, N., Taoka, H., Fujiwara, D., Schreiner, E. P., Wolff, B., et al. (1999) Leptomycin B inactivates CRM1/exportin 1 by covalent modification at a cysteine residue in the central conserved region. *Proc. Natl. Acad. Sci. USA* **96,** 9112–9117.

32. Wolff, B., Sanglier, J. J., and Wang, Y. (1997) Leptomycin B is an inhibitor of nuclear export: inhibition of nucleo-cytoplasmic translocation of the human immunodeficiency virus type 1 (HIV-1) Rev protein and Rev-dependent mRNA. *Chem. Biol.* **4,** 139–147.
33. Richards, S. A., Carey, K. L., and Macara, I. G. (1997) Requirement of guanosine triphosphate-bound ran for signal-mediated nuclear protein export. *Science* **276,** 1842–1844.
34. Arts, J., Lansink, M., Grimbergen, J., Toet, K. H., and Kooistra, T. (1995) Stimulation of tissue-type plasminogen activator gene expression by sodium butyrate and trichostatin A in human endothelial cells involves histone acetylation. *Biochem. J.* **310,** 171–176.
35. Shibasaki, F., Price, E. R., Milan, D., and McKeon, F. (1996) Role of kinases and the phosphatase calcineurin in the nuclear shuttling of transcription factor NF-AT4. *Nature* **382,** 370–373.
36. Klein, P. S. and Melton, D. A. (1996) A molecular mechanism for the effect of lithium on development. *Proc. Natl. Acad. Sci. USA* **93,** 8455–8459.
37. Kutay, U., Izaurralde, E., Bischoff, F. R., Mattaj, I. W., and Görlich, D. (1997) Dominant-negative mutants of importin-beta block multiple pathways of import and export through the nuclear pore complex. *EMBO J.* **16,** 1153–1163.

20

Characterization of the Effects of RanGTP on the Microtubule Cytoskeleton

Sofia B. Lizarraga, Yixian Zheng, and Andrew R. Wilde

1. Introduction

Ran is a GTPase of the Ras superfamily. Until recently, its primary role was believed to be in interphase, regulating nucleocytoplasmic transport *(1)*. However, recent studies have identified additional roles for Ran in regulating spindle assembly and maintenance in mitosis *(2–10)*. In addition, Ran regulates nuclear-membrane assembly upon exit from mitosis *(11,12)*. This chapter focuses on the regulatory role of Ran in spindle assembly and maintenance, outlining techniques used to study the effects of Ran on the microtubule cytoskeleton.

The initial indication that Ran was involved in regulating the microtubule cytoskeleton came with the identification of a RanGTP-binding protein—RanBPM—which localizes to the centrosome and induces ectopic aster formation when overexpressed in tissue-culture cells *(13)*. Subsequently, the role of Ran in spindle assembly was directly tested in cytostatic factor (CSF)-arrested *Xenopus* egg extract using activated alleles of Ran, which mimic Ran in the guanosine triphosphate (GTP)-bound state, and dominant-negative alleles of Ran, which mimic Ran in the guanosine diphosphate (GDP)-bound state. The activated alleles of Ran, RanG19V *(2,4)*, RanL43E *(3)*, or RanQ69L *(5)*, induced the formation of microtubule asters and bipolar spindles in CSF-arrested *Xenopus* egg extract in the absence of chromatin and centrosomes, whereas the dominant-negative alleles did not.

Chromatin exerts stabilizing and organizational effects on microtubules *(14–16)*. Recently, a potential mechanism for the chromatin-derived effects was revealed when microtubule spindle formation by demembranated sperm

From: *Methods in Molecular Biology, vol. 189: GTPase Protocols: The Ras Superfamily*
Edited by: E. J. Manser and T. Leung © Humana Press Inc., Totowa, NJ

was found to be dependent upon the activity of the Ran nucleotide-exchange factor RCC1 *(4,5)*. Since RCC1 is bound to the chromatin, it was hypothesized that the generation of RanGTP around the mitotic chromatin was required for the assembly of the bipolar spindle in the vicinity of the chromatin. Therefore, the effect of the chromatin on microtubule production and organization now appears to be at least partially caused by the generation of RanGTP by chromatin-associated RCC1.

Recently, RanGTP was shown to induce microtubule production and organization by affecting microtubule dynamics, microtubule nucleation, microtubule motor activities ***(9,10)***, and spindle-pole assembly ***(6–8)***. The mechanism by which Ran mediates these effects on the microtubules parallels the mechanism by which Ran regulates nucleocytoplasmic transport. In general, the targeting of proteins to the nucleus requires them to interact with a nuclear-trafficking receptor, either importin α or importin β *(1)*. These receptors then facilitate the trafficking of the proteins to the nucleus. Once in the nucleus, binding of RanGTP to the receptor dissociates the cargo protein-receptor complex by reducing the affinity of the receptor for the cargo, thereby releasing the cargo into the nucleoplasm. Several proteins involved in spindle assembly are localized to the nucleus in interphase ***(17–20)***. It is possible that upon nuclear-envelope breakdown, these proteins may be free to once again interact with the nuclear-import receptors, an interaction that could inhibit their activity. Thus, it was hypothesized that the generation of RanGTP around the chromatin prevents the interaction between proteins involved in spindle assembly and nuclear-transport receptors, allowing spindle assembly to proceed in the vicinity of the chromatin ***(6–8)***.

This chapter describes various techniques that can be used to study the microtubule cytoskeleton in *Xenopus* egg extracts, including techniques that have been used to study the role of Ran in regulating spindle assembly through the modulation of microtubule dynamics and microtubule motor activities.

2. Materials

2.1. Buffers, Reagents, and Specialized Equipment

1. XB: 10 m*M* HEPES, 1 m*M* $MgCl_2$, 0.1 m*M* $CaCl_2$, 100 m*M* KCl, 50 m*M* sucrose, 5 m*M* EGTA, pH 7.7 with KOH.
2. BRB80: 80 m*M* PIPES, 1 m*M* $MgCl_2$, 1 m*M* EGTA, pH 6.8 with KOH.
3. VALAP: A 1:1:1 w/w/w mixture of vaseline: lanolin: beeswax. The VALAP is melted in a warm water bath prior to use.
4. Oxygen-scavenger system: A 1:1:1 v/v/v mix of glucose oxidase (G-1760 Sigma), catalase (C-40 Sigma) (both made as 10 mg/mL stocks in 50% glycerol in BRB80) and glucose (1 *M* in water). Solutions can be stored at –20°C for up to 1 wk, and should be defrosted immediately before use. The mix should be

kept on ice, and used within 1 h after being mixed. In addition, a freshly made saturated solution of hemoglobin (H-2500 Sigma ~40 mg/mL), can be added to a final concentration of ~4 mg/mL.

5. ATP regeneration system (x20): 20 m*M* MgATP, pH 7.4, 200 m*M* creatine phosphate, and 1 mg/mL creatine kinase.
6. Spindown tubes described in **refs.** ***21*** and ***22***: 15 mL Corex (Corning Glass Co.) tubes containing custom-made plexiglass removable inserts.
7. Cover slip spinner: Custom-made, for details on how to construct a cover slip spinner, *see* http://blocks.fhcrc.org/~kinesin/Methods/SpinnerBox.html.

2.2. Purification of Tubulin

We purify tubulin from bovine brains using the method described by Hyman et al ***(23)***. Three bovine brains yield 0.5–1 g of tubulin. A similar protocol can be used to purify tubulin from porcine brains. Tubulin can be stored at –80°C for up to 1 yr.

2.3. Purification of Centrosomes

Centrosomes are purified from mammalian tissue-culture cells as previously described (*see* **ref.** ***24***). Centrosomes are frozen and stored at -80°C at a final concentration of ~1.5–2× 10^7 centrosomes/mL. Centrosomes are competent for microtubule nucleation for at least 1 yr.

2.4. Bacterially Expressed Proteins in the Ran Pathway

Ran alleles can be expressed and purified from bacteria either free of tags ***(25)***, fused to glutathione-S-transferase (GST) ***(26)***, or fused to 6 histidines ***(27)***. Alternatively, small quantities of protein of the various Ran alleles can be purchased from Cytoskeleton (www.cytoskeleton.com). Other proteins in the Ran pathway can also be expressed and purified from bacteria, including RCC1 ***(25)***, RanBP1 ***(28)***, importin α, and importin β ***(29)***.

2.5. Fluorescently-Labeled Stabilized Microtubules

To visualize microtubules by fluorescence microscopy, purified tubulin can be labeled with different fluorophores without significantly affecting the activity of the tubulin. We commonly use rhodamine or fluorescein to label purified tubulin as described (*see* **ref.** ***23***). Alternatively, labeled tubulin can be purchased from Cytoskeleton (www.cytoskeleton.com). Labeled tubulin can be used to generate further reagents such as fluorophore-labeled taxol-stabilized microtubules, microtubule seeds, and polarity-marked microtubules (where the minus end of the microtubule is brightly labeled and the plus end is dimly labeled). The production of these reagents is described in **refs.** ***9*** and ***23***. Fluorophore-labeled tubulin is aliquoted, flash-frozen in liquid nitrogen, and stored at –80°C. Tubulin should be thawed quickly and kept on ice for no longer than 30 min prior to use.

2.5.1. Taxol-Stabilized Microtubules

The protocol we use is based on the one described at http://iccbweb.med.harvard.edu/mitchisonlab/.

1. Dilute purified tubulin solution (1:7 ratio of rhodamine-labeled tubulin to unlabeled tubulin) to 1 mg/mL final concentration, in BRB80 containing 1 m*M* dithiothreitol (DTT) and 1 m*M* GTP. Centrifuge in a Beckman TLA100 rotor at 360,000*g* for 5 min at 4°C to remove tubulin aggregates.
2. Add an equal vol of BRB80 containing 1 m*M* DTT, 1 m*M* GTP, and 20% dimethyl sulfoxide (DMSO).
3. Incubate at 37°C for 30 min to allow the tubulin to polymerize.
4. Pellet the microtubules over a 150-µL cushion of 40% glycerol in BRB80 (kept at room temperature) in a TLA100 Beckman rotor at 110,000*g* rpm for 15 min at 25°C.
5. Resuspend the microtubule pellet in 10 µ*M* taxol in BRB80 and store at room temperature. We recommend using fresh taxol-stabilized microtubules for each experiment.

2.5.2. Stabilized Microtubule Seeds

The method presented here is based on a previously described protocol ***(30)***. In addition, polarity-marked seeds can be made by combining this method with that described in **ref. *31***.

1. Remove tubulin aggregates as in **Subheading 2.5.1., item 1**.
2. Prepare a 200-µL reaction of 50% glycerol, 1X BRB80 (final), 5 m*M* $MgCl_2$, 1 m*M* GTP, and 3 mg/mL tubulin (1:7 ratio of rhodamine-labeled tubulin to unlabeled tubulin).
3. Incubate at 37°C for 20 min.
4. Pass solution through a 26-gauge needle to shear the microtubules into 2–3 µm fragments (this must be done by trial and error).
5. Incubate for 5 min at 37°C.
6. Add 20 µL of 37.5 m*M* ethylene glycol-*bis*-succinimidylsuccinate (EGS, Calbiochem) dissolved in DMSO.
7. Incubate at 37°C for 15 min.
8. Dilute the reaction into 1 mL of BRB80 containing 50% sucrose, 10 m*M* potassium glutamate, and 1 m*M* GTP. Collect seeds by centrifugation at 100,000*g* for 30 min.
9. Resuspend the pellet in BRB80 containing 1 m*M* GTP, aliquot, flash-freeze in liquid nitrogen, and store at –80°C.

2.6. Purification of Demembranated Xenopus Sperm

Demembranated sperm were purified as described (*see* **ref.** ***32***), flash-frozen in liquid nitrogen, and stored at –80°C at a final concentration of 50–100 sperm/nL.

2.7. Preparation of CSF-Arrested Xenopus Egg Extract

The protocol we use has been previously described in detail (*see* **ref.** ***32***). Prior to use, add the ATP regeneration system (1X final concentration, *see* **Subheading 2.1.**), to the extracts. Extracts should be handled with wide bore tips. For most experiments, we use extracts that have been frozen in liquid nitrogen and stored at –80°C. However, for bipolar spindle assembly and microtubule dynamics experiments, we use fresh egg extracts. To ensure the mitotic state of our extracts, we incubate extracts with recombinant nondegradable Cyclin-B (Cyclin B-Δ90, at ~0.1 mg/mL final concentration) for 30 min at room temperature, in the presence of an ATP regeneration system.

3. Methods

3.1. Microtubule Polymerization Assay

Several studies have shown that RanGTP can stimulate microtubule production in *Xenopus* egg extract ***(2–5)***. One possible explanation is that RanGTP directly stimulates tubulin to polymerize. However, we found that RanGTP did not directly stimulate microtubule polymerization from pure tubulin ***(3)***, suggesting that Ran regulates other factors that stimulate microtubule production. The assay described here can be used to test potential downstream factors of Ran for microtubule nucleating and polymerizing activity.

1. Before starting the experiment, prepare the spindown tubes (*see* **Subheading 2.1.**) by placing a cover slip on top of the insert. Add 2 mL BRB80 and underlay with 2 mL 10% glycerol in BRB80.
2. Prepare 5-μL reactions on ice containing 0.5× BRB80, 500 μ*M* GTP, 4 mg/mL tubulin (consisting of a 1:7 ratio of rhodamine-labeled tubulin to unlabeled tubulin), and various concentrations of test proteins. We used 25 μ*M* Ran, the concentration used to stimulate aster and spindle assembly in *Xenopus* egg extract ***(3)***.
3. Initiate the reactions at 30-s intervals by placing in a 37°C water bath.
4. Remove the reaction after 5–10 min from the water bath and gently mix with 45 μL of 1% glutaraldehyde in BRB80 (prepared fresh prior to use from a 25% glutaraldehyde stock stored at –20°C). Incubate for 3 min at room temperature.
5. Dilute the reaction in 30 vol of BRB80 and mix gently by inversion. Layer 10 μL of the diluted reaction on top of the BRB80 in the spindown tubes (*see* **Subheading 2.1.**).

6. Centrifuge in a swinging bucket rotor at 12,000*g* for 2 h at 20°C. Suitable rotors include the HB-4 and HB-6 (Sorvall) or JS13.1 (Beckman).
7. Aspirate most of the liquid from the tube, remove cover slip, and postfix in –20°C MeOH for 5 min. Rehydrate in TBS and mount in antifade medium as described (*see* **ref.** ***33***).

3.2. Microtubule Aster and Spindle-Formation Assay

To address the effect of various Ran alleles on microtubule aster and spindle formation, we used a previously described assay (*see* **refs.** ***3***, ***34***, and ***45***) with either purified centrosomes or demembranated sperm to initiate microtubule production and organization in *Xenopus* egg extract. We found that the addition of an activated allele of Ran, RanL43E (final concentration 25 µ*M*), to *Xenopus* egg extract stimulated aster and spindle assembly, whereas a dominant-negative allele of Ran, RanT24N did not ***(3)***. Similar results have been obtained with other activated alleles of Ran, including RanG19V ***(2,4)*** and RanQ69L ***(5)***. The assay is carried out in a method similar to that described in **Subheading 3.1.**; however, the centrifugation time, cushion, and temperature conditions are modified as follows:

1. Prepare the reaction as follows: 10 µL *Xenopus* egg extract containing 0.1 mg/mL rhodamine-labeled tubulin, 1 µL centrosomes (final concentration of 1.5×10^6 centrosomes/mL), or sperm (diluted to yield approx 200 sperm per cover slip), 1 µL of test protein (final concentration of 25 µ*M* for Ran proteins).
2. Place the reactions at room temperature at 30-s intervals and incubate either for 10 min, to observe aster formation, or 30–60 min to observe spindle formation.
3. Stop the reaction by adding 1 mL of 30% glycerol in BRB80 (at room temperature) and mix gently by inverting the tubes.
4. Overlay the diluted reaction onto a 5-mL cushion of 40% glycerol in BRB80 in a spindown tube as described in **Subheading 3.1.** Centrifuge for 10 min at 15,000*g* in a HB-4, HB-6 (Sorvall) rotor, or a JS 13.1 rotor (Beckman), or 20 min at 9,000*g* in a JS 7.5 rotor (Beckman).
5. Sperm chromatin is visualized by staining with 4′,6-diamidino-2-phenylindole dilactate (DAPI). Centrosomes are visualized by immunofluorescence using an antibody against acetylated tubulin, which primarily stains the centrioles.

3.3. The Measurement of Microtubule Dynamics

Microtubules are highly dynamic polymers that alternate between phases of polymerization and depolymerization, a property defined as dynamic instability. For a detailed discussion of microtubule dynamics, *see* **refs.** ***36–38***. The dynamic behavior of microtubules is defined by four parameters: the growth rate, the shrinkage rate, the frequency of catastrophe (transition from growth to shrinkage), and the frequency of rescue (transition from shrinkage to growth). These parameters are regulated within the cell by various microtubule-associated proteins (MAPs) at different points of the cell cycle. In order to study the effect of RanGTP on microtubule dynamics, we used time-lapse fluorescence microscopy on microtubules nucleated from purified mammalian centrosomes incubated in Xenopus CSF-arrested egg extract in the presence and absence of RanGTP *(9)*. We found that RanGTP stabilizes microtubules by increasing the rescue frequency, which could at least partially explain how RanGTP stimulates microtubule production in CSF-arrested egg extracts.

3.3.1. Assay for Microtubule Dynamics

1. Prepare the reaction as follows: use 6.5 μL fresh egg extract, 0.5 μL rhodamine-labeled tubulin (final concentration 0.3 μ*M*), 0.32 μL oxygen-scavenger mix (*see* **Subheading 2.1.**), 1 μL of a 40 mg/mL hemoglobin solution, 1 μL centrosomes (final concentration 1.5×10^6 centrosomes/mL), 1 μL of 10 mg/mL RanL43E, or 1 μL XB. The reaction can be kept on ice for up to 4 h.
2. Prior to use, clean the glass slides and cover slips, first in ethanol, then in deionized water. The glass slides are dried using Kimwipes, and the cover slips are dried with a cover slip spinner (*see* **Subheading 2.1.**). Remove any residual dust using a compressed air spray (Control Company).
3. Spot 1 μL of the reaction onto a clean microscope slide and use forceps to overlay a clean cover slip onto the reaction. Seal the cover slip with VALAP and place immediately on a 25°C temperature-controlled microscope stage (Physitemp Instruments Inc.).
4. Locate a microtubule aster and collect time-lapse images within 5 min of spotting the reaction onto the cover slip.
5. We collected images using 250-ms exposures at 2-s intervals with a 100x lens and a Princeton instruments CCD camera operated by Metamorph software (Universal Imaging Systems).

3.3.2. Analysis of Microtubule Dynamics

Various programs for the analysis of microtubule dynamics have been previously described (*see* **refs. *39–42***). We measured the length of individual microtubules throughout the time series of collected images using Metamorph software. We specified that a microtubule would be considered to have grown or shrunk only if a microtubule exhibited a change in length of at least 0.5 μm over four consecutive frames (~6 s). From these measurements, microtubule lifetime graphs were constructed by plotting microtubule length vs time, using Microsoft Excel. Microtubule growth and shrinkage rates were obtained from these graphs using the least-squares linear-regression function in Microsoft Excel. Catastrophe and rescue frequencies were obtained by dividing the total number of each of these events by the total time microtubules spent growing (for catastrophe) or shrinking (for rescue).

3.4. Methods Used to Analyze Microtubule Motor Activity

Microtubule motor proteins utilize energy derived from ATP hydrolysis to move microtubules or cargo, and play prominent roles in regulating the organization of microtubules, causing them to bundle and to form asters and bipolar spindles. **Subheading 3.4.1.** describes different assays to measure and observe microtubule motor activities in response to RanGTP.

3.4.1. Microtubule Gliding Assay

To determine whether RanGTP could directly affect microtubule motors, we used an assay that has been previously described (*see* **ref. *43***). This assay is performed in a flow chamber, and utilizes the fact that microtubule motors remain active when bound to glass surfaces. The glass-bound motors bind microtubules and can push them over the surface of the glass. When this moving microtubule is picked up by another adjacent motor, the microtubule appears to glide smoothly over the surface of the glass. The microtubule movements can be visualized by either differential interference (DIC) microscopy or by fluorescence microscopy (if fluorophore-labeled microtubules are used). We use rhodamine-labeled taxol-stabilized microtubules.

1. Create a flow chamber by placing 2 parallel strips of double-sided sticky tape, 7–10 mm apart on a glass slide. A cleaned glass cover slip is then placed on the tape (*see* **Subheading 3.3.1.** for cover slip cleaning) to form the chamber.
2. Pipet a solution containing purified motor proteins (ranging from 10–50 μg/mL in BRB80 and containing 3 m*M* ATP) into the chamber. The motors should adhere to the glass within 5 min.
3. Wash the chamber with 80 μL of BRB80 containing 3 m*M* ATP. Flow through the chamber is achieved by pipetting the solution at one side of the chamber and placing a piece of Whatman paper at the other end to act as a wick to draw the solution through the chamber.
4. To prevent microtubules from adhering to the glass directly, pipet a BRB80 solution containing 3 m*M* ATP and either 5 mg/mL casein or BSA into the chamber and incubate for 5 min. Wash the chamber as decribed in **step 3**.
5. Flow into the chamber a solution containing taxol-stabilized rhodamine-labeled microtubules and an oxygen-scavenging system (1:100 dilution of the stock, *see* **Subheading 2.1.**). Incubate in the chamber for a time period that allows a sufficient number of microtubules to become bound to the motor proteins (5 min is a good starting point).
6. Once enough microtubules have attached to the glass cover slip, wash the chamber as in **step 3** with a solution containing taxol and the oxygen-scavenger system.

7. Place the chamber on the microscope stage and visualize microtubules using a 100X lens. Take a series of time-lapse images using similar parameters to those described in **Subheading 3.3.1.**
8. Once the images have been acquired, we track the position of microtubule ends throughout the course of the time-lapse series using the trackpoint function in the Metamorph software. These data reveal the velocity of the microtubule movements.

This assay can be modified to use polarity-marked microtubules (*see* **Subheading 2.5.**), in order to assess the direction the motor moves microtubules.

3.4.2. Seed Translocation Assay

The effect of RanGTP on microtubule motor activities can also be studied on microtubule asters and spindles *(9,44)* formed in *Xenopus* egg extract using short stabilized microtubule seeds (*see* **Subheading 2.4.2.**). These assays provide insight into the predominant motor activities altered in response to RanGTP within the astral microtubules—information that is valuable in understanding how these structures are formed and maintained.

1. Prepare the assay as described in **Subheading 3.2.**, but using fluorescein-labeled tubulin to visualize the microtubules. The reaction is allowed to proceed in a microfuge tube until the desired structures have formed. The reaction time must be determined prior to the experiment. In our hands, Ran-induced asters formed in egg extract by 7–10 min of initiating the experiment, and spindles began to form after about 30 min.
2. Once the desired microtubule structures have formed, add the rhodamine-labeled EGS-stabilized seeds (*see* **Subheading 2.4.2.**) and the oxygen-scavenging system (*see* **Subheading 2.1.**). Pipet 2 µL of the reaction mix onto a glass slide and cover it with a cleaned glass cover slip (*see* **Subheading 3.3.1.**).
3. Seal the cover slip with VALAP and immediately view the microtubule structures using a 60X lens.
4. Once a suitable structure is found, take a time-lapse series of images using the parameters described in **Subheading 3.3.1.** This can be done in one of two ways. A dual-wavelength time-series can be collected, taking successive images in both the fluorescein channel (to view the microtubule aster or spindle) and the rhodamine channel (to view the seeds). However, if the microscope does not have this feature, the first and last images can be taken in the fluorescein channel, and the rest can be taken in the rhodamine channel with the filter blocks moved manually.

From the time-lapse images, it is then possible to measure the velocity of seed movement and the number of seeds moving in different directions. We found that RanGTP induced a greater number of seeds to move away from the astral microtubule center (to the plus end of microtubules) than RanGDP.

3.4.3. Monitoring the Movement of Microtubule Motors on Microtubule Structures

The analysis described in **Subheading 3.4.2.** predominantly follows the dominant motor activity on a microtubule structure. However, to study one particular motor, it would be necessary to label an individual motor to observe it alone. It is possible to label a motor in one of two ways, either directly by labeling a purified recombinant motor protein with a fluorophore or by using a fluorophore labeled, noninhibitory antibody added into the egg-extract reaction. In **steps 1–3**, we describe the approach we used to study Eg5, a plus-end-directed bipolar kinesin in response to RanGTP, in *Xenopus* egg extract. We used a rhodamine-labeled antibody *(9)* which binds to the central "stalk" domain of Eg5, which, at 20 μ*M*, did not inhibit spindle assembly.

1. Prepare the reaction in the same way as described for the aster-formation assay in the presence of a rhodamine-labeled, noninhibitory anti-motor antibody and fluorescein-labeled tubulin.
2. Once asters are formed, pipet 1.5 μL of the reaction onto a glass slide cover with a cleaned glass cover slip and view using a 100X lens as described in **Subheading 3.4.2.**
3. Collect time-lapse series as described in **Subheading 3.4.2.**

We found that the labeled antibody, when bound to Eg5, appeared as a bright speckle. The speckles could be tracked using the same protocol used to track the microtubule seeds. Using this analysis, we found that the velocity of Eg5 on the microtubules did not change in response to RanGTP. However, RanGTP did induce an increase in the amount of Eg5 molecules being directed to the plus end of the microtubules *(9)*.

4. Notes

1. Microtubules are susceptible to shearing. Therefore, use either wide-bore or "cut off" pipet tips when manipulating them.
2. At room temperature, high concentrations of tubulin (2 mg/mL and above) will self-polymerize. Therefore, keep solutions containing tubulin on ice to prevent microtubule polymerization. In addition, to prevent microtubules from depolymerizing prior to fixation, keep solutions containing microtubules at room temperature.
3. Image analysis: We use Metamorph software to operate the microscope, the CCD camera, and to conduct all the image analysis. However, there are many

other software packages that can be used to perform similar analyses. For detailed descriptions of how to process the image data and prepare the data for presentation (especially the production of movies), refer to **refs. *45*** and ***46***.

4. Other useful resources for studying microtubules. Below are listed two websites that are directly relevant to the methods described in this chapter. Both websites have extensive links to further sites which describe additional protocols that are useful for studying the microtubule cytoskeleton. (A) http://iccbweb.med.harvard.edu/mitchisonlab/. This is the home page web address for Dr. Tim Mitchison's lab. It lists numerous detailed methods for studying both the microtubule and actin cytoskeleton. (B) http://blocks.fhcrc.org/~kinesin/. This website is an excellent source for protocols and general information on motor proteins.

References

1. Mattaj, I. W. and Englmeier, L. (1998) Nucleocytoplasmic transport: the soluble phase. *Annu. Rev. Biochem.* **67,** 265–306.
2. Kalab, P., Pu, R. T., and Dasso, M. C. (1999) The Ran GTPase regulates mitotic spindle assembly. *Curr. Biol.* **9,** 481–484.
3. Wilde, A. and Zheng, Y. (1999) Stimulation of microtubule aster and spindle assembly by the small GTPase Ran. *Science* **284,** 1359–1362.
4. Ohba, T., Nakamura, M., Nishitani, H., and Nishimoto, T. (1999) Self-organization of microtubule asters induced in Xenopus egg extracts by GTP-bound Ran. *Science* **284,** 1356–1358.
5. Carazo-Salas, R. E., Guarguaglini, G., Gruss, O. J., Segref, A., Karsenti, E., and Mattaj, I. W. (1999) Generation of GTP-bound Ran by RCC1 is required for chromatin-induced mitotic spindle formation. *Nature* **400,** 178–181.
6. Gruss, O. J., Carazo-Salas, R. E., Schatz, C., Guarguaglini, G., Kast, J., Wilm, M., et al. (2001) Ran induces spindle assembly by reversing the inhibitory effect of importin alpha on TPX2 activity. *Cell* **104,** 83–93.
7. Nachury, M. V., Maresca, T. J., Salmon, W. C., Waterman-Storer, C. M., Heald, R., and Weis, K. (2001) Importin beta is a mitotic target of the small GTPase Ran in spindle assembly. *Cell* **104,** 95–106.
8. Wiese, C., Wilde, A., Moore, M.S., Adam, S.A., Merdes, A., and Zheng, Y. (2001) Role of importin beta in coupling Ran to downstream targets in microtubule assembly. *Science* **291,** 653–656.
9. Wilde, A., Lizarraga, S. B., Zhang, L., Wiese, C., Gliksman, N. R., Walczak, C. E., et al. (2001) Ran stimulates spindle assembly by altering microtubule dynamics and the balance of motor activities. *Nat. Cell Biol.* **3,** 221–227.
10. Carazo-Salas, R. E., Gruss, O. J., Mattaj, I. W., and Karsenti, E. (2001) Ran-GTP coordinates regulation of microtubule nucleation and dynamics during mitotic-spindle assembly. *Nat. Cell Biol.* **3,** 228–234.
11. Zhang, C. and Clarke, P. R. (2000) Chromatin-independent nuclear envelope assembly induced by Ran GTPase in *Xenopus* egg extracts. *Science* **288,** 1429–1432.

12. Hetzer, M., Bilbao-Cortez, D., Walther, T. C., Gruss, O. J., and Mattaj, I. W. (2000) GTP hydrolysis by Ran is required for nuclear envelope assembly. *Mol. Cell.* **5,** 1013–1024.
13. Nakamura, M., Masuda, H., Horii, J., Kuma, K., Yokoyama, N., Ohba, T., et al. (1998) When overexpressed, a novel centrosomal protein, RanBPM, causes ectopic microtubule nucleation similar to gamma-tubulin. *J. Cell Biol.* **143,** 1041–1052.
14. Karsenti, E., Newport, J., Hubble, R., and Kirschner, M. W. (1984) Interconversion of metaphase and interphase microtubule arrays as studied by the injection of centrosomes and nuclei in eggs. *J. Cell Biol.* **98,** 1730–1745.
15. Dogterom, M., Felix, M., Guet, C., and Leibler, S. (1996) Influence of M-phase chromatin on the anisotropy of microtubule asters. *J. Cell Biol.* **133,** 125–140.
16. Heald, R., Tournebize, R., Blank, T., Sandaltzopoulos, R., Becker, P., Hyman, A., et al. (1996) Self-organization of microtubules into bipolar spindles around artificial chromosomes in *Xenopus* egg extracts. *Nature* **382,** 420–425.
17. Andersen, S. S. and Karsenti, E. (1997) XMAP310: a Xenopus rescue-promoting factor localized to the mitotic spindle. *J. Cell Biol.* **139,** 975–983.
18. Walczak, C. E., Verma, S., and Mitchison, T. J. (1997) XCTK2: a kinesin-related protein that promotes mitotic spindle assembly in Xenopus laevis egg extracts. *J. Cell Biol.* **136,** 859–870.
19. Wittmann, T., Wilm, M., Karsenti, E., and Vernos, I. (2000) TPX2, A novel *Xenopus* MAP involved in spindle pole organization. *J. Cell Biol.* **149,** 1405–1418.
20. Compton, D. A. and Cleveland, D. W. (1993) NuMA is required for the proper completion of mitosis. *J. Cell Biol.* **120,** 947–957.
21. Evans, L., Mitchison, T., and Kirschner, M. (1985) Influence of the centrosome on the structure of nucleated microtubules. *J. Cell Biol.* **100,** 1185–1191.
22. Mitchison, T. J. and Kirschner, M. W. (1984) Dynamic instability of microtubule growth. *Nature* **312,** 237–242.
23. Hyman, A., Drechsel, D., Kellogg, D., Salser, S., Sawin, K., Steffen, P., et al. (1991) Preparation of modified tubulins. *Methods Enzymol.* **196,** 478–485.
24. Mitchison, T. J. and Kirschner, M. W. (1986) Isolation of mammalian centrosomes. *Methods Enzymol.* **134,** 261–268.
25. Klebe, C., Nishimoto, T., and Wittinghofer, F. (1993) Functional expression in Escherichia coli of the mitotic regulator proteins p24Ran and p45Rcc1 and fluorescence measurements of their interaction. *Biochemistry* **32,** 11,923–11,928.
26. Dasso, M., Seki, T., Azuma, Y., Ohba, T., and Nishimoto, T. (1994) A mutant form of the Ran/TC4 protein disrupts nuclear function in Xenopus laevis egg extracts by inhibiting the RCC1 protein, a regulator of chromosome condensation. *EMBO J.* **13,** 5732–5744.
27. Izaurralde, E., Kutay, U., von Kobbe, C., Mattaj, I. W., and Gorlich, D. (1997) The asymmetric distribution of the constituents of the Ran system is essential for transport into and out of the nucleus. *EMBO J.* **16,** 6535–6547.
28. Lounsbury, K. M. and Macara, I. G. (1997) Ran-binding protein 1 (RanBP1) forms a ternary complex with Ran and karyopherin beta and reduces Ran GTPase-

activating protein (RanGAP) inhibition by karyopherin beta. *J. Biol. Chem.* **272,** 551–555.

29. Kutay, U., Izaurralde, E., Bischoff, F.R., Mattaj, I.W., and Gorlich, D. (1997) Dominant-negative mutants of importin beta block multiple pathways of import and export through the nuclear pore complex. *EMBO J.* **16,** 1153–1163.
30. Fanara, P., Oback, B., Ashman, K., Podtelejnikov, A., and Brandt, R. (1999) Identification of MINUS, a small polypeptide that functions as a microtubule nucleation suppressor. *EMBO J.* **18,** 565–577.
31. Hyman, A. A. (1991) Preparation of marked microtubules for the assay of the polarity of microtubule-based motors by fluorescence. *J. Cell Sci. Suppl.* **14,** 125–127.
32. Murray, A. W. (1991) Cell cycle extracts. *Methods Cell Biol.* **36,** 581–605.
33. Oegema, K., Wiese, C., Martin, O. C., Milligan, R. A., Iwamatsu, A., Mitchison, T. J., et al. (1999) Characterization of two related Drosophila gamma-tubulin complexes that differ in their ability to nucleate microtubules. *J. Cell Biol.* **144,** 721–733.
34. Felix, M. A., Antony, C., Wright, M., and Maro, B. (1994) Centrosome assembly in vitro: role of γ-tubulin recruitment in Xenopus sperm aster formation. *J. Cell Biol.* **124,** 19–31.
35. Martin, O., Gunawardane, R., Iwamatsu, A., and Zheng, Y. (1998) Xgrip109: a gamma tubulin-associated protein with an essential role in gamma tubulin ring complex (gammaTuRC) assembly and centrosome function. *J. Cell Biol.* **141,** 675–687.
36. Desai, A. and Mitchison, T. J. (1998) Microtubule polymerization dynamics. *Annu. Rev. Cell Dev. Biol.* **13,** 83–117.
37. Andersen, S. S. (1999) Balanced regulation of microtubule dynamics during the cell cycle: a contemporary view. *Bioessays* **21,** 53–60.
38. Cassimeris, L. (1999) Accessory protein regulation of microtubule dynamics throughout the cell cycle. *Curr. Opin. Cell Biol.* **11,** 134–141.
39. Gliksman, N. R., Parsons, S. F., and Salmon, E. D. (1992) Okadaic acid induces interphase to mitotic like microtubule dynamic instability by inactivating rescue. *J. Cell Biol.* **119,** 1271–1276.
40. Tournebize, R., Andersen, S. S. L., Verde, F., Doree, M., Karsenti, E., and Hymman, A. A. (1997) Distinct roles of PP1 and PP2A-like phosphatases in control of microtubule dynamics in mitosis. *EMBO J.* **16,** 5537–5549.
41. Belmont, L. D., Hyman, A. A., Sawin, K. E., and Mitchison, T. J. (1990) Real-time visualization of cell cycle-dependent changes in microtubule dynamics in cytoplasmic extracts. *Cell* **62,** 579–589.
42. Dhamodharan, R. and Wadsworth, P. (1995) Modulation of microtubule dynamic instability in vivo by brain microtubule associated proteins. *J. Cell Sci.* **108,** 1679–1689.
43. Kapoor, T. M. and Mitchison, T. J. (1999) Allele-specific activators and inhibitors for kinesin. *Proc. Natl. Acad. Sci. USA* **96,** 9106–9111.
44. Heald, R., Tournebize, R., Habermann, A., Karsenti, E., and Hyman, A. (1997) Spindle assembly in Xenopus egg extracts: respective roles of centrosomes and microtubule self-organization. *J. Cell Biol.* **138,** 615–628.

45. Waterman-Storer, C. M., Shaw, S. L., and Salmon, E. D. (1998) Production and presentation of digital movies. *Trends Cell Biol.* **7,** 503–506.
46. Waterman-Storer, C. M., Desai, A., and Salmon, E. D. (1999) Fluorescent speckle microscopy of spindle microtubule assembly and motility in living cells, in *Mitosis and Meiosis* (Rieder, C. L., ed.), vol. 61, Academic Press, San Diego, CA, pp. 156–172.

Index

A

ADP ribosylation factor, *see* Arfs
Allophycocyanin, 213
Antennapedia, *see* penetratin
Antibodies,
  anti-Aic2 for panning, 123–125
  anti-PAK, 81
  anti-RhoA, Rac-1 and Cdc42 for western blot, 101–105
Arf1,
  recombinant myristoylated protein, 171–174
  nucleotide–binding, kinetics of, 174–176
Arf6,
  aluminium flouride activation, 192,196
  ARNO involvement, 192
  cortical actin arrangement, involvement in, 192,196
  membrane protrusions/ruffling, 192–193
  nucleotide-binding kinectics of, 174–176
  recombinant myristoylated protein from *E. coli* , 171–174
  relationship to Rac, 193, 195
Arf GTPase activating protein 1 (GAP1),
  cysteine zinc-finger motif, 199
  expression and purification of, 200–201
  GAP assay with, 204–205
ATP regenerating system, 220, 232
  phopholipids in GTP loading, requirement of, 200–201
  renaturation of recombinant protein, 200
Arf guanine nucleotide exchange proteins (GEPs or GEFs),
  activity assay of, 185–186
  cytohesin family, 182–184
  brefeldin A (BFA) sensitivity, 181–182
  GTPγS binding, 184
  protein–protein interactions of, 186–184
  sec7 domain, 182–183

B

Blot overlays,
  Nck SH3 domain, by, 89–90,
  radio-labeling with [γ^{32}P]GTP, 91–92
  Ras-GTP module, 87–89
Bovine serum albumin,
  modification of, 212–215, 235
Bound nucleotide, *see* guanine nucleotide, bound
Brefeldin A, 181

C

Carcinoma *see* transformed cells
Cytohesin, *see* Arf GEFs
Cdc42,
  ERM phosphorylation, involvement in, 114
  filopodial formation, role, 114
  inhibition of signaling by, 76, 80

quantitation of GTP state, pull-down assay 100–104
Cdc42-Rac interaction/binding (CRIB) domain, 76, 100–105, 129
cDNA-representational difference analysis (cDNA-RDA)
adaptors for, 34–35
cDNA preparation, 32
enzyme and buffers, 29–30
hybridization protocol, 33–35
microarray analysis, 39–40
PCR primers for, 30–31,
single-colony-PCR, 38–38
Southern blot analysis, 37–37
Cell–cell adhesion, 121
E-cadherin-mediated, 121–123
effects of RhoGTPases on, 127
Cell-growth and anchorage, 80
Cell lines, cultured,
Chinese hamster ovary (CHO), 197
EL, 123–124
GFP NFAT reporter, 233–233
LAP86, 123–124
nEaCL cell, 123–124
Rat1 fibroblast expressing Rho, 130–131
v-Ha-Ras transformed NIH3T3, 80
Cell microinjection, *see* Microinjection of cells
Cell motility and migration assays,
Boyden–chamber based, 133–134
collagen invasion assay, 132, 135–136
Dunn chamber, use of, 136
wound healing assay, 136
Cell transformation (soft agar) assay, 8
Centrosomes, 249
Chromatin, 247–248
Collagen matrix, *see* cell motility
Cytochalasin D, 194

D

Deoxyribose nucleic acid (DNA), *see* transfection of cDNA *see* expression vectors
Diaminobenzidine (DAB), 68–71
Digitonin, for permiabilization of cells, 221–222
Dominant inhibitory mutants of GTPases, 3–8
Dynamics, of microtubules, 252–253

E

Endosomal membranes, 191
Epidermal growth factor (EGF), 78
ErbB1/B2, 78
Expression vectors,
cytomegalovirus (CMV) promoter, 192
penetratin peptide, 77–80
pGEX, cleavage problems, 15
pGEX-TK-Ras. 89
pLMM (light meromyosin), 49
ptac, 48
pRSET, 154
pS65T-C1 (Clontech)

F

Fluorescence-based GTPase protein assays,
dissociation from Rho, measurement of, 51–53
GAP-stimulated Ras GTPase, 55, 57
hGBP1 activity assays, 50–53
methyanthraniloyl (mant) guanine nucleotide based, 50–53

p190RhoGEF effects, 51–53
Raf RBD and Ras binding, 53–54
Ras intrinsic GTPase activity, 54–56
tryptophan excitation of, 51, 53
Fluorescence resonance energy transfer (FRET), 55
Fluorescent activated cell sorting (FACS), 238–242
Fluorescein-5-isothiocyanate (FITC), 71, 212, 249

G

GAP, *see* GTPase activating protein
GDI, *see* guanine nucleotide dissociation inhibitor
Gerenylgerenylation, 157–164
G-proteins, *see* under specific GTP binding protein
Green fluorescent protein (GFP)
Nuclear import assay, 212
GFP-NFAT, 232–233
Golgi complex, 19
Guanidinium hydrochloride, for protein refolding, 200–201
Guanine nucleotide dissociation assay, 149
Guanine nucleotide dissociation inhibitor (GDI), 143–144, 146–148
Guanine nucleotide exchange,
analysis of bound nucleotide, 20
alkaline phosphatase, use of, 17
radiolabelling protocol with γ32P–GTP, 88, 91–92
Ras GEFs in promotion of, 6–7
Rho GTPases, 17
Ras, preparation of free and bound states, 48–49
Rab GEP (GEF), 148–149
Guanine nucleotide exchange factor (GEF) *see* specific GTPase
Guanine nucleotide exchange protein (GEP) *see* specific GTPase
Guanylate binding protein 1(GBP1), recombinant protein, 48
binding kinectics to mant-GDP, 50–52

I

Image analysis software,
microscopy, 256
microarray, 29
Invasion assay, *see* cell motility
Importins, *see* nuclear protein import
Imidazole, purity of, 16
Ionomycin, 238

L

Latrunculin, 194
Leptomycin B, 236, 239–240
Lipid modification *see* Prenylation and myristoylation
Liposome preparation, 202

M

Mant, *see* N-methylanthraniloyl nucleotides
Mass spectrometry, 15, 175
Membrane traffic, 170, 191–192
Micro-array, *see* cDNA-RDA
Microinjection of cells, 69–70, 116
Microtubules,
aster and spindle-formation assays, 251–252
dynamics by time-lapse imaging, 252
fluorescent-labeled stabilized microtubules, 249–250

gliding assays, 253
movement on, 255–256
polymerization assay of, 250
purification of tubulin, 249
seed translocation assay, 254–255
taxol stabilized seeds, 250
Migration assay, *see* cell motility
Multimeric protein complexes,
myotonic dystrophy related Cdc42 binding kinase (MRCK), 114
RhoA-binding kinase (ROK), 11
Myristoylation,
membrane trafficking, 170
nucleotide-binding kinectics, effects of, 169
vesicle transport and membrane trafficking, 170

N

N-methylanthraniloyl (mant) nucleotides, 46–47
N-myristoyl transferase (NMT), 169–174
NBT *see* Nitro blue tetrazolium
Nerve growth factor,
PC12 cells, effect on 113–116
laminin seeded cells, 116
neurite outgrowth assays, for, 113–116
Neurofibromin,
recombinant protein, 48
GAP activity of, 56–58
Nitro blue tetrazolium (NBT), 68–70
Nitrous oxide, 8
Nuclear export signal (NES), 231
Nuclear factor of activated T-cells (NFAT), 232
Nuclear localization signal (NLS), 209–210, 231, 235
Nuclear pore complex (NPC), 209, 231
Nuclear protein transport,
analysis by flow cytometry, 237–241
analysis by fluorescent microscopy, 237–238
in vitro assay of, 236
preparation of cytosol for, 233
Ran and Crm1 recombinant proteins, 234–235
specific inhibitors of, 239–240
stable cell line with reporter GFP–NFAT, use of, 233
Nuclear protein import,
cell extracts for soluble import factors, 214–215
digitonin permeablilized cell assays, 209–211
fluorescent import substrates, 209–211
fluorescent NLS–conjugated substrate, 212–214
importin α and β, 210–211, 248
inhibition of, 211
nuclear localization signal (NLS), 209, 231
Ran GTPase, roles of, 210
RCC1 and RanGAP, roles of, 210–211
requirement of soluble factors, 210–211
Nuclear protein export,
assays in-vitro, 221–224
Crm, involvement in, 232–233, 239, 241
fluorescent reporters, 212

nuclear export signal (NES), 211–212
permeablilized–cell assay, 211–212, 236
PKI, biotinylated, 212, 217–218
quantitation by fluorescent microscopy, 224–225, 236–238
Ran.GTP, role of, 212, 239, 241
streptavidin-NLS, fluroescent-labeled (FITC-STV-NLS) 218–219
Nucleotide exchange, *see* guanine nucleotide exchange

O

Oncogenic Ras signaling,
ACK CRIB peptide, effects on, 76–77, 83
blocking by cell-permeable small peptides, 76
PAK CRIB peptide, effects on, 76–77
PAK proline–rich domain fragment, effects on, 77
Overlays, *see* blot overlays

P

p21s, *see* specific GTPases
p21-activated kinases (PAKs), 76, 100
activation assays of, 81
CRIB domain (PAK-CD), 100–102
Rac/Cdc42 quantitation, CRIB domain 100
p190 RhoGEF, 53
PAK, *see* p21-activated kinase
Penetratin (PEN) 77–80
Permiabilization of cells, 221
Phalloidin, 117, 194
Phosphatidylinositol-3-OH (PI-3) kinase, 4, 76
Phospholipase D (PLD), 169–170, 191
Pinocytosis, 68
Pleckstrin homology (PH) domain, 112, 114, 182
Polymerization of tubulin, *see* microtubules
Prenylation of GTPases,
involvement in signaling, 4
purification of prenylated Rab3A, 145
gerenylgerenyl transferase, 157–164
Probes for blot overlays, 87–88, 91–94
Pull-down assays for Rho-like GTPases, 100–104, 106

R

Rab1a, recombinant protein, 159–160
Rab3,
isoforms of, 143–144
regulatory proteins, 144
lipid-modified form, 145
Rab Escort Protein (REP), 157–159
Rab Escort Protein:Rab complexes,
prenylation of, 157
membrane targeting by, 158
purification of, 158, 160
REP:GG-Rab comlex, 158–159, 160–162
Rab GAP,
assays of, 151–152
purification from rat brain, 153–154
recombinant p130 and p150 subunits, 154
Rab geranylgeranyl transferase,
α and β subunits, 157

recombinant protein from Sf9 cells, 159–160
Rab guanine nucleotide dissociation inhibitor (GDIa) , 143–144, 146–148
Rab guanine nucleotide exchange protein (GEP or GEF)
assays of, 148–149
purification from rat brain, 149–151
recombinant protein, purification of, 151
Rac,
Arf6, involvement with, 192
bound GTP quantitation, pull-down assay 100–104
Raf kinases, 4
Ras binding domain, 49, 56
Ran,
microtubule aster and bipolar spindles, induction of, 248
microtubule organization, 247–248
nuclear protein transport role, 210–212
spindle assembly and maintenance, 248
Ras,
blocking Ras-signaling, 78
dominant inhibitory mutants of, 3
fusion protein for g32P labeling, 87–89
GAP-stimulated GTPase activity, 55–56
GDI interaction, kinetics of, 54
intrinsic GTPase activity, 55
nucleotide exchange reaction, 48
Raf –Ras binding domain, 49
mutant A15 and N17, 4–8
GTPase cycle, 4
Ras GEF binding to Ras mutants, 6–8
Ras-homology (*see* Rho)
RCC1, 248
Reactive oxygen species (ROS),
Ras and Rac as activators, 67–68
nitroblue tetrazolium (NBT) assay, 69–71
diaminobenzidine (DAB) assay, 69, 71
REP, *see* Rab Escort Protein
RhoA.GTP quantitation, pull-down assay 100–104
Rho family GTPases,
bound nucleotide, analyses of, 17–18, 20–21
cell–cell adhesion, involvement in, 127
detection of GTP-bound forms, assays for, 100–104, 106
expression of, 15–16, 18–19
fluorescent assays of, 50, 54
hexahistidine-tagged version, 15–16
nucleotide exchange reactions, 17, 20
p190 RhoGEF catalyzed nucleotide dissociation from, 53–54
purification of, 16–17, 19–20
Rho-associated kinase (ROK),
focal adhesion formation, involvement in, 111–114
stress fibers, involvement in, 111–114
Rhotekin, Rho binding domain, 100–104
ROS, *see* reactive oxygen species

S

Signal transduction,
therapy of human cancers, 75–76

oncogenic Ras, 76–7
Streptavidin, modification of, 218–219
Stopped-flow, for kinetic analysis, 55
Superoxide, *see* reactive oxygen species

T

Taxol, for stable microtubules, 250
Tetracyclin-regulated expression (tTA), 130
Transfection of cDNA into cells,
Arf6 mutants, 194–195
calcium phosphate method, 68–70
interleukin receptor as marker, 124
liposome–mediated, 115–116, 131
Rho family mutants, 131, 194
Transformed cells, 67, 75–78, 129
assays in nude mice, 81
Transwell, *see* cell motility
Trichostatin A, 236
Tryptophan fluorescence, 55–57

V

v-Ha-RAS-induced sarcoma, 81

W

Wheat-germ agglutinin, 239–240

X

Xenopus,
egg extract, 248, 251–254
sperm, demembranated, 251

FSC
www.fsc.org
MIX
Papier | Fördert
gute Waldnutzung
FSC® C083411

Kolibri 360 GmbH
Ferdinand-Jühlke-Straße 7
99095 Erfurt, Deutschland
produktsicherheit@kolibri360.de